Local Geometry of the Fermi Surface

Springer-Science+Business Media, LLC

Nataliya A. Zimbovskaya

Local Geometry of the Fermi Surface

And High-Frequency Phenomena in Metals

With 37 Illustrations

 Springer

Nataliya A. Zimbovskaya
Department of Physics, J419
City College of New York
138 Street and Convent Avenue
New York, NY 10031
USA

Library of Congress Cataloging-in-Publication Data
Zimbovskaya, Nataliya A.
    Local geometry of the Fermi surface: and high-frequency phenomena in metals /
Nataliya A. Zimbovskaya.
        p.   cm.
    Includes bibliographical references and index.
    ISBN 978-1-4612-6557-3   ISBN 978-1-4613-0193-6 (eBook)
    DOI 10.1007/978-1-4613-0193-6
    1. Fermi surfaces.   2. Fermi liquid theory.   I. Title.
    OC176.8.F4 Z56 2001
    530.4′1—d21   00-061862

Printed on acid-free paper.

© 2001 Springer Science+Business Media New York
Originally published by Springer-Verlag New York, Inc. in 2001
Softcover reprint of the hardcover 1st edition 2001

Production managed by Michael Koy; manufacturing supervised by Joe Quatela.
Typeset by The Bartlett Press, Inc., Marietta, GA.

9 8 7 6 5 4 3 2 1

ISBN 978-1-4612-6557-3   SPIN 10711946

Preface

The electronic properties of normal metals have been under active study for several decades. In the 1950s and 1960s most investigations sought to describe the Fermi surfaces of metals. These studies were based on experimental data obtained as a result of numerous observations of particular phenomena responsive to the structure of the electronic spectra of metals and thus to band-structure calculations [1]. The high-frequency properties of metals were also actively studied. These investigations were initiated by the development of the theory of the anomalous skin effect [2]. Later, significant achievements were reached in studies of the high-frequency properties of metals in the presence of an applied magnetic field. Cyclotron resonance in a parallel magnetic field [3], electromagnetic waves in metals [4]–[6], size effects [7], [8], Doppler shifted cyclotron resonance, and dopplerons [9], [10] were predicted in theoretical studies and repeatedly observed in experiments. This offered new scope for analysis of the properties of the electron system of metals. The main results of theoretical and experimental studies of the electronic characteristics of metals performed during this period are expounded in several books and review articles (see, e.g., [11]–[14], [16]).

Great progress was also achieved in studies of the interaction between an electron system and the ultrasonic waves propagating in metals. At low temperatures ($T < 10$ K) the electrons produce a strong effect on the dispersion and attenuation of ultrasound waves. Studies in this field were first stimulated by the work of Pippard [15] who proposed using magnetoacoustic effects to determine the shape of the Fermi surfaces. Magnetoacoustic oscillations became one of the most important tools to study the geometry of the Fermi surfaces of various metals [16]–[19]. One more effect promoted further development of investigations of the effects arising when an ultrasound wave propagates in metal: the discovery of giant quantum oscillations of the ultrasound attenuation rate in the presence of a quantizing magnetic field [20].

In the 1970s significant progress was achieved in investigations of the possible manifestations of the electron–electron interaction in observables. It follows from the theory of an electron Fermi liquid [4], [21] that a response of the electron system of a metal to high-frequency and spatially inhomogeneous disturbance can exhibit some qualitative difference compared to the response of the noncorrelated electron gas under the same conditions.

An experimental observation of the collective excitations, which arise due to the electron–electron correlation (so-called Fermi liquid spin and cyclotron waves) in alkali metals [22]–[24] and which gave information about the parameters of the Fermi-liquid interaction in sodium and potassium, was an important success in this field. It generated further development of the phenomenological theory of the electron Fermi liquid. The theory was extended to be applicable to research on a broad class of phenomena arising in the presence of a strong (quantizing) magnetic field. These new methods were first applied to study the spectra of Fermi liquid quantum waves in metals [25], [26]. A systematic exposition of the theory of electron Fermi liquid in a quantizing magnetic field was first presented in [27], [28].

At the same time investigations proceeded to develop a microscopic theory of correlated electron systems [29]–[31]. One of the main purposes of these treatments was to derive the basic equations of the phenomenological theory by means of the microscopic approach. Such a justification of the basic points of the phenomenological theory of electron liquid was achieved.

Those effects which are determined by the fundamental geometric characteristics of Fermi surfaces (connectivity, presence or absence of open orbits, and so on) were studied rather well by the mid-1970s. The major manifestations of electron correlation in the electronic plasma of metals were also investigated in detail for simple metals whose Fermi surfaces can be supposed to be spherical. At the same time, new problems and objects to be studied were put forward. Both theorists and experimentalists started to do research on the electronic properties of thin metallic films. Semiclassical and quantum size effects [13], [32], [33], collective excitations [27], [34]–[37], and transport [38] in these metallic films were analyzed in detail. For further development of these studies it appeared to be necessary to perform an in-depth study of the process of electron scattering from a surface of the film [39]–[43]. The influence of the surface layer of the metal on the electron wave functions had also been studied in detail [44]. Other important fields of research, concerning the electron properties of metals in the 1970s, include studies of magnetic breakdown, nonlinear effects in metals, and some others. Further treatment of these leads was continued for the next decade. However, major attention moved to the field of investigations of the electron properties of low-dimensional conductors. The phenomenon of high-temperature superconductivity was also intensively treated.

The subject of the present book is closely connected with the principal trends of studies in the field of the electron theory of metal developed in the 1970s and up to the present. The main subject-matter of the book is the local geometry of Fermi surfaces and its influence on high-frequency phenomena in metals and other materials of a metal-like type of conductivity. The author's interest in this subject was first stimulated by the results of several experiments carried out in the 1960s [45]–[49] which revealed cyclotron resonance in a magnetic field directed along a normal to the surface of a metallic sample. In these experiments, the cyclotron resonance in a normal magnetic field was observed in cadmium [45]–[47], zinc [48], and potassium [49]. A detailed theoretical analysis showed that this resonance arose due to a local flattening of the Fermi surfaces of these metals in the vicinities of some points. This conclusion was a starting point for further research whose results are presented here. These results were published in part in several papers [50]–[75].

The first chapter of this book is of an introductory character. It covers material which can be found in most of the books on the theory of metals. However, a brief summary of some basic concepts of the theory of metals seems to be of importance for a better comprehension of the major portion of the book. Here the concept of the electronic liquid of a metal is introduced; the definition of the Fermi surface is given and some geometric features of the Fermi surfaces are described. These are points of flattening and parabolic points of the Fermi surface where its Gaussian curvature becomes zero, as well as lines of zero or extremely large curvature which can be found on the Fermi surfaces of some metals. The influence of such points or lines upon the observables comes from the change in the electron density of states in their neighborhood. The manifestations of this influence are studied in detail in the main body of the book. Besides, the first chapter contains a brief description of several phenomena in metals which can be affected due to the local geometry of the Fermi surfaces, notably: the skin effect, the cyclotron resonance, the magnetoacoustic oscillations of the attenuation rate, and the velocity shift of the ultrasound waves propagating in metals.

It is known that the phenomenological theory of the Fermi liquid in metals is well developed within the framework of an isotropic model of a metal where the Fermi surface is supposed to be a sphere. However, real metals, as a rule, have complicated shape anisotropic Fermi surfaces. This gives rise to significant difficulties in studies of the Fermi-liquid effects. The point of Chapter 2 of this book is to propose some asymptotic expansions of the kernel of the Fermi–liquid interaction (Landau F-function) to accommodate the theory of the electron liquid to a more systematic study of the Fermi-liquid effects in real metals. These asymptotic expansions are based on the symmetry properties of the Fermi surfaces. Thus Chapter 2 contains a body of mathematical techniques which permit us to study the response of an electron liquid of real metals to external disturbances.

In subsequent chapters these techniques are applied to analyze concrete problems. Chapter 3 deals with the anomalous skin effect in metals whose Fermi surfaces have local anomalies of their curvature. The analysis carried out in this chapter leads to the conclusion that in the presence of points or lines of zero, or anomalously large curvature on a Fermi surface, the magnitude and frequency dependence of the surface impedance of a semi-infinite metal can be significantly changed. It is shown that under certain conditions a new kind of weakly damping electromagnetic wave can propagate in a metal.

The content of Chapter 4 is a systematic theory of a cyclotron resonance in metals, in a geometry where an external magnetic field is directed along a normal to the surface of the metal. Unlike the well-known Azbel–Kaner cyclotron resonance [3] this effect can exhibit itself only in the presence of locally flattened or nearly cylindrical segments on the effective parts of the Fermi surfaces. When the Fermi surface of a considered metal everywhere has a finite and nonzero curvature, the resonance feature corresponding to the cyclotron resonance is smeared out until it is scarcely detectable. The proposed theory of the cyclotron resonance in metals in a normal magnetic field gives good agreement with the results of the experiments of [47]–[49].

Geometric oscillations of the attenuation rate of the ultrasonic waves, propagating in metals perpendicularly to an applied magnetic field, are very sensitive to the local geometry of the Fermi surface at the points corresponding to the stationary points of a cyclotron orbit of an electron. The characteristic features of the geometric oscillations of the ultrasonic attenuation, and the velocity shift caused by the local geometry of the Fermi surface, are analyzed in Chapter 5.

Chapter 6 includes a theory of a special kind of collective excitations, the so-called Fermi-liquid cyclotron dopplerons. These collective excitations are doppleron-like extensions of the well-known Fermi-liquid cyclotron waves predicted by V.P. Silin [4]. Cyclotron dopplerons can propagate in metals only under the condition that their Fermi surfaces have a paraboloid-like shape. Thus these collective excitations can be referred to as a manifestation of the Fermi-liquid interaction among the electrons in metals with nonspherical Fermi surfaces.

The presence of points and lines of zero curvature affects the density of states of electrons on the Fermi surface and, consequently, the magnitude and shape of the quantum oscillations of the observables in a strong magnetic field. Some manifestations of this influence in metals are analyzed in Chapter 7.

The topics covered by the final chapter are related to those of the preceding chapters. This chapter deals with the possible manifestations of the local geometry of the Fermi surface in low-dimensional conductors with a metal-like type of conductivity. The skin effect, the cyclotron resonance, and the quantum oscillatory phenomena in organic metals are discussed in the first three sections of the chapter. The last two sections include the theory of the magnetoacoustic response of modulated two-dimensional electron systems in a quantum Hall regime near half-filling of the lowest Landau level. It is shown that as well as in conventional metals, the local geometry of the Fermi surfaces can strongly influence the response of such systems to the electromagnetic or ultrasonic disturbance. The theoretical analysis corroborates the results of recent experiments on the cyclotron resonance in layered conductors and on the velocity shift and attenuation of the surface acoustic waves propagating in the modulated GaAs/AlGaAs heterostructures.

An acquaintance with this text should prepare the reader for a more detailed study of particular areas of the theory of metals and other metal-like systems especially of the "Fermiology" which is not exhausted to the present. It is assumed that the essentials of electrodynamics, and quantum and statistical mechanics are a sufficient theoretical background for reading this book. As far as possible the author has tried to present the necessary calculations thoroughly. Some special details are considered in the Appendices. The book is intended to appeal to theoretical and experimental physisists, whose field of research concerns magnetotransport in metals and low-dimensional systems or the neighboring areas of solid state theory, and to graduate students.

During the development of the English edition some alterations were introduced into the text. New material (Chapter 1, and Sections 8.1, 8.2, 8.4, and 8.5) and a number of new figures have been added, and the list of references has been changed substantially.

I am pleased to acknowledge the hospitality of the Physics Department of City College, City University of New York, which made computer and other facilities available to me during the preparation of the manuscript for the English edition of this book.

I am sincerely grateful to all my colleagues with whom I collaborated to make these studies. This collaboration was a honor for me. It is my pleasant duty to thank Professor J.L. Birman for his proposal to have the book published. I take this opportunity to express my deep gratitude to my husband, Dr. G.M. Zimbovsky. Without his enthusiastic help and support this book would never have been written.

New York   Nataliya A. Zimbovskaya

Contents

CHAPTER 1

The Electronic Liquid of Metals

1.1 The Quasi Particle Concept

Any metal can be considered as a Fermi sea of electrons which moves within a crystalline lattice. Each of these conduction or "free" electrons strongly interacts with ions of the lattice. Besides, there exists a strong interaction among electrons. Actually, the potential energy of the conduction electrons in metals is of the same order as their kinetic energy. However, it appears that this strongly interacting system of conduction electrons (electron liquid) in many aspects behaves like a system of noninteracting particles (electron gas) affected by the electric field of the crystalline lattice. This similarity is the basis of a phenomenological theory of Fermi liquids. This theory was first proposed by Landau for neutral Fermi liquids in [77], [78] and then generalized by Silin for charged Fermi liquids like the electron liquid of metals [4].

The relevant quasi particles of the Landau–Silin theory of the electron Fermi liquid are electrons surrounded by time-dependent polarization clouds which accompany them. The interaction between the quasi particles is thus dynamically screened and is weaker than the interaction among electrons. Like electrons themselves, these quasi particles are also described by Fermi–Dirac statistics and obey the Pauli exclusion principle. Any quasi particle bears an electric charge equal to the charge of electrons and their total number equals the total number of conduction electrons.

Within the framework of the Landau theory of Fermi liquids it is assumed that there exists a one-to-one correspondence between the eigenstates of the system of interacting quasi particles (liquid) and noninteracting quasi particles (gas). Certainly, energy levels of the interacting systems are shifted compared to those for the system of noninteracting electrons. However, they are specified by the same quantum numbers as those for the noninteracting system and there is no gap between the ground state and the first excited level.

This characteristic of the eigenstates cannot be applied to a general interacting system. When the interaction is strong enough it can lead to radical changes in the ground state of the system and in the character of the current carrying states. These changes can give rise to a transition to a superconducting state, or to some other phase transitions, for example, a metal–insulator transition or structural changes in the lattice. However, a system of conduction electrons of the usual nonsuperconductive metal may usually be regarded as a normal Fermi liquid.

Let us first assume that the potential of the crystalline lattice $V(\mathbf{r})$ is constant in space ($V(\mathbf{r}) = V_0$). Under this assumption we have an isotropic Fermi liquid of conduction electrons. Neglecting at first the interaction between quasi particles we can write their equilibrium distribution function; it is the well-known Fermi function

$$f(E) = \left(\exp \left[\frac{E - \zeta(T)}{T} \right] + 1 \right)^{-1}. \tag{1.1}$$

Here E is the energy of the quasi particle, ζ is the chemical potential, and T is the temperature expressed in energy units. When $T = 0$ the Fermi function is a step function $f = 1$ for $E < \zeta(0)$ and $f = 0$ for $E > \zeta(0)$. This quantity $\zeta(0)$ is usually called the Fermi energy and denoted as ζ.

For reasons of symmetry, surfaces of constant energy for the isotropic system are spheres in quasi-momentum space. The radius of the Fermi sphere p_F corresponding to the Fermi energy ζ is determined by the following relation:

$$N = \frac{2V}{(2\pi\hbar)^3} \frac{4\pi}{3} p_F^3, \tag{1.2}$$

where N is the total number of quasi particles and V is the occupied volume in a coordinate space.

When the Fermi liquid is in equilibrium at $T = 0$, all states inside the Fermi sphere are filled by the quasi particles whose quasi momenta $\mathbf{p}$ are less in magnitude than the Fermi momentum p_F, while the states outside the Fermi sphere are empty. The ground state of the Fermi liquid for nonzero temperatures, as well as its excited states, can be constructed by transferring some quasi particles from inside the Fermi sphere to outside. Interaction between quasi particles of an isotropic Fermi liquid does not change their total number, therefore the radius of the Fermi sphere remains unchanged. However, the Fermi energy can be shifted and we can take this into account by introducing an effective mass of the quasi particles m^* which is modified due to the interaction among quasi particles, $\zeta = p_F^2/2m^*$.

When the state of the Fermi liquid is not a pure stationary state it has to decay in time. Therefore the quasi particles can be characterized by a finite lifetime. Consider a simple case when we introduce a single quasi particle with quasi momentum $\mathbf{p}$ into the electron liquid in the equilibrium state at $T = 0$. In this case, the probability of finding this quasi particle decreases in time as $\exp(-t/\tau_p)$. The lifetime τ_p for our case can be estimated as $\tau_p^{-1} = (a/\hbar\zeta)(E(\mathbf{p}) - \zeta)^2$ where the dimensionless coefficient "a" is of the order of unity [12], [29]. We can interpret the decay as originating from an imaginary part of the energy of the quasi particle

$$i\Gamma_p = \frac{i\hbar}{2\tau_p} = \frac{ia}{2\zeta}(E(\mathbf{p}) - \zeta)^2. \tag{1.3}$$

In the vicinity of the Fermi surface we have $|E_p - \zeta| \ll \zeta$, and the imaginary part Γ_p is negligible compared to the free energy of the quasi particle $E_p - \zeta$. On the contrary, the quantity Γ_p has the same order as the free energy of the quasi particles when $|E_p - \zeta| \sim \zeta$. This means that the quasi particles are rather stable when their energies are close to the Fermi energy but they tend to be dissipated being far away from the Fermi surface. When the temperature is nonzero we can suppose that $|E_p - \zeta| \sim T$ and the quasi particle concept makes sense under the condition $T/\zeta \ll 1$.

For simplicity we will call these quasi particles electrons or conduction electrons. The position of the Fermi level for the electrons in a real metal is determined by the equation $E(\mathbf{p}) = \zeta$ which describes the Fermi surface in the quasi momentum space. A variety of experimental techniques, together with the methods of band-structure calculations, give information which allows us to determine the most important characteristics of the Fermi surfaces of metals. Usually Fermi surfaces have a very complicated shape and contain several sheets [1]. Some of them can be associated with different partly filled energy bands. The Fermi surface symmetry is completely determined by the symmetry of the crystalline lattice.

To analyze, in general, the geometry of the Fermi surfaces of metals we can start from the structure of the constant energy surfaces within an energy band. The energy reaches its minima and maxima at certain values of the quasi momentum. In the vicinities of these points the energy momentum relation can be represented in the form

$$E(\mathbf{p}) = E(\mathbf{p}_0) + \left(\frac{1}{m}\right)^{\alpha\beta}_{\mathbf{p}=\mathbf{p}_0} (p_\alpha - p_{0\alpha})(p_\beta - p_{0\beta}). \tag{1.4}$$

Here $(1/m)_{\alpha\beta} = \partial^2 E/\partial p_\alpha \partial p_\beta$ is a tensor of the inverse effective masses. The components of this tensor have the dimensions of the reciprocal of mass.

It follows from expression (1.3) that near the points of extrema the constant energy surfaces are ellipsoids. Any such surface near a point of the minimum of energy encloses a region in which the energy takes on values less than that on the surface. Correspondingly, the velocity vector $\partial E/\partial\mathbf{p}$ is directed along the outward normal to the surface of constant energy. When the constant energy surface surrounds the region in which the values of energy are larger than on the surface itself, the velocity vector is directed along the inward normal. It would be impossible to proceed continuously from the surfaces surrounding the points of minima of energy to those surrounding its maxima. Therefore these topologically simple ellipsoid-like surfaces have to be interspersed by more complex open surfaces of constant energy. These open surfaces extend throughout the whole of the reciprocal lattice. Following [13] we consider as an example the surfaces of a constant energy which correspond to a following energy-momentum relation for

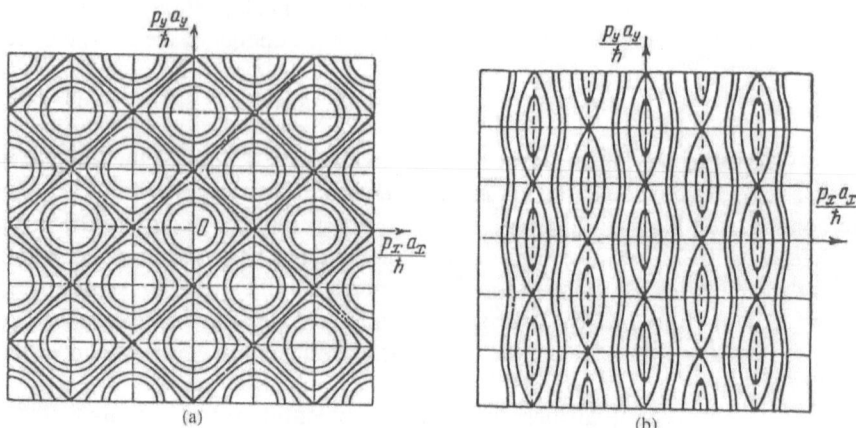

FIGURE 1.1. Constant-energy curves.

a two-dimensional system of electrons

$$E(\mathbf{p}) = A_1 \cos \frac{p_x a_x}{\hbar} + A_2 \cos \frac{p_y a_y}{\hbar},$$

where A_1, A_2 are constants of the dimensions of energy. The constant energy curves are shown in Figure 1.1. When $A_1 = A_2$ we only have one "open surface" (the system of straight lines in Figure 1.1(a)). For $A_1 \neq A_2$ there appears to be a layer of such surfaces as is shown in Figure 1.1(b). Some examples of open surfaces are shown in Figure 1.2.

The topology of surfaces of a constant energy (and the Fermi surface which is one of these surfaces) determines the nature of their planar sections. The curves formed by planar sections of the Fermi surface can be closed or open, and the presence of open sections can essentially change the dynamics of conduction electrons in the presence of an external magnetic field. It is clear that open sections can only exist when the Fermi surface itself is open. However, some open surfaces (like the surface "b" in Figure 1.2) do not contain any open planar sections. In more typical cases, open sections exist for a one-dimensional (dihedral angle) or even a two-dimensional (solid angle) set of directions of normals to the secant plane. Examples of the corresponding surfaces (surfaces "d" and "e") are presented in Figure 1.2.

1.2   Local Geometry of the Fermi Surface and High-Frequency Properties of Metals

Fermi surfaces of real metals are complex and intricate in shape and this essentialy influences observables. Phenomena which are determined by the main geometrical characteristics of the Fermi surfaces, i.e., their connectivity, are well studied.

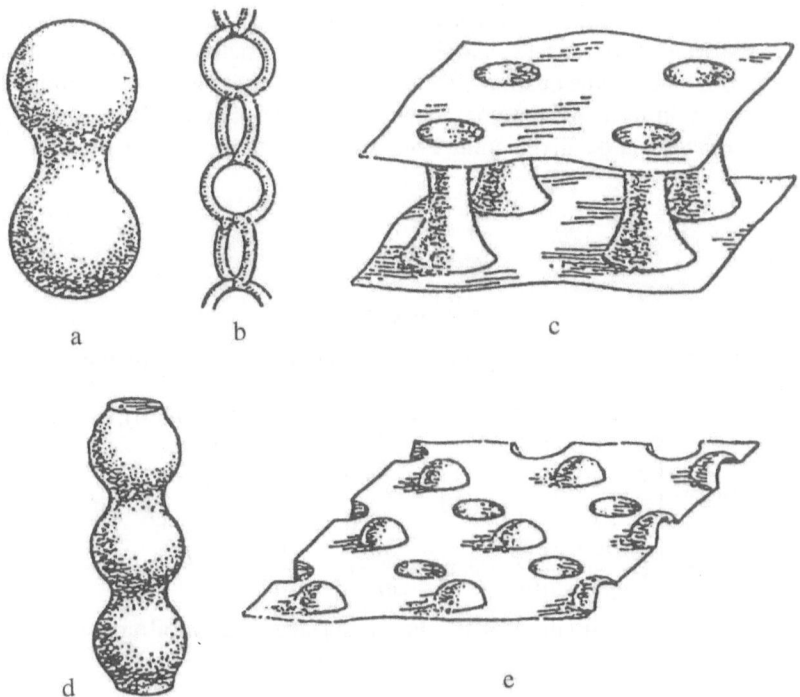

FIGURE 1.2. Various types of constant-energy surfaces.

However, those effects, which appear due to the local geometry of the Fermi surface, have not been investigated in detail at present. This influence of the local geometry of the Fermi surface on observable properties of a metal becomes essential for high frequencies of the electromagnetic or ultrasonic perturbation. In this high-frequency range only those electrons which are moving in synchronism with the propagating perturbation strongly participate in the absorption of the energy. These quasi particles concern are concentrated on the so-called "effective" parts of the Fermi surface where

$$\omega = \mathbf{q}\mathbf{v}. \qquad (1.5)$$

Here ω, and $\mathbf{q}$ are the frequency and the wave vector of the perturbation, $\mathbf{v}$ is the velocity of the electron.

For sound waves, and also for electromagnetic waves under conditions of strong spatial dispersion (large q), (1.5) is satisfied for electrons which are moving nearly transversely to the direction of propagation of the wave. The local geometry of the effective parts of the Fermi surface can strongly influence the high-frequency properties of metal. The most essential characteristic of the local geometry of Fermi surfaces is their Gaussian curvature $K(\mathbf{p})$:

$$K(\mathbf{p}) = \frac{LN - M^2}{EG - F^2} \equiv \frac{1}{R_1(\mathbf{p})R_2(\mathbf{p})}. \qquad (1.6)$$

Here E, G, F and L, M, N represent coefficients of the first and second differential quadratic forms of the surface, and $R_{1,2}(\mathbf{p})$ are the principal radii of curvature.

The coefficients of the quadratic forms at an arbitrary point of the Fermi surface are expressed in terms of the components of the velocity of electrons and their partial derivatives at the given point. For coefficients of the first quadratic form we obtain

$$
\begin{aligned}
E &= 1 + v_x^2/v_z^2, \\
G &= 1 + v_y^2/v_z^2, \\
F &= v_x v_y/v_z^2.
\end{aligned}
\tag{1.7}
$$

The coefficients of the second quadratic form are accordingly equal to

$$
\begin{aligned}
L &= \left[\left(\frac{\partial v_z}{\partial p_x} + \frac{\partial v_x}{\partial p_z}\right) v_x v_z - \frac{\partial v_z}{\partial p_z} v_x^2 - \frac{\partial v_x}{\partial p_x} v_z^2\right] \Big/ (v_z^2 v), \\
M &= \left[\left(\frac{\partial v_z}{\partial p_x} v_y + \frac{\partial v_z}{\partial p_y} v_x\right) v_z - \frac{\partial v_z}{\partial p_z} v_x v_y - \frac{\partial v_y}{\partial p_x} v_z^2\right] \Big/ (v_z^2 v), \\
N &= \left[\left(\frac{\partial v_z}{\partial p_y} + \frac{\partial v_y}{\partial p_z}\right) v_y v_z - \frac{\partial v_z}{\partial p_z} v_y^2 - \frac{\partial v_y}{\partial p_y} v_z^2\right] \Big/ (v_z^2 v).
\end{aligned}
\tag{1.8}
$$

Thus the Gaussian curvature of the Fermi surface at an arbitrary point can be represented as

$$
K(\mathbf{p}) = v^{-4} \det\left(\frac{1}{m}\right) v_\alpha \left(\frac{1}{m}\right)_{\alpha\beta}^{-1} v_\beta.
\tag{1.9}
$$

All points of a surface, whose Gaussian curvature is finite, can be separated into three classes—elliptic, hyperbolic, and parabolic points, which correspond to positive, negative, or zero values of the second differential form at the considered point. For the elliptic points we have $LN - M^2 > 0$, for the hyperbolic points $LN - M^2 < 0$, and for the parabolic points this differential form and, consequently, the Gaussian curvature of the surface tends to zero.

Points of zero curvature on a smooth surface are arranged on lines which separate the "hyperbolic" parts of the surface from its "elliptic" parts. Such lines of zero curvature exist on the Fermi surfaces of most metals. Arrangements of lines of zero curvature on surfaces of several elementary types, which are elements of the Fermi surfaces of real metals, are represented in Figure 1.3. The line of zero curvature does not certainly divide the elliptic and hyperbolic parts of the Fermi surface. It can belong to such parts of the surface where the second differential quadratic form keeps its sign. The shape of the surface, near to its extreme cross-sections, can be close to cylindrical. To illustrate, we consider an axial-symmetric Fermi surface which corresponds to the following energy momentum relation:

$$
E(\mathbf{p}) = \frac{\mathbf{p}_\perp^2}{2m_\perp} + \tfrac{1}{4} p_m v_m(x)^4.
\tag{1.10}
$$

Here $x = p_z/p_m$, a constant $m_\perp$ has dimensions of mass, v_m, p_m are maximum values of components of the velocity, and the quasi momentum of electrons along

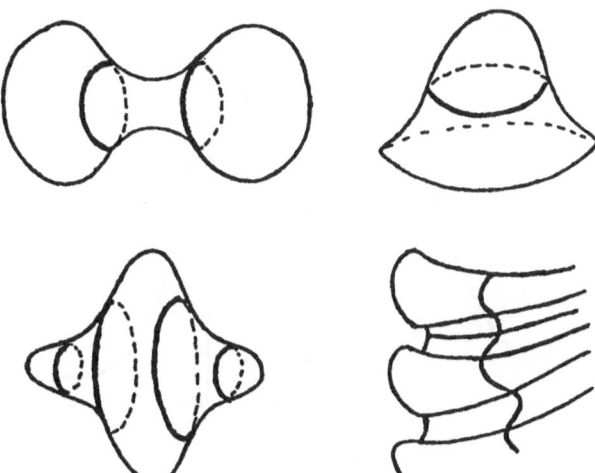

FIGURE 1.3. Examples of lines of zero curvature on the Fermi surfaces of metals [82].

the z-axis in the selected coordinate system which is the symmetry axis of the surface. The Gaussian curvature of the axial-symmetric Fermi surface is given by the expression

$$K(p_z) = m \left(v_z^2 + \frac{p_\perp^2}{m_\perp} \frac{\partial v_z}{\partial p_z} \right) \Big/ (p_\perp^2 + m^2 v_z^2)^2. \tag{1.11}$$

It follows from this expression (1.11) that, in the vicinity of the central cross-section of the Fermi surface by a plane perpendicular to the symmetry axis, $K(p_z)$ equals $12\pi m_\perp v_m p_z^2 / (S(0) p_m^3)$ where $S(0)$ is the area of the central cross-section. For all points of the cross-section ($p_z = 0$) the curvature equals zero, but for the points belonging to the remaining part of the surface it takes positive values, i.e., all points of the surface are of elliptic type and the line of zero curvature is not a line of inflection of the surface (see Figure 1.4).

The influence of lines of zero curvature of the Fermi surface on the observables in metals comes from the enhancement of the electron density of states in the vicinities of such lines. The electron density of states on the Fermi surface is determined by the formula

$$N_\zeta = \frac{V}{(2\pi\hbar)^3} \int \frac{dS}{v}. \tag{1.12}$$

Here the integration is performed over the Fermi surface and dS is the element of the surface area. We can transform the integral over the Fermi surface to the integral over the angles which determine the direction of an electron velocity vector $\mathbf{v}$, i.e., the direction of the normal to the surface. The surface element is shown in Figure 1.5. The element of the surface area dS equals $R_1 R_2 d\theta_1 d\theta_2$. Let us choose

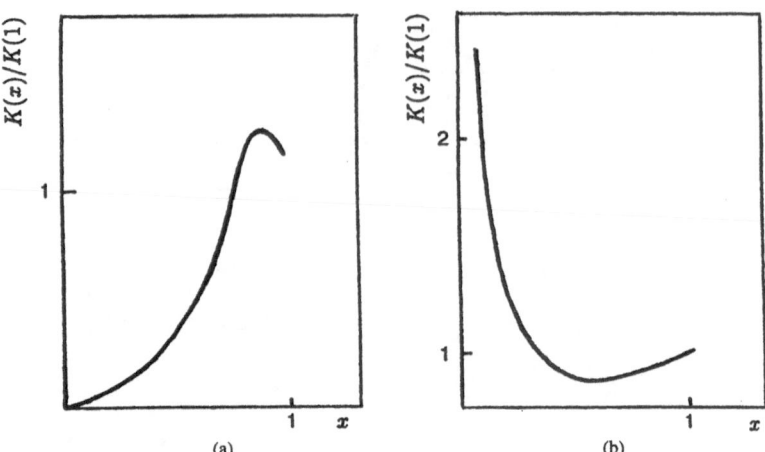

FIGURE 1.4. The curvature of the Fermi surfaces corresponding to (a) the energy-momentum relations (1.10) and (b) (1.15) versus p_z/p_m.

polar coordinates θ, φ in such way that $d\theta_1 = d\theta$ and $d\theta_2 = \sin\theta d\varphi$. Then

$$dS = R_1 R_2 \sin\theta d\theta d\varphi \equiv \frac{d\Omega}{K(\theta, \varphi)}, \qquad (1.13)$$

where $K(\theta, \varphi)$ is the curvature of the Fermi surface. We can now write

$$N_\zeta = \frac{2V}{(2\pi\hbar)^3} \int \frac{d\Omega}{K(\theta, \varphi)}. \qquad (1.14)$$

It follows from this formula that contributions to the electron density of states, from those parts of the Fermi surface which correspond to the vicinities of lines of zero curvature ($K(\theta, \varphi) = 0$), are larger than contributions of the remaining parts of the Fermi surface. In other words, when the curvature of the Fermi surface tends to zero at some point (or line) it gives an enhancement of electrons associated with the vicinity of this point (or line) on the Fermi surface.

The presence of lines of zero curvature on the Fermi surface affects high-frequency phenomena in metals when such lines intersect effective belts on the surface, or partially or completely coincide with them. In the vicinity of a point of zero curvature, the Fermi surface resembles the shape of a cylinder (if one of the principal radii of curvature goes to infinity) or a plane (so-called points of flattening where both principal radii of curvature go to infinity). Therefore, when a line of zero curvature coincides with an effective belt or even intersects it, the number of effective electrons increases. In turn, this leads to changes in the observable properties for certain directions of propagation of the external disturbance. The dependence of the attenuation rate of a sound wave on the direction of its propagation, when the Fermi surface is plane or cylindrical in shape, was predicted and analyzed in [79]. The influence of lines and points of zero curvature on frequency, and the angular dependence of the dispersion and attenuation of ultra-sound, conductivity, and surface impedance of metal at the abnormal skin effect,

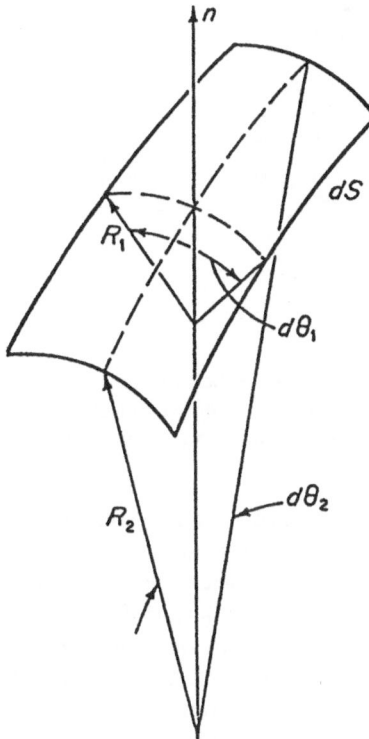

FIGURE 1.5. The surface elements of the Fermi surface. The principal radii of curvature of the shown elements are R_1 and R_2; **n** is a normal vector to the surface element.

was treated in [80]–[84]. In all these works the analysis was carried out within the framework of some simple models of the Fermi surface shape near to the points of zero curvature. Correspondingly, only terms smaller than $(p_\alpha - p_{0\alpha})^3$ ($p_{0\alpha}$ are the coordinates of the point of zero curvature) were kept in the expansions of the velocity components in powers of $(p_\alpha - p_{0\alpha})$. Some results of [79]–[84] were confirmed in experiments concerning the propagation of ultrasound waves in metals (see [85], [86]).

The influence of points and lines of zero curvature on the high-frequency properties of metals arises, due not only to the increase in the number of effective electrons which was discussed above. The presence of the lines of inflection, separating the "hyperbolic" parts of the Fermi surface from the "elliptic" parts, can lead to changes in the topology of the effective belt on the surface (change of connectivity, collapse of the belt, etc.) for certain directions of propagation of the disturbance (Figure 1.6). For instance, it also causes anomalies in the angular dependence of the attenuation rate of the sound waves (see [80], [87]). It also results in modification of the spectrum of the geometrical oscillations of ultrasonic attenuation when the sound wave propagates in metal in the presence of an external magnetic field [80].

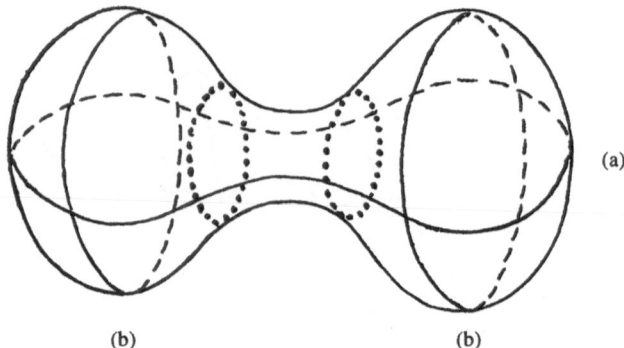

(a)

(b)   (b)

FIGURE 1.6. Geometry of the effective belt on the Fermi surface for various directions of propagation of the disturbance: (a) a simply connected effective belt and (b) a multiply connected effective belt. The dotted curves are lines of inflection on the Fermi surface.

Along with the points and lines of zero curvature there can exist points and lines where the curvature of the surface has singularities. A very simple illustration for such situations is a fracture line on a piecewise smooth surface. If the Fermi surface has fracture lines, then the effective belts can be missed for the appropriate directions of propagation of electromagnetic or sound waves, which causes anomalies in high-frequency properties of the metal. It is shown in [88], [89] that under these conditions electromagnetic waves of some special kind can propagate in metals.

The lines of singular (infinite) curvature can be arranged on edges of narrow lenses or needle-shaped sheets, which are elements of the Fermi surfaces of some metals. These lines, where the curvature has singularity, are not certainly the fracture lines. To be convinced, consider an axial-symmetric Fermi surface appropriate to the energy momentum relation

$$E(\mathbf{p}) = \frac{\mathbf{p}_\perp^2}{2m} + \tfrac{2}{3} p_m v_m |x|^{3/2}. \tag{1.15}$$

In this case, the velocity of electrons on the Fermi surface varies continuously and its longitudinal component on the central cross-section reduces to zero. However, the curvature of the surface near the central cross-section is

$$K(p_z) = \frac{3\pi m v_m}{4 S(0) p_m} \sqrt{\left|\frac{p_m}{p_z}\right|}. \tag{1.16}$$

When p_z tends to zero the curvature tends to infinity (see Figure 1.4). The absence of a jump in the longitudinal velocity of electrons on the line of "infinite curvature," which coincides with the central cross-section, means that the effective belt does not disappear. However, there is a considerable decrease in the number of the effective electrons when the disturbance propagates along the symmetry axis. It can influence some high-frequency properties of the metal, in particular, frequency dependence of the surface impedance.

Under conditions of a weak spatial dispersion (small $\mathbf{q}$), relation (1.5) is satisfied for electrons with the maximum value of the projection of their velocity on the direction of propagation of the perturbation. Therefore, the vicinities of limiting points of the Fermi surface and other segments, where the appropriate component of the velocity of electrons reaches its maxima, are "effective" segments of the surface. In this case, the local geometry of these effective segments can also essentially change the number of electrons strongly interacting with a perturbing field. The greatest increase in the number of effective electrons is achieved, when these effective segments are close in shape to axisymmetric paraboloids whose axes are parallel to the direction of propagation of the perturbing wave. Fermi surfaces of some metals include nearly paraboloidal pieces. For example, the paraboloidal model was repeatedly used to describe the electronic lens in cadmium (see [90]–[92]). When such nearly paraboloidal segments exist on the Fermi surface it can significantly influence some observables. In particular, the frequency range, where Fermi liquid cyclotron waves can propagate, is considerably extended which changes their contribution to the surface impedance of the metal.

Thus the local geometry of the Fermi surfaces can considerably influence the high-frequency phenomena in metals. Some manifestations of this influence were studied already and observed in experiments. However, the research in this field is still not complete. The following sections of the present work contain further development of this topic.

1.3   Semiclassical Dynamics of Electrons in a Magnetic Field and Magnetotransport in Metals

Consider an electron moving in an applied magnetic field $\mathbf{B}$. When quantum effects are negligible we can describe its motion with a Lorentz force equation

$$\frac{d\mathbf{p}}{dt} = \frac{e}{c}[\mathbf{v} \times \mathbf{B}].$$ (1.17)

It follows from this equation that electrons move in a plane normal to the magnetic field $\mathbf{B}$. At the same time, the field $\mathbf{B}$ does not change the energy of electrons

$$\frac{\partial E}{\partial t} = \frac{\partial E}{\partial \mathbf{p}}\frac{d\mathbf{p}}{dt} = \frac{e\mathbf{v}}{c}[\mathbf{v} \times \mathbf{B}] = 0.$$ (1.18)

It means that the electron orbit in quasi momentum space is a cross-section of the Fermi surface by the plane normal to the magnetic field (Figure 1.7). For a closed Fermi surface all these orbits are closed. When a Fermi surface is extending continuously throughout the quasi momentum space (in the extended zone scheme), some electron orbits are not closed curves but they extend through the reciprocal lattice.

From (1.18), by an elementary time integration,

$$\mathbf{p} = \frac{e}{c}[\mathbf{B} \times \mathbf{r}].$$ (1.19)

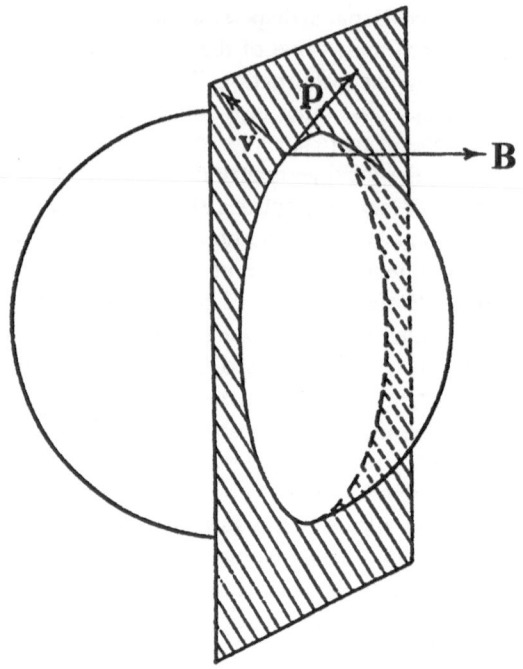

FIGURE 1.7. Orbit of an electron in a magnetic field.

Assume that the magnetic field **B** is directed along the "z"-axis of the chosen coordinate system. We can conclude from (1.16) that the orbit of an electron in the momentum space has the same shape as the projection of the electron path in the coordinate space on the plane perpendicular to **B** $((x, y)$-plane) except that the latter is rotated through a right angle about **B**. The size of the electron–cyclotron orbit in the coordinate space differs from the size of the corresponding orbit in **p** space by a factor of $c/|e|B$.

When the orbit in the quasi-momentum space is closed, the motion of the electron in the (x, y)-plane is periodic and the period T can be calculated as follows:

$$T = \frac{c}{|e|B} \oint \frac{d\lambda}{v_\perp}. \tag{1.20}$$

Here $\lambda = \sqrt{dp_x^2 + dp_y^2}$ is the element of length along the orbit in quasi-momentum space and $v_\perp = \sqrt{v_x^2 + v_y^2}$.

We can re-express the cyclotron period T in terms of a so-called cyclotron mass of electron. Consider a cross-section of the Fermi surface by a plane perpendicular to the magnetic field $p_z = $ const. The enclosed cross-sectional area is

$$S = \int\int dp_x \, dp_y = \int dE \oint \frac{d\lambda}{v_\perp} \tag{1.21}$$

which gives us a formula

$$\frac{\partial S}{\partial E} = \oint \frac{d\lambda}{v_\perp}.$$

The cyclotron mass $m_\perp$ is proportional to this derivative $\partial S/\partial E$:

$$m_\perp = \frac{1}{2\pi}\frac{\partial S}{\partial E} \equiv \frac{1}{2\pi}\oint \frac{d\lambda}{v_\perp}. \tag{1.22}$$

Comparing (1.22) with (1.20) we see that the cyclotron period T can be written as

$$T = \frac{2\pi c m_\perp}{|e|B}. \tag{1.23}$$

Correspondingly, the cyclotron frequency Ω equals $2\pi/T = |e|B/m_\perp c$. It is obvious that the cyclotron mass can only be defined for closed orbits. In the isotropic model the cyclotron mass at the Fermi surface coincides with the effective mass of a quasi particle m^*.

When the external magnetic field is strong enough we cannot neglect the effects arising due to the quantization of the electron motion in the presence of the magnetic field. We will analyze the quantum effects within the framework of a semiclassical quantization approach which is valid at large quantum numbers. For this limit we can use the Bohr phase-integral formula

$$\oint \mathbf{P} \cdot d\mathbf{r} = 2\pi\hbar(n + \gamma), \tag{1.24}$$

where n is an integer, γ is a phase correction which takes values from an interval (0, 1), $\mathbf{P} = \mathbf{p} - (e/c)\mathbf{A}$ is a canonical momentum, and $\mathbf{A}$ is the vector potential of the external magnetic field $\mathbf{B}$. To calculate $\mathbf{r}$, which describes the position of the electron on its path in real coordinate space, we can use (1.19). After straightforward calculation we arrive at the well-known formula

$$S(E, p_z) = \frac{2\pi\hbar|e|B}{c}(n + \gamma). \tag{1.25}$$

Within the framework of an isotropic model near the Fermi surface we have $E = p^2/2m^*$ and $S(E, p_z) = \pi(p^2 - p_z^2)$. Substituting this expression into (1.25) we arrive at the result

$$E = \frac{p_z^2}{2m^*} + \hbar\Omega(n + \gamma). \tag{1.26}$$

This formula gives the energies of the Landau levels of electrons in a strong magnetic field which are labeled with the quantum number "n". However, it does not take into account that each electron has a spin $\hbar s$ and intrinsic magnetic momentum $2\mu s$ where μ is the Bohr magneton. Accordingly, the expression for the quasi-particle energy (1.26) has to be modified as follows:

$$E = \frac{p_z^2}{2m^*} + \hbar\Omega(n + \gamma) + \sigma\mu B, \tag{1.27}$$

where σ is the spin quantum number.

Returning to (1.25) we can see that the difference ΔS between cross-sectional areas corresponding to adjacent Landau levels is

$$\Delta S = \frac{2\pi |e| B\hbar}{c}. \tag{1.28}$$

Multiplying this result by a factor $c^2/e^2 B^2$ we obtain the change in the area in real space. Then, multiplying by B, we get the corresponding change in magnetic flux in passing from the nth to the $(n + 1)$st Landau level

$$\Delta\Phi = \frac{2\pi\hbar c}{|e|} = \Phi_0. \tag{1.29}$$

This quantity Φ_0 is known as the magnetic flux quantum [93]. So a general principle is that the Landau levels have to be arranged in a such way that the difference between the magnetic flux through the associated orbits in real coordinate space is equal to an increasing integer number of magnetic flux quanta.

Conduction electrons in metals can be affected by external fields and by temperature gradients. They are also scattered from impurities, lattice waves, and so on. Assume that all applied fields and temperature gradients are slowly varying over a lattice distance, and changes in a quasi-particle energy caused by interactions with these fields are small compared to the Fermi energy. Under these assumptions we can analyze the transport properties of a system of conduction electrons within the framework of a semiclassical approach. To proceed we define an electron distribution function $n(\mathbf{p}, \mathbf{r}, t)$ in a phase space. The total number of electrons in any given region of the phase space can be obtained simply by integration over the corresponding momentum and coordinate ranges:

$$N = \frac{2}{(2\pi\hbar)^3} \int n(\mathbf{p}, \mathbf{r}, t)\, d^3r\, d^3p. \tag{1.30}$$

The total time rate of change in this distribution function is due to collisions among conduction electrons and due to electron scattering from the crystalline lattice. Therefore we can write

$$\frac{dn}{dt} = I[n], \tag{1.31}$$

where $I[n]$ is a collision integral which describes the scattering processes. The total derivative of the distribution function can be represented in the form

$$\frac{dn}{dt} = \frac{\partial n}{\partial t} + \frac{\partial n}{\partial \mathbf{r}}\frac{d\mathbf{r}}{dt} + \frac{\partial n}{\partial \mathbf{p}}\frac{d\mathbf{p}}{dt}. \tag{1.32}$$

Now we can note the meaning of some parameters included in this expression for the total derivative of the distribution function. The quantity $d\mathbf{r}/dt$ is the rate of change of position of a quasi particle moving along its trajectory. This is the quasi-particle velocity $\mathbf{v}$. The term $d\mathbf{p}/dt$ is equal to the applied force at that point and time. When both an electric field $\mathbf{E}$ and a magnetic field $\mathbf{B}$ are applied to the system of conduction electrons, we have

$$\frac{d\mathbf{p}}{dt} = \frac{e}{c}[\mathbf{v} \times \mathbf{B}] + e\mathbf{E}. \tag{1.33}$$

Combining the expressions (1.31)–(1.33) we arrive at the Boltzmann transport equation

$$\frac{\partial n}{\partial t} + \frac{\partial n}{\partial \mathbf{r}}\mathbf{v} + \frac{\partial n}{\partial \mathbf{p}}\left(\frac{e}{c}[\mathbf{v} \times \mathbf{B}] + e\mathbf{E}\right) = I[n]. \tag{1.34}$$

It is known that if the distribution function was the equilibrium distribution function there would be no change in this function due to scattering. Let us take the distribution function in the form

$$n(\mathbf{r}, \mathbf{p}, t) = f(E) + g(\mathbf{r}, \mathbf{p}, t). \tag{1.35}$$

Here $f(E)$ if the Fermi function (1.1) which corresponds to the equilibrium distribution of electrons, and the correction $g(\mathbf{r}, \mathbf{p}, t)$ describes deviations from this equilibrium distribution. These deviations are supposed to be small. It follows from (1.34) that we can write the transport equation for the nonequilibrium correction $g(\mathbf{r}, \mathbf{p}, t)$ in the form

$$\frac{\partial g}{\partial t} + \frac{\partial g}{\partial \mathbf{r}}\mathbf{v} + \frac{e}{c}\frac{\partial g}{\partial \mathbf{p}}[\mathbf{v} \times \mathbf{B}] + \frac{\partial f}{\partial E}e\mathbf{E}\mathbf{v} = I[g]. \tag{1.36}$$

Actually only this kinetic equation describing the flow of excited quasi particles close to the Fermi surface can be used to study their magnetotransport. Equation (1.34) which refers to the total distribution function of all quasi particles cannot be justified within the framework of the Landau theory of Fermi liquid because those quasi particles which are far away from the Fermi surface are not stable.

Let us consider the Fourier transform of the function $g(\mathbf{p}, \mathbf{r}, t)$ with respect to space and time. Within the assumption of linear response we may treat each Fourier component independently. Thus we can assume that the nonequilibrium correction to the distribution function can be written as

$$g(\mathbf{p}, \mathbf{r}, t) = g_{q\omega}(\mathbf{p}) \exp(i\mathbf{q} \cdot \mathbf{r} - i\omega t). \tag{1.37}$$

Here, $\mathbf{q}$ and ω are the wave vector and frequency of a plane wave perturbation produced by the electric field $\mathbf{E}(\mathbf{r}, t) = \mathbf{E}_{q\omega} \exp(i\mathbf{q} \cdot \mathbf{r} - i\omega t)$. Within the framework of the Fermi liquid theory possible values of q and ω are restricted with inequalities $\hbar q v_F \ll \zeta$, $\hbar\omega \ll \zeta$.

Using this expression (1.37) we can transform the kinetic equation (1.36) to the form

$$(i\mathbf{q} \cdot \mathbf{v} - i\omega)g_{q\omega}(\mathbf{p}) + \frac{e}{c}\frac{\partial g_{q\omega}}{\partial \mathbf{p}}[\mathbf{v} \times \mathbf{B}] + \frac{\partial f_p}{\partial E_p}e\mathbf{E}_{q\omega} \cdot \mathbf{v} = I[g]. \tag{1.38}$$

The simplest approximation for the collision integral is the so-called relaxation time approximation, $I[g] = -g/\tau$ where τ is the relaxation time. This approximation is based on the assumption that the nonequilibrium distribution function decays exponentially in time to the equilibrium form. It gives good results when we study high-frequency phenomena which are not strongly affected by electron scattering, but in the low-frequency range where the electron scattering has to be taken into account in detail, the relaxation time approximation can be used only to obtain some qualitative estimations.

It is convenient to convert the transport equation introducing new variables instead of p_x and p_y. These new variables are the energy E and the time t measured along a particular orbit [12]. The latter is

$$t = \frac{c}{|e|B} \int \frac{d\lambda}{v_\perp}. \tag{1.39}$$

Here t is a function of p_x and p_y defined by the Lorentz force equation (1.17). Since $dt = (c/|e|B)(d\lambda/v_\perp)$ we can perform a change in variables in integrals over quasi-momentum space

$$\int dp_x\, dp_y\, dp_z = \frac{|e|B}{c} \int dE\, dt\, dp_z. \tag{1.40}$$

Using these new variables we can rewrite the Boltzmann transport equation (1.38) in the relaxation time approximation as follows:

$$\frac{\partial g_{q\omega}}{\partial t} + (i\mathbf{q}\cdot\mathbf{v} - i\omega)g_{q\omega} + \frac{\partial f}{\partial E}e\mathbf{E}_{q\omega}\cdot\mathbf{v} = -\frac{g_{q\omega}}{\tau}. \tag{1.41}$$

The nonequilibrium correction g obviously has to be periodic in t when an orbit is closed. For an open orbit, g need not be periodic in t but anyway it has to be bounded everywhere. These boundary conditions have to be used to find a desired particular solution of the linearized transport equation (1.41).

The Boltzmann transport equation can be applied to calculate the electron current. We can compute it by adding the contributions from each occupied state. It is obvious that no current would be obtained with the equilibrium distribution function so we have to use the nonequilibrium correction g. As a result we have the following expression for the electric current density:

$$\mathbf{J}_{q\omega} = \frac{2e}{(2\pi\hbar)^3} \int d^3p\, \mathbf{v} g_{q\omega} \tag{1.42}$$

which gives us information concerning the magnetotransport properties of a metal.

Let us consider, for example, an isotropic metal and assume that a plane wave perturbation has a long wavelength ($ql \ll 1$, where l is a mean free path of electrons). Under this assumption we can rewrite the transport equation (1.41) in the form

$$\frac{\partial g_{q\omega}}{\partial t} + \left(\frac{1}{\tau} - i\omega\right) g_{q\omega} = -\frac{\partial f}{\partial E}e\mathbf{E}_{q\omega}\cdot\mathbf{v}. \tag{1.43}$$

This gives us

$$g_{q\omega} = -\frac{\partial f}{\partial E} \int_{-\infty}^{t} e\mathbf{E}_{q\omega}\mathbf{v}(t')\exp\left[(t' - t)\left(\frac{1}{\tau} - i\omega\right)\right] dt'. \tag{1.44}$$

Substituting this expression into formula (1.42) and using (1.40) we obtain

$$\mathbf{J}_{q\omega} \equiv \sigma \mathbf{E}_{q\omega} = \frac{2e^2 m^* \Omega}{(2\pi\hbar)^3} \int dp_z \int_0^T dt\, \mathbf{v}(t) \int_{-\infty}^t \mathbf{v}(t')\mathbf{E}_{q\omega} \exp\left[(t'-t)\left(\frac{1}{\tau} - i\omega\right)\right] dt'.$$

(1.45)

Here Ω is the cyclotron frequency of electrons and $T = 2\pi/\Omega$ is their cyclotron period. It follows from the Lorentz force equations (1.17) that $v_x(t) = v_\perp \cos\Omega t$ and $v_y(t) = v_\perp \sin\Omega t$. The component of the velocity along the direction of the magnetic field ($v_z = p_z/m^*$) does not depend on the variable t defined by (1.39). Using these expressions for the components of the velocity of electrons we arrive at the following expressions for the elements of the tensor of electron conductivity:

$$\sigma_{xx} = \sigma_{yy} = \tfrac{1}{2}\sigma_0 \left[\frac{1}{1 - i(\omega - \Omega)\tau} + \frac{1}{1 - i(\omega + \Omega)\tau}\right],$$

$$\sigma_{xy} = -\sigma_{yx} = -\frac{\sigma_0 \Omega\tau}{2(1 - i\omega\tau)}\left[\frac{1}{1 - i(\omega - \Omega)\tau} + \frac{1}{1 - i(\omega + \Omega)\tau}\right], \quad (1.46)$$

$$\sigma_{zz} = \frac{\sigma_0}{1 - i\omega\tau}, \qquad \sigma_{xz} = \sigma_{zx} = \sigma_{yz} = \sigma_{zy} = 0.$$

Here $\sigma_0 = Ne^2\tau/m^*$ is the Drude conductivity. In the absence of the external magnetic field all off-diagonal elements of the conductivity tensor are zero and its diagonal components are equal to $\sigma_0/(1 - i\omega\tau)$.

1.4   Skin Effect

In this section we consider the penetration of a high-frequency electromagnetic field into a metal in the absence of an external magnetic field. Let us assume that an isotropic metal occupies the half-space $z < 0$. The electromagnetic wave is supposed to be normally incident at the surface of the metal with the electric field $\mathbf{E}$ polarized along the x-axis. It follows from Maxwell's equations that

$$\left[\nabla \times [\nabla \times \mathbf{E}]\right] = -\frac{4\pi}{c^2}\frac{\partial \mathbf{J}}{\partial t}.$$

(1.47)

When we deal with an electromagnetic wave with a long wavelength ($ql \ll 1$), we can use the current–field relationship $\mathbf{J} = \sigma\mathbf{E}$ where $\sigma = \sigma_0/(1 - i\omega\tau)$. Substitution into (1.47) gives

$$\left[\nabla \times [\nabla \times \mathbf{E}]\right] = -\frac{4\pi\sigma}{c^2}\frac{\partial \mathbf{E}}{\partial t}.$$

(1.48)

We look for a solution to this equation which is proportional to $\exp(-iqz - i\omega t)$. Inserting this dependence into (1.48) we arrive at the result

$$q^2 = \frac{4\pi i\omega\sigma}{c^2}.$$

(1.49)

Correspondingly, we obtain

$$q \equiv q' + iq'' = \left(\frac{2\pi\omega\sigma}{c^2}\right)^{1/2}(1+i) \equiv \delta^{-1}(1+i). \qquad (1.50)$$

Substituting (1.50) into $\exp(-iqz - i\omega t)$ we get the result

$$E(z,t) = E(0)\exp\left[\frac{z}{\delta}(1-i) - i\omega t\right], \qquad z < 0. \qquad (1.51)$$

Thus the field decays in the bulk of the metal as $\exp(z/\delta)$, and it practically vanishes at distances of the order of δ from the surface. This phenomenon is called the "skin effect." The parameter δ is the penetration or skin depth. For good metals at frequencies $\omega \sim 10^{10}$–10^{11} s^{-1} the skin depth is of the order of 10^{-5} cm.

For pure metals at low temperatures the conductivity σ enhances, which results in a decrease of the skin depth δ. The skin depth can be further decreased as a result of the increase in the frequency ω. At high frequencies ($\omega\tau \gg 1$) the skin depth δ passes to the limit c/ω_p where $\omega_p = (4\pi Ne^2/m^*)^{1/2}$ is the so-called plasma frequency of the electrons. For good metals $\omega_p \sim 10^{16}$ s^{-1} which gives, for the skin depth, the estimate $\delta \sim 10^{-6}$ cm. On the other hand, the mean free path of electrons increases at low temperatures. It leads us to the conclusion that at sufficiently low temperatures and sufficiently high frequencies l will exceed δ. Under the condition $l > \delta$ the electric field significantly changes in magnitude over the distances of the order of the mean free path. In this case we cannot use our expression (1.50) for the skin depth because it is valid only in the local regime of propagation of the electromagnetic wave when the electric field is nearly uniform over the distances of the order of l. In the case when $l > \delta$ the skin effect is referred to as the "anomalous skin effect."

The systematic theory of the anomalous skin effect in metals was developed in the work of Reuter and Zondheimer [2]. However, we can estimate the skin depth of a strongly anomalous skin effect following the qualitative approach proposed by Pippard. The point of this approach is that when an electromagnetic field penetrates into a metal to a depth δ, which is small compared to the free path length, only the electrons moving nearly in parallel to the surface of the metal strongly interact with the field and absorb some energy. Other electrons, which move at larger angles to the surface, spend little time in the electric field and therefore they cannot receive a perceptive amount of energy from the field.

In the isotropic model of metal the number of electrons whose velocities are within a solid angle $d\Omega$ is proportional to the angle. Therefore we can estimate the number of electrons which remain near the surface over their whole mean free path as follows (see Figure 1.8):

$$\frac{\delta n}{n} \sim \frac{d\Omega}{4\pi} \sim \frac{\sin\theta d\theta}{4\pi} \sim \frac{\delta}{l}. \qquad (1.52)$$

According to Pippard we can replace the conductivity σ by the "effective" conductivity σ_{eff} which contains the extra factor δ/l. We can write this "effective"

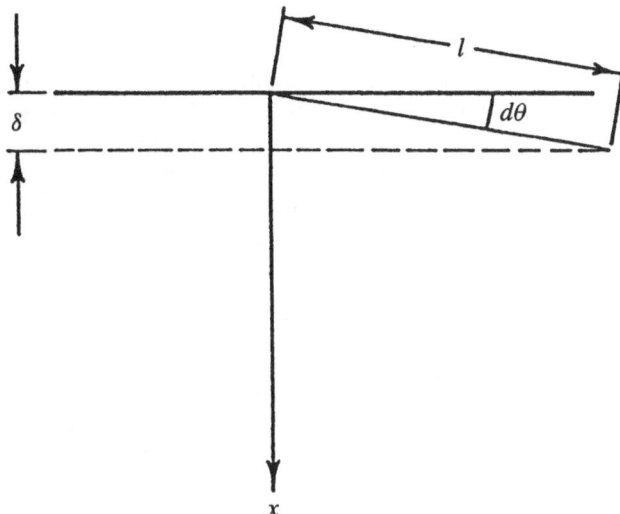

FIGURE 1.8. Effective electrons in an anomalous skin effect.

conductivity in the form (see [12]):

$$\sigma_{\text{eff}} = \frac{ia\sigma}{ql},$$ (1.53)

where a is a real dimensionless factor of the order of unity. Substituting this expression into (1.49) we arrive at the result

$$q = \left(\frac{4\pi\omega a\sigma}{c^2 l}\right)^{1/3}\left(\tfrac{1}{2} + i\frac{\sqrt{3}}{2}\right).$$ (1.54)

Skin depth δ is determined by the imaginary part of q. Therefore, the skin depth for the anomalous skin effect equals

$$\delta = \left(\frac{c^2 l}{4\pi\omega a\delta}\right)^{1/3}\frac{2}{\sqrt{3}}.$$ (1.55)

Comparison of expressions (1.50) and (1.55) shows that the frequency dependence of the skin depth changes in the nonlocal regime ($l > \delta$). Under the conditions of the normal skin effect (local regime, $l \ll \delta$) the skin depth is proportional to $\omega^{-1/2}$ whereas the skin depth of the anomalous skin effect is proportional to $\omega^{-1/3}$. In the order of magnitude the skin depth of the anomalous skin effect is larger than the skin depth of the normal skin effect for the same value of conductivity σ. The ratio of magnitudes is of the order of $\sqrt{l/\delta}$. Within the low-frequency region ($\omega\tau < 1$) the conductivity σ is close to the d.c. conductivity σ_0. Replacing σ by σ_0 in the expression (1.55) for the skin depth we obtain the approximation $\delta \sim (c^2 p_F/Ne^2\omega)^{1/3}$. Using the condition $l > \delta$ and this estimation we can find the frequency requirements necessary to observe the anomalous skin effect, $\omega > c^2 p_F/Ne^2 l^3$. For good metals we have $N \sim 10^{22}$ cm^{-3}, $p_F \sim 10^{-19}$ gm cm

s^{-1}, which give us $\omega > 10^{-2}l^{-3}$ s^{-1} cm^3. When the mean free path is of the order of 10^{-4} cm we find $\omega > 10^{10}$ s^{-1}.

To analyze the behavior of a metal placed in a high-frequency electromagnetic field one can introduce a new quantity—the so-called surface impedance Z. The surface impedance is a generalized characteristic of high-frequency properties of the metal. The elements of the tensor of the surface impedance for a metal filling the half-space $z < 0$ can be defined as follows:

$$Z_{\alpha\beta} = E_\alpha(0) \left/ \int_0^{-\infty} J_\beta(z)\,dz \right. , \tag{1.56}$$

where $\alpha, \beta = x, y$.

In the considered geometry when the electric field of the incident wave is directed along the "x"-axis (its magnetic field is directed, correspondingly, along the "y"-axis of the chosen coordinate system) the only nonzero element of the surface impedance is Z_{xx} which equals

$$Z_{xx} = E_x(0) \left/ \int_0^{-\infty} J_x(z)\,dz \right. .$$

It results from the relevant Maxwell's equations that

$$J_x(z) = -\frac{c}{4\pi}\frac{dB_y}{dz} \quad \text{and} \quad B_y(z) = \frac{c}{i\omega}\frac{dE_x}{dz}.$$

Therefore we can transform the expression for the surface impedance as follows:

$$Z_{xx} = -\frac{4\pi}{c}\frac{E_x(0)}{B_y(z)\big|_0^{-\infty}} = \frac{4\pi}{c}\frac{E_x(0)}{B_y(0)} = \frac{4\pi i\omega}{c^2}\frac{E_x(0)}{E_x'(0)}. \tag{1.57}$$

This quantity Z_{xx} is a complex quantity, $Z_{xx} = R - iX$. Both real and imaginary parts of the surface impedance can be determined in experiments from the change in amplitude and phase shift of an electromagnetic wave reflected from the surface of the metal.

To proceed with the calculation of the surface impedance we have to make assumptions about the character of the reflection of electrons from the metal boundary. It was shown in [39] that in the extreme anomalous limit electrons are reflected from the surface of the metal essentially specularly. This result has a simple physical interpretation. Suppose that an electron moves at a small angle to the metal surface ($\theta \sim \delta/l \ll 1$) and that the characteristic scale of the surface irregularities is of the order of the interatomic distance $d(d \sim \hbar/p_F)$. The time spent by the electron in the corresponding diffuse layer is $d/v_F\theta$ and the associated uncertainty in the momentum can be estimated as $\Delta p/p_F \sim \theta \sim \delta/l \ll 1$. Most of the effects discussed in this book can be displayed under the conditions of the strongly anomalous skin effect ($l \gg \delta$). We suppose, correspondingly, that the electrons undergo specular reflection and we calculate the surface impedance under this assumption.

We can symmetrically extend the electric field $\mathbf{E}(z)$ into the region $z > 0$ (Figure 1.9). This implies that the derivative $E_x'(z)$ has to undergo a jump from $E'(0)$ to $-E'(0)$ at the boundary of the metal $z = 0$. Correspondingly, the second derivative

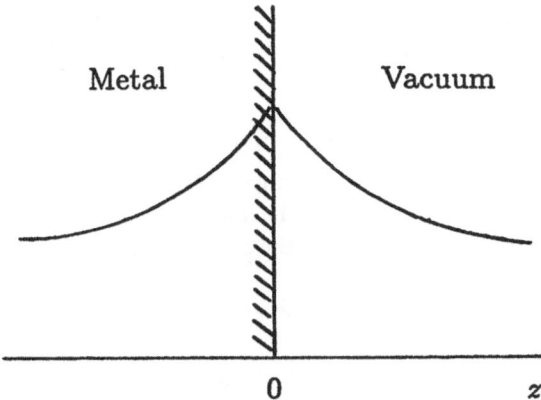

FIGURE 1.9. Under the assumption that electrons are specularly reflected at the boundary of the metal the electric field can be symmetrically extended in the half-space $z > 0$.

$d^2 E_x/dz^2$ contains a contribution proportional to the δ-function at $z = 0$. Taking this into account we can rewrite (1.47) as

$$\frac{d^2 E_x}{dz^2} - 2E_x'(0)\delta(z) = -\frac{4\pi i \omega}{c^2} J_x(z). \qquad (1.58)$$

Taking the Fourier transform of this equation in the space variable z we obtain

$$-q^2 E_q - 2E_x'(0) = -\frac{4\pi i \omega}{c^2}\sigma_{xx}(q, \omega). \qquad (1.59)$$

This gives us the following expression for the Fourier component of the electric field E_q:

$$E_q = 2E_x'(0) \left/ \left(\frac{4\pi i \omega}{c^2}\sigma_{xx}(q, \omega) - q^2 \right) \right. \qquad (1.60)$$

For the electric field itself we get

$$E_x(z) = \frac{1}{\pi} \int_0^\infty E_q \cos qz \, dq = \frac{2c^2}{\pi} E_x'(0) \int_0^\infty \frac{\cos qx \, dq}{4\pi i \omega \sigma_{xx}(q, \omega) - c^2 q^2}. \qquad (1.61)$$

Correspondingly,

$$\frac{E_x(0)}{E_x'(0)} = \frac{2c^2}{\pi} \int_0^\infty \frac{dq}{4\pi i \omega \sigma_{xx}(q, \omega) - c^2 q^2}. \qquad (1.62)$$

So, under specular reflection of the electrons from the boundary of the metal, we get the result

$$Z_{xx} = \frac{8i\omega}{\pi} \int_0^\infty \frac{dq}{4\pi i \omega \sigma_{xx}(q, \omega) - c^2 q^2}. \qquad (1.63)$$

We can easily calculate the surface impedance of the metal under the conditions of the normal skin effect ($l \ll \delta$). In this case, the conductivity σ_{xx} does not depend on q. In the low-frequency region, $\sigma_{xx} = \sigma_0$. Substituting this into the expression for the surface impedance (1.63) we arrive at the result

$$Z_{xx} = R - iX = \left(\frac{2\pi\omega}{\sigma_0 c^2}\right)^{1/2} (1 - i). \tag{1.64}$$

Thus, both the active part of the impedance R and its reactive part X are equal to each other and proportional to $\omega^{1/2}$. In the high-frequency region ($\omega\tau \gg 1$) $\sigma_{xx} \approx i\sigma_0/\omega\tau$. Using this approximation for the conductivity we obtain

$$Z_{xx} = -iX = -\frac{4i}{c}\frac{\omega}{\omega_p}, \tag{1.65}$$

here ω_p is the plasma frequency for the electrons.[1]

When the skin effect is of anomalous character we can estimate the surface impedance using the approximation (1.53) for the conductivity. As a result in the low-frequency region we get

$$Z_{xx} \equiv R - iX = \left(\frac{2}{a}\right)^{1/3} \left(\frac{\pi\omega}{c^2}\right)^{2/3} \left(\frac{l}{\sigma_0}\right)^{1/3} (1 - i\sqrt{3}). \tag{1.66}$$

We conclude that the impedance is proportional to $\omega^{2/3}$ and $X = \sqrt{3}R$.

1.5 Cyclotron Resonance

Now we proceed to a description of the high-frequency properties of metals in the presence of an external uniform magnetic field. A periodical motion of electrons in the magnetic field can cause a resonance with the electric field of an incident electromagnetic wave. The oscillating electric field penetrates into the skin layer nearly to the surface of the metal. When the cyclotron frequency Ω of an electron coincides with the frequency of the electromagnetic wave ω the electron is accelerated by the electric field every time it passes through the skin layer and, in this manner, it absorbs the electromagnetic energy of the wave. The phenomenon of the resonance absorption of the energy of the incident electromagnetic wave, whose frequency ω matches the cyclotron frequency Ω, is called cyclotron resonance. The same phenomenon occurs when ω is an integer multiple of Ω. The cyclotron resonance can exhibit itself in a strong magnetic field when $\Omega\tau > 1$ and an electron should make at least one circuit of its cyclotron orbit before being scattered.

To analyze the cyclotron resonance we consider the response of the system of electrons of a metal to a circularly polarized electromagnetic wave which travels

---

[1]It is necessary to remark here that the skin effect in metals exhibits itself when the frequency of the incident electromagnetic field ω is smaller than the plasma frequency ω_p. For $\omega > \omega_p$ the effect of suppression of the incident electromagnetic field inside the metal vanishes: the field penetrates into the metal practically without decreasing.

in parallel to the magnetic field. For simplicity, we neglect spatial variations of the electromagnetic field and we carry out the analysis within the framework of the isotropic model of a metal. It is convenient to go over to circular components of the electric current and electric field

$$J_\pm = J_x \pm i J_y, \qquad E_\pm = E_x \pm i E_y. \qquad (1.67)$$

For the isotropic metal the tensor of conductivity diagonalizes in circular components. So we get

$$J_\pm = \sigma_\pm E_\pm. \qquad (1.68)$$

Using expressions (1.46) for the components of the electron conductivity, we can write the circular components $\sigma_\pm = \sigma_{xx} \pm i\sigma_{yx}$ in the form

$$\sigma_\pm = \sigma_0(1 - i(\omega \mp \Omega)\tau)^{-1}. \qquad (1.69)$$

We can calculate the mean electric power absorbed in the unit volume (Q) by means of the Joule–Lenz law. Using a complex representation for the field and the current:

$$\mathbf{J} \to \tfrac{1}{2}[\mathbf{J}\exp(-i\omega t) + \mathbf{J}^*\exp(i\omega t)],$$
$$\mathbf{E} \to \tfrac{1}{2}[\mathbf{E}\exp(-i\omega t) + \mathbf{E}^*\exp(i\omega t)],$$

where $\mathbf{J}^*$ and $\mathbf{E}^*$ are the conjugates of $\mathbf{J}$ and $\mathbf{E}$, we arrive at the result

$$Q = \mathbf{JE} = \tfrac{1}{2}\,\mathrm{Re}\,(\mathbf{JE}^*). \qquad (1.70)$$

To obtain this result we neglected rapidly oscillating terms proportional to $\exp(\pm 2i\omega t)$, therefore expression (1.70) corresponds to the absorbed power averaged over a time interval that is larger than the period of the variation of the field. In terms of the circular components we have

$$Q = \tfrac{1}{2}\,\mathrm{Re}\,\{\sigma_+|E_+| + \sigma_-|E_-|\}. \qquad (1.71)$$

According to (1.69), the real parts of the quantities $\sigma_\pm$ take the form

$$\mathrm{Re}\,\sigma_\pm = \sigma_0[1 + (\omega \mp \Omega)^2\tau^2]^{-1}. \qquad (1.72)$$

In a strong magnetic field ($\Omega\tau > 1$) both the real part of the conductivity σ_+ and the absorbed power Q exhibit a pronounced maximum at $\omega = \Omega$. This maximum corresponds to the cyclotron resonance. The cyclotron resonance line has a width of $1/\tau$. The cyclotron resonance can also exhibit itself in the magnetic field or frequency dependencies of the reflected power and the surface impedance. It is more conveniently observed by varying the magnetic field with the frequency ω fixed.

In semiconductors the cyclotron resonance is observed for any direction of the magnetic field relative to the surface of the sample. For metals, however, there is a difficulty which impedes the observation of this phenomenon. The thing is that we can resolve the resonance line only at high frequencies, when $\omega\tau \gg 1$. Under these conditions we arrive at the region of the anomalous skin effect ($l \gg \delta$). On

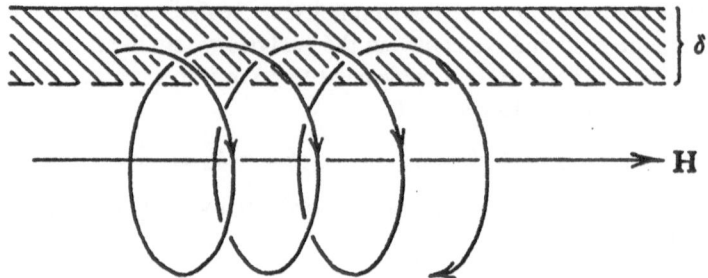

FIGURE 1.10. Azbel–Kaner cyclotron resonance.

the other hand, in the presence of a strong magnetic field $\Omega\tau \gg 1$, the maximum radius of the cyclotron orbit $R = v_F/\Omega$ is much smaller than the mean free path l ($R \ll l$). Comparison of the parameters R and δ for good metals leads us to the conclusion that in magnetic fields ~ 10 kG, which are typical for observations of the cyclotron resonance, the ratio R/δ is of the order of 10^2, i.e., we have the inequality $l \gg R \gg \delta$. That is, spiraling in the magnetic field, electrons rapidly leave the skin layer and cease the opportunity to absorb the energy of the incident electromagnetic wave.

There are two geometries which provide the display of the cyclotron resonance in metals. At first, strong resonance arises when the magnetic field is parallel to the surface of the metal (Figure 1.10) or is oriented so that the angle of inclination of the field is smaller than δ/τ. The theory of this effect was proposed by Azbel and Kaner [3]. The mechanism of this Azbel–Kaner resonance is similar to the mechanism of the acceleration of charged particles in a cyclotron. The electron spiralls around the axis parallel to the surface of a metal. At each revolution it returns to the skin layer and there it can gain energy from the electric field.

Another exception, basically, is a magnetic field oriented strictly perpendicular to the surface of the metal sample. In this geometry the resonating electrons can remain within the skin layer for a long time and absorb the energy. It appears, however, that for conventional metals a resonance feature in observables at $\omega = \Omega$ corresponding to this effect is smeared out until it is scarcely detectable. The reason is that in good metals the skin layer is very thin at high frequencies and the relative number of the electrons which move within the skin layer in parallel to the surface of the metal is too small to produce a distinguishable resonance feature at $\omega = \Omega$ in the surface impedance of the metal or in the absorbed power. This conclusion can be verified by the results of the calculation of the surface impedance of an isotropic metal carried out for the corresponding geometry when the external magnetic field is directed along the normal to the surface of the metal (see e.g., [111]).

Nevertheless, the resonance features at $\omega = \Omega$ were observed in the magnetic field dependencies of the surface impedance of several metals (potassium, cadmium, and zinc [47]–[49]) in the geometry where the magnetic field is directed perpendicularly to the metal surface. The theory of this effect is proposed in Chapter 4 of this book. The point of the analysis is that some special features of the local geometry of the Fermi surface may provide a significant enhancement of the relative number of electrons which can participate in the resonance absorption of the energy of the electromagnetic wave. It gives rise to the appearance of the resonance line in the real part of the surface impedance or its derivative with respect to the magnetic field. The results of the theoretical analysis are in agreement with the experiments of [47]–[49].

1.6   Ultrasonic Attenuation in Metals: Geometric Resonances

When an ultrasound wave propagates in a metal the crystalline lattice is periodically deformed. It gives rise to electric fields which influence the electrons. Besides, the periodical deformations of the lattice cause changes in the electronic spectrum. In the presence of the sound wave the energy of an electron gets an extra contribution ΔE of the form

$$\Delta E = \Lambda_{ij}(\mathbf{r}, \mathbf{p})u_{ij}(\mathbf{r}, t). \tag{1.73}$$

Here u_{ij} is the strain tensor

$$u_{ij} = \frac{1}{2}\left(\frac{\partial u_i}{\partial x_j} + \frac{\partial u_j}{\partial x_i}\right), \tag{1.74}$$

where $\mathbf{u}$ is the vector of the displacement of the lattice and the tensor Λ_{ij} is called the deformation potential. We can consider (1.73) as a definition of the deformation potential which was proposed in [96], [97] and used in numerous works concerning problems of the propagation of ultrasonic waves in metals.

The effect of the ultrasound wave upon an electron can be described by means of the force equation

$$\frac{d\mathbf{p}}{dt} = -\frac{\partial}{\partial \mathbf{r}}E + e\mathbf{E}, \tag{1.75}$$

where the first term arises due to the direct effect of deformation of the electron spectra in the presence of the sound and is proportional to the deformation potential Λ_{ij}. The second term originates from the interaction of the electron with the electric field accompanying the sound wave.

Interaction of the ultrasound waves with the conduction electrons of a metal results in the electron contribution to the ultrasonic attenuation and the velocity shift. Let us consider the ultrasonic attenuation in the presence of a moderately strong external magnetic field ($\Omega\tau \gg 1$, $qR \gg 1$). The second inequality means that the radii of cyclotron orbits of electrons are large compared to the wavelength of the sound. Suppose that the sound wave propagates perpendicularly to the magnetic field and consider the projection of the cyclotron path of electrons on the plane perpendicular to the magnetic field (Figure 1.11). Here dashed lines indicate planes of the equal phase of the electric field of the sound wave. When the sound frequency

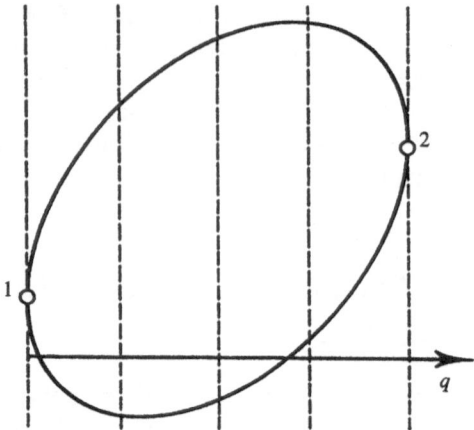

FIGURE 1.11. Geometry for a geometric resonance in the ultrasonic attenuation. The magnetic field is directed perpendicularly to the plane of the figure and to the direction of propagation of the sound wave. The dashed lines indicate planes of equal phase of the electric field of the wave.

ω is small compared to the cyclotron frequency Ω we can treat this electric field as a stationary field inhomogeneous in the real space. It is evident from Figure 1.11 that the electric field can be arranged so that it accelerates the electron in the same sense round the orbit at both stationary points of the orbit where the velocity vector $\mathbf{v}$ is directed along the planes of equal phase (points 1 and 2 in Figure 1.11). Under this condition we will observe a maximum in the ultrasonic attenuation. For a fixed wavelength of the sound the absorption will oscillate as the magnetic field is varied. This phenomenon is known as a geometric resonance. The theory of this effect was proposed by Pippard [15].

Consider an isotropic metal where the projection of a cyclotron path to the plane perpendicular to the magnetic field is a circle. Assume that the transverse ultrasonic wave propagates along the "x"-axis of the chosen coordinate system with a polarization vector along the "y"-axis. The magnetic field as usual is supposed to be directed along the "z"-axis. It follows from the Lorentz force equation (1.17) that for the electron moving along the cyclotron orbit of the maximum diameter, $v_y = v_F \sin \Omega t$ and $x = R \sin \Omega t$. The energy of the electric field of the sound wave $E_y \sin(qx)$ absorbed by the electron per one cycle equals

$$|e| \int_0^T \mathbf{E} \cdot \mathbf{v} \, dt = |e| E_y v_F \int_0^T \sin(qR \sin \Omega t) \sin \Omega t \, dt. \qquad (1.76)$$

To proceed we expand the factor $\sin(qR \sin \Omega t)$ in the integrand in Bessel functions

$$\sin(qR \sin \Omega t) = 2 \sum_{k=0}^{\infty} J_{2k+1}(qR) \sin[(2k+1)\Omega t]. \qquad (1.77)$$

Substitution of this expansion to (1.76) gives the result

$$|e| \int_0^T \mathbf{E} \cdot \mathbf{v} \, dt = 2\pi |e| E_y R J_1(qR). \tag{1.78}$$

The geometric resonances occur when the Bessel function $J_1(qR)$ gets its maxima. For a moderately strong magnetic field ($qR \gg 1$) we can replace $J_1(qR)$ by its asymptotic expression

$$J_1(qR) = \sqrt{\frac{2}{\pi qR}} \cos\left(qR - \frac{3\pi}{4}\right). \tag{1.79}$$

In this limit expression (1.79) describes the oscillations of the absorbed energy periodical in $1/B$. The geometric resonances are obseved in the attenuation and the velocity shift of the ultrasound waves (longitudinal as well as transverse) propagating perpendicularly to the external magnetic field. The magnitude of the geometric oscillations can be enhanced when the stationary points of the cyclotron orbit are points of zero curvature, because in this case an electron spends more time moving in parallel to the planes of equal phase and therefore it can absorb a larger amount of energy of the electric field accompanying the sound wave. The cyclotron trajectory of the electron in the real space is similar in shape to its orbit on the Fermi surface. Therefore the local geometry of the Fermi surface can strongly influence the magnitude of the geometric oscillations. This effect is analyzed in detail in Chapter 5 of the present book.

1.7 Fermi-Liquid Interaction between Quasi Particles

Any quasi particle of the electron liquid of metals interacts with other quasi particles. This Fermi-liquid interaction between quasi particles is weaker than the interaction among conduction electrons themselves, however it does play an important role in a real electron Fermi liquid. In the framework of Landau theory this Fermi-liquid interaction can be represented as a self-consistent field which arises due to the surrounding electrons which act upon a given electron. Because of this self-consistent field the quasi-particle energy depends on the state of other quasi particles of the system and can be represented as a functional of the distribution function of quasi-particles.

The energy of a "bare" quasi particle $E(\mathbf{p})$ has to be replaced by the renormalized energy $\tilde{E}_\sigma$ which is determined by the relation

$$\tilde{E}_\sigma(\mathbf{p}, \mathbf{r}, t) = E(\mathbf{p}) + \sum_{\mathbf{p}'\sigma'} F(\mathbf{p}, \sigma; \mathbf{p}', \sigma') g(\mathbf{p}', \mathbf{r}, t). \tag{1.80}$$

Here the function $F(\mathbf{p}, \sigma; \mathbf{p}', \sigma')$ is the Landau F-function or the kernel of the Fermi-liquid interaction. The F-function is a second variational derivative of the total energy of the system of quasi particles with respect to their distribution function. It can be defined in the vicinity of the Fermi surface and we assume it to be continuous when $\mathbf{p}$ or $\mathbf{p}'$ crosses the Fermi surface. Actually, we only need values of this F-function on the Fermi surface. Then it depends on quasi momenta $\mathbf{p}$ and $\mathbf{p}'$ and on the spin indices σ and σ'. It follows from the definition of the F-function that it is invariant under the permutation of $\mathbf{p}$, σ with $\mathbf{p}'$, σ'.

In the absence of the external magnetic field the system is invariant under time reversal and we have

$$F(\mathbf{p}, \sigma; \mathbf{p}', \sigma') = F(-\mathbf{p}, -\sigma; -\mathbf{p}', -\sigma'). \tag{1.81}$$

If, furthermore, the Fermi surface is invariant under reflection, $\mathbf{p} \to -\mathbf{p}$, (1.81) gives us

$$F(\mathbf{p}, \sigma; \mathbf{p}', \sigma') = F(\mathbf{p}, -\sigma; \mathbf{p}', -\sigma'). \tag{1.82}$$

It follows from expression (1.80), defining the renormalized electron energy, that this energy depends on $\mathbf{r}$. This dependence $\tilde{E}$ upon $\mathbf{r}$ arises due to an inhomogeneity of the nonequilibrium correction to the distribution function of quasi particles g in real space. As a result $\tilde{E}_\sigma$ may be considered as a local energy, perturbed by a distortion of the surrounding medium. The gradient of $\tilde{E}_\sigma$ in the coordinate space

$$\frac{\partial \tilde{E}}{\partial \mathbf{r}} = \frac{\partial}{\partial \mathbf{r}} \left(\sum_{\mathbf{p}', \sigma'} F(\mathbf{p}, \sigma; \mathbf{p}', \sigma') g(\mathbf{p}, \mathbf{r}, t) \right). \tag{1.83}$$

describes an averaged force exerted by the surrounding medium on the given electron. This new force equals $-\partial \tilde{E}/\partial \mathbf{r}$. It is equivalent to a kind of "diffusion" force, which tends to push the quasi particle toward regions of minimum energy. The "diffusion" force has to be included into the net force acting on an electron. Therefore expression (1.33) has to be replaced by the new expression

$$\frac{d\mathbf{p}}{dt} = \frac{e}{c}[\mathbf{v}_p| \times |\mathbf{B}] + e\mathbf{E} - \sum_{\mathbf{p}'\sigma'} F(\mathbf{p}, \sigma; \mathbf{p}', \sigma') \frac{\partial g}{\partial \mathbf{r}}. \tag{1.84}$$

Here $\mathbf{v_p} = \partial \tilde{E}_p / \partial \mathbf{p}$.

Substituting expression (1.84) into the transport equation of (1.36) we arrive at the following result:

$$\frac{\partial g}{\partial t} + \frac{\partial g^e}{\partial \mathbf{r}} \mathbf{v}_p + \frac{e}{c}[\mathbf{v}_p \times \mathbf{B}] \frac{\partial g^e}{\partial \mathbf{p}} + \frac{\partial f}{\partial E} e\mathbf{v}_p \mathbf{E} = I[g]. \tag{1.85}$$

Here the distribution function g, as before, describes a deviation from the ground equilibrium state of the system of quasi particles and the function g^e is defined by the relation

$$g^e(\mathbf{p}, \mathbf{r}, t) = g(\mathbf{p}, \mathbf{r}, t) - \frac{\partial f_\mathbf{p}}{\partial E_\mathbf{p}} \sum_{\mathbf{p}'\sigma'} F(\mathbf{p}, \sigma; \mathbf{p}', \sigma') g(\mathbf{p}', \mathbf{r}, t). \tag{1.86}$$

This function corresponds to a deviation from the "local equilibrium" state (see [29]). The latter is described with the Fermi distribution function (1.1) where the quasi-particle energy E is replaced by the "renormalized" energy $\tilde{E}$.

Let us again consider the simplest case of a particular plane wave perturbation, of a wave vector $\mathbf{q}$ and frequency ω and use a relaxation time approximation for the collision integral. Then we can rewrite the kinetic equation (1.85) in the form

$$\left(-i\omega + \frac{1}{\tau}\right)g_{q\omega} + i\mathbf{q}\cdot\mathbf{v}g_{q\omega}^e + \frac{e}{c}[\mathbf{v}\times\mathbf{B}]\frac{\partial g_{q\omega}^e}{\partial\mathbf{p}} + \frac{\partial f_\mathbf{p}}{\partial E_\mathbf{p}}e\mathbf{v}\mathbf{E}_{q\omega} = 0. \quad (1.87)$$

Replacing p_x and p_y by E and t where the variable t (time measured along the cyclotron orbit) is defined by relation (1.39), we can write the transport equation (1.87) in form similar to (1.41):

$$\left(-i\omega + \frac{1}{\tau}\right)g_{q\omega} + i\mathbf{q}\cdot\mathbf{v}_pg_{q\omega}^e + \frac{\partial g_{q\omega}^e}{\partial t} + \frac{\partial f_\mathbf{p}}{\partial E_\mathbf{p}}e\mathbf{v}\mathbf{E}_{q\omega} = 0. \quad (1.88)$$

Finally we can find the correct expression for the electrical current density. We will start from the kinetic equation (1.85). In the absence of external forces this equation has the form

$$\frac{\partial g}{\partial t} + \frac{\partial g^e}{\partial\mathbf{r}}\cdot\mathbf{v} = I[g]. \quad (1.89)$$

Let us sum (1.89) over $\mathbf{p}$ and σ. The contribution from the collision integral is zero, since collisions do not change the total number of particles. Therefore we obtain

$$\frac{\partial\rho}{\partial t} + \frac{\partial}{\partial\mathbf{r}}\sum_{p\sigma}g^e(\mathbf{p},\mathbf{r},t)\mathbf{v}_p = 0. \quad (1.90)$$

Here $\sum_{p\sigma}g(\mathbf{p},\mathbf{r},t)$ is the electron density. The total electron current density $\mathbf{J}$ is related to their material density ρ by the well-known conservation law

$$e\frac{\partial\rho}{\partial t} + \operatorname{div}\mathbf{J} = 0. \quad (1.91)$$

Comparing (1.90) and (1.91) we see that

$$\mathbf{J} = e\sum_{p\sigma}g^e(\mathbf{p},\mathbf{r},t)\mathbf{v}_p = e\sum_{p\sigma}g(\mathbf{p},\mathbf{r},t)\left(\mathbf{v}_p - \sum_{p'\sigma'}\frac{\partial f_{\mathbf{p'}}}{\partial E_{\mathbf{p'}}}F(\mathbf{p},\sigma;\mathbf{p}',\sigma')\mathbf{v}_{p'}\right). \quad (1.92)$$

On the other hand, we have

$$\mathbf{J} = e\sum_{p\sigma}g(\mathbf{p},\mathbf{r},t)\mathbf{v}_p^0. \quad (1.93)$$

where $\mathbf{v}_p^0 = \partial E/\partial\mathbf{p}$. Summarizing the obtained results we can write

$$\mathbf{J} = e\sum_{p\sigma}g(\mathbf{p},\mathbf{r},t)\mathbf{v}_p^0 \equiv e\sum_{p\sigma}g^e(\mathbf{p},\mathbf{r},t)\mathbf{v}_p, \quad (1.94)$$

where $\mathbf{v}_p^0$ and $\mathbf{v}_p$ are related as follows:

$$\mathbf{v}_p = \mathbf{v}_p^0 + \sum_{p'\sigma'}\frac{\partial f_{\mathbf{p'}}}{\partial E_{\mathbf{p'}}}F(\mathbf{p},\sigma;\mathbf{p}',\sigma')\mathbf{v}_p. \quad (1.95)$$

This relation gives us a connection between an unrenormalized quasi-particle velocity $\mathbf{v}_p^0$ and the renormalized velocity $\mathbf{v}_p$. The difference $\mathbf{v}_p - \mathbf{v}_p^0$ can be interpreted as a backflow current, arising from the interaction of a moving conduction electron with the surrounding electron liquid [29].

A similar analysis may be used to obtain the expressions for the momentum and energy currents in the electron liquid. Again we start from the kinetic equation (1.85) and arrive at formulas similar to (1.94). These formulas again define renormalized momentum and energy currents whose relation to the corresponding unrenormalized currents can be described by the equations similar to (1.95). The derivation of these formulas in more detail is given in the next chapter.

CHAPTER 2

Main Equations of the Theory of the Electron Fermi Liquid of Anisotropic Metal

2.1 Main Relations of the Quantum Theory of an Electronic Liquid of Metals

Originally Landau–Silin Fermi liquid theory described the electron characteristics of metals in the framework of a semiclassical approach. However, later the main equations of the theory were generalized to study quantum phenomena. The fundamentals of the quantum theory of the electron liquid are expounded in [27]–[29], [94], [95]. The detailed justification of the equations given below is contained in [27], [95]. Here we will give a brief summary.

The phenomenological theory of an electronic liquid is based on the assumption that weakly excited states of the electronic system of normal metal are of a single-particle character. It proposes the existence of a single-particle Hamiltonian H_0 with a set of quantum numbers ν and energies E_ν. The weakly excited states have energies close to the Fermi energy ζ, and are featured by a single-particle density matrix, which represents the sum of an equilibrium part $f(f_{\nu\nu'} = f_\nu \delta_{\nu\nu'})$, f_ν is the Fermi distribution function for the energy E_ν and the small nonequilibrium correction ρ. This density matrix ρ satisfies the equation

$$i\hbar \frac{\partial \rho}{\partial t} = [H, \rho]. \tag{2.1}$$

Here the Hamiltonian H includes the term H_0 and corrections describing the interactions of quasi particles with external disturbances (e.g., electromagnetic fields), their scattering on the defects of the crystalline lattice, and the interaction among the quasi particles. For a disturbance harmonic in time ($\sim \exp(-i\omega t)$) the matrix elements $\rho_{\nu\nu'}$ satisfy the kinetic equation

$$-i\omega\rho_{\nu\nu'} + \frac{1}{i\hbar}(E_{\nu'} - E_\nu)\rho_{\nu\nu'} + \frac{1}{i\hbar}(f_\nu - f_{\nu'})W_{\nu\nu'}^* = I_{\nu\nu'}[\rho], \tag{2.2}$$

where $I[\rho]$ is the quantum collision integral, and the operator W^* describes the interaction of an electron with the disturbance which causes the deviation from the equilibrium state.

Within the framework of the Fermi liquid theory W^* is a linear functional of ρ:

$$W^*_{\nu\nu'} = W_{\nu\nu'} + \sum_{\nu_1\nu_2} F^{\nu_1\nu_2}_{\nu\nu'} \rho_{\nu_1\nu_2},\tag{2.3}$$

where $W_{\nu\nu'}$ represents a matrix element of the interaction with the perturbing fields, and the integral term describes the interaction among the quasi particles. For disturbances whose frequencies are small compared to $\zeta/\hbar$ and whose wavelengths are larger than a wavelength of the electron on the Fermi surface, such as electromagnetic fields or ultrasound waves propagating in a metal, matrix elements of the Fermi-liquid kernel $F^{\nu_1\nu_2}_{\nu\nu'}$ represent parameters characterizing the ground state of the electron system.

To study the electromagnetic response of the system of electrons we can write this operator W as follows:

$$W_{\nu\nu'} \equiv W^e_{\nu\nu'} = \int d\mathbf{r} \left(en_{\nu\nu'}(\mathbf{r})\Phi(\mathbf{r}) - \frac{1}{c}\mathbf{j}_{\nu\nu'}(\mathbf{r})A(\mathbf{r}) \right).\tag{2.4}$$

Here $\Phi(\mathbf{r})$, $\mathbf{A}(\mathbf{r})$ are the scalar and vector potentials of the alternating self-consistent electromagnetic field and $en(\mathbf{r})$, $\mathbf{j}(\mathbf{r})$ are the operators of a charge and current density of the electrons. When the disturbance of the electronic system is caused by the displacement of an ionic lattice from its equilibrium state (e.g., as a result of the propagation of a sound wave), the form of the Hamiltonian W is determined by the character of the effect of alternating in time deformation of the lattice on the conduction electrons. First, the ion displacement described by a vector $\mathbf{u}(\mathbf{r})\exp(-i\omega t)$ generates a nonequilibrium density of charge $Q\,\mathrm{div}\,\mathbf{u}(\mathbf{r})$ (Q is an equilibrium density of the charge of the lattice) and an ionic current with the density $-i\omega Q\mathbf{u}(\mathbf{r})$, which together with the induced electron current create a self-consistent electromagnetic field accompanying the sound wave. Thus the operator W includes the term of the form (2.3), describing interaction with this field. Second, the deformation of the lattice leads to redistribution of the nonuniform part of the ionic charge at short (about interatomic) distances. It gives rise to the other part of the interaction of electrons with the oscillating lattice—so-called deformation interaction, which is of local character. The corresponding contribution to W is proportional to the strain tensor and has the form

$$W^d_{\nu\nu'} = \int d\mathbf{r}\Lambda^{\nu\nu'}_{\alpha\beta}(\mathbf{r})\frac{\partial u_\beta}{\partial r_\alpha},\tag{2.5}$$

where $\Lambda(\mathbf{r})$ is the tensor operator of a density of deformation potential whose classical analog was defined by formula (1.73).

Thus, for the disturbance of the electronic system of metals caused by deformations of the ionic lattice, the Hamiltonian W is the sum of terms (2.4) and (2.5):

$$W_{\nu\nu'} = W^e_{\nu\nu'} + W^d_{\nu\nu'}.\tag{2.6}$$

The kinetic equation (2.2) where the Hamiltonian W has the form (2.4), together with Maxwell's equations for the self-consistent electromagnetic field, gives a basis to study the electromagnetic response of an electronic liquid in metals. The self-consistent field is completely determined by the average current density of electrons

$$\mathbf{J}(\mathbf{r}) = -\frac{e^2 N}{m^* c} \mathbf{A}(\mathbf{r}) + \sum_{\nu\nu'} \rho_{\nu\nu'} \mathbf{j}_{\nu'\nu}(\mathbf{r}). \qquad (2.7)$$

Here $N = \text{Tr}\{f, n(\mathbf{r})\}$ is the electron density and m^* is the mass of a "free" electron.

In the case when the perturbed state of the electronic system is due to oscillations of the lattice, the kinetic equation (2.2) with the Hamiltonian W of the form (2.6) should be supplemented with Maxwell's equations where the current density is a sum of the eletronic and ionic contributions, and the equation describing the displacement of the lattice

$$- \omega^2 \rho_m \mathbf{u}(\mathbf{r}) = \mathbf{F}(\mathbf{r}). \qquad (2.8)$$

Here ρ_m is a mass density of the lattice and $\mathbf{F}(\mathbf{r})$ represents the amplitude of a density of force, originating due to the displacement of ions from equilibrium positions. The force $\mathbf{F}(\mathbf{r})$ contains a term corresponding to the contribution from the self-consistent electromagnetic field

$$\mathbf{F}^e(\mathbf{r}) = Q\mathbf{E}_\omega(\mathbf{r}) + \frac{Q}{c}[\mathbf{u}(\mathbf{r}) \times B]. \qquad (2.9)$$

Besides, the force contains a term arising due to the inhomogeneous strain of the system of ions

$$\mathbf{F}^d(\mathbf{r}) = \frac{\partial}{\partial r_\beta} \left(\lambda_{\alpha\beta\gamma\delta} \frac{\partial u_\gamma}{\partial r_\delta} + \overline{\Lambda_{\alpha\beta}(\mathbf{r})} \right). \qquad (2.10)$$

Here λ is a lattice tensor of elastic moduli averaged with the equilibrium density matrix of electrons and $\overline{\Lambda(\mathbf{r})}$ is the tensor of the deformation potential averaged with the nonequilibrium part of the electronic density matrix.

Finally, the force $\mathbf{F}(\mathbf{r})$ contains one more contribution which appears because electrons are scattered by defects of the crystalline lattice which move together with the lattice. The electrons produce an additional force acting upon the lattice due to the scattering by moving defects. The density of this force equals

$$\mathbf{F}^{dr}(\mathbf{r}) = -\sum_{\nu\nu'} I_{\nu\nu'}[\rho] \mathbf{P}_{\nu\nu'}(\mathbf{r}). \qquad (2.11)$$

The equation of motion of the lattice obeys the conservation law for the total momentum of the system of electrons and the lattice.[1]

The kinetic equation (2.2) can be transformed to a form more convenient for later calculations (especially in the low-frequency range) and for comparison with the

---

[1] Here, for simplicity, we do not consider the scattering of electrons by that kind of defect which breaks the translation symmetry and, consequently, gives breaking to this momentum conservation.

semiclassical kinetic equation. To transform the equation we introduce an operator of effective current density $\mathbf{j}^*(\mathbf{r})$ which determines the form of the correction ρ_A to the electronic density matrix in a weak static magnetic field $\mathbf{b}(\mathbf{r}) = \text{rot } \mathbf{A}(\mathbf{r})$:

$$\rho_{\nu\nu'}^A = -\frac{1}{c}\frac{f_\nu - f_{\nu'}}{E_\nu - E_{\nu'}}\int d\mathbf{r}\,\mathbf{j}_{\nu\nu'}^*(\mathbf{r})\mathbf{A}(\mathbf{r}). \qquad (2.12)$$

Here the matrix elements $\mathbf{j}_{\nu\nu'}^*(\mathbf{r})$ are defined as follows:

$$\mathbf{j}_{\nu\nu'}^*(\mathbf{r}) = \mathbf{j}_{\nu\nu'}(\mathbf{r}) + \sum_{\nu_1\nu_2}\frac{f_{\nu_1} - f_{\nu_2}}{E_{\nu_1} - E_{\nu_2}}F_{\nu\nu'}^{\nu_1\nu_2}\mathbf{j}_{\nu_1\nu_2}^*(\mathbf{r}). \qquad (2.13)$$

In the semiclassical limit this expression turns into (1.95), where $\mathbf{v}_p = (1/e)\mathbf{j}_p^*$ and $\mathbf{v}_p^0 = (1/e)\mathbf{j}_p$. Thus our definition (2.13) is a generalization of the relation between "bare" and the renormalized velocity of a quasi particle derived in the semiclassical limit. We can separate the contribution ρ_A from a nonequilibrium part of a density matrix

$$\rho_{\nu\nu'} = \rho_{\nu\nu'}^A + g_{\nu\nu'}. \qquad (2.14)$$

When the disturbance arises due to the external electromagnetic field, the matrix elements $g_{\nu\nu'}$ are defined by an equation of the form

$$\frac{1}{i\hbar}(\hbar\omega + E_{\nu'} - E_\nu)g_{\nu\nu'} \qquad (2.15)$$

$$+\frac{f_\nu - f_{\nu'}}{E_\nu - E_{\nu'}}\left\{\int d\mathbf{r}\,\mathbf{j}_{\nu\nu'}^*(\mathbf{r})\mathbf{E}_\omega(\mathbf{r}) + \frac{1}{i\hbar}(E_\nu - E_{\nu'})\sum_{\nu_1\nu_2}F_{\nu\nu'}^{\nu_1\nu_2}g_{\nu_1\nu_2}\right\} = I_{\nu\nu'}[g].$$

Here $\mathbf{E}_\omega(\mathbf{r}) = (i\omega/c)\mathbf{A}(\mathbf{r}) - \text{grad}\,\Phi(\mathbf{r})$ is the amplitude of the electric field. Equation (2.14) is a quantum analog of the semiclassical kinetic equation of [21], [96]:

$$-i\omega g(\mathbf{p}) + i q \mathbf{v}_p g^e(\mathbf{p}) + \frac{1}{c}[\mathbf{j}^* \times \mathbf{B}]\frac{\partial}{\partial\mathbf{p}}g^e(\mathbf{p}) + e\frac{\partial f_p}{\partial E_p}\mathbf{E}_{q\omega}\mathbf{v}_p = I[g],$$

where

$$g_{q\omega}^e(\mathbf{p}) = g_{q\omega}(\mathbf{p}) - \frac{\partial f_p}{\partial E_p}\frac{1}{(2\pi\hbar)^3}\sum_{\sigma'}\int d\mathbf{p}'\,F(\mathbf{p},\sigma;\mathbf{p}',\sigma')g(\mathbf{p}'),$$

$$\mathbf{j}^*(\mathbf{p}) = \int d\mathbf{r}\,\mathbf{j}_{pp}^*(\mathbf{r}). \qquad (2.16)$$

Naturally, this result agree as with the Boltzmann kinetic equation (1.87) which is derived within the framework of the semiclassical theory of the Fermi liquids.

We remark that (2.15) does not include matrix elements of any operators non-renormalized due to the Fermi-liquid interaction: the nonuniform part of the equation includes only the operator of the effective current density $\mathbf{j}^*(\mathbf{r})$. The kinetic equation can be similarly transformed, when the disturbance of the state of the electronic liquid is caused by the vibrations of the lattice. Introducing an operator of the effective density of the deformation potential $\Lambda^*(\mathbf{r})$, connected with $\Lambda(\mathbf{r})$, by the relation similar to (2.13):

$$\Lambda^*_{\nu\nu'}(\mathbf{r}) = \Lambda_{\nu\nu'}(\mathbf{r}) + \sum_{\nu_1\nu_2} \frac{f_{\nu_1} - f_{\nu_2}}{E_{\nu_1} - E_{\nu_2}} F^{\nu_1\nu_2}_{\nu\nu'} \Lambda^*_{\nu_1\nu_2}(\mathbf{r}), \qquad (2.17)$$

we can present $\rho_{\nu\nu'}$ in the form

$$\rho_{\nu\nu'} = -\frac{f_\nu - f_{\nu'}}{E_\nu - E_{\nu'}} \int d\mathbf{r} \left(\frac{1}{c} \mathbf{j}^*_{\nu\nu'}(\mathbf{r}) \mathbf{A}(\mathbf{r}) - \Lambda^{*\alpha\beta}_{\nu\nu'}(\mathbf{r}) \frac{\partial u_\beta}{\partial r_\alpha} \right) + \rho'_{\nu\nu'}. \qquad (2.18)$$

Substituting (2.18) into (2.2) we can see that the matrix elements $\rho'_{\nu\nu'}$ satisfy the equation

$$\frac{1}{i\hbar}(\hbar\omega + E_{\nu'} - E_\nu)\rho'_{\nu\nu'} + \frac{f_\nu - f_{\nu'}}{E_\nu - E_{\nu'}} \left\{ \int d\mathbf{r} \left(\mathbf{j}^*_{\nu\nu'}(\mathbf{r})\mathbf{E}_\omega(\mathbf{r}) - i\omega\Lambda^{*\alpha\beta}_{\nu\nu'}(\mathbf{r}) \frac{\partial u_\beta}{\partial r_\alpha} \right) \right.$$

$$\left. + \frac{1}{i\hbar} \sum_{\nu_1\nu_2} F^{\nu_1\nu_2}_{\nu\nu'} \rho'_{\nu_1\nu_2}(E_\nu - E_{\nu'}) \right\} = I_{\nu\nu'}[\rho']. \qquad (2.19)$$

This equation can be used as a starting point to develop a quantum Fermi liquid theory describing the interaction between the conduction electrons and ultrasonic waves.

The range of frequencies of ultrasonic waves can be divided into the intervals of low and high frequencies depending on a relation between ω and a characteristic collision frequency $1/\tau$. The low-frequency range is more typical for real experimental conditions. Therefore it appears to be necessary to transform the kinetic equation to the form which allows us to simplify it in the low-frequency range. At $\omega\tau < 1$ the state of the system of electrons is close to a local equilibrium, which is described by the locally equilibrium density matrix ρ^L:

$$\rho^L_{\nu\nu'} = -\frac{f_\nu - f_{\nu'}}{E_\nu - E_{\nu'}} \int d\mathbf{r} \left[\frac{1}{c} \mathbf{j}^*_{\nu\nu'}(\mathbf{r}) \mathbf{A}(\mathbf{r}) \right.$$

$$\left. - \Lambda^{*\alpha\beta}_{\nu\nu'}(\mathbf{r}) \frac{\partial u_\beta}{\partial r_\alpha} + n^*_{\nu\nu'}(\mathbf{r})\zeta'(\mathbf{r}) - i\omega\mathbf{P}^*_{\nu\nu'}(\mathbf{r})\mathbf{u}(\mathbf{r}) \right]. \qquad (2.20)$$

Here $n^*(\mathbf{r})$ and $\mathbf{P}^*(\mathbf{r})$ are the operators of the effective density of the electrons and their canonical momentum connected with the corresponding nonrenormalized "bare" operators $n(\mathbf{r})$ and $\mathbf{P}(\mathbf{r})$ by relations similar to (2.13), (2.17). The value $\zeta'(\mathbf{r})$ represents the nonequilibrium correction to the chemical potential of electrons in the state of the local equilibrium. This correction can be found from the condition of a local electroneutrality of the system. The last term in (2.20) occurs because the local equilibrium is realized in a coordinate system bound to the moving lattice.

Presenting the density matrix ρ as the sum of ρ^L and the correction g circumscribing deviations from the local equilibrium we can derive from (2.19) the following equation for g:

$$-i\omega g_{\nu\nu'} + \frac{1}{i\hbar}(E_{\nu'} - E_\nu)g^e_{\nu\nu'} + \frac{f_\nu - f_{\nu'}}{E_\nu - E_{\nu'}} \int d\mathbf{r}[\mathbf{j}^*_{\nu\nu'}(\mathbf{r})\mathbf{E}'_\omega(\mathbf{r})$$

$$+ i\omega n^*_{\nu\nu'}(\mathbf{r})\zeta'(\mathbf{r}) - i\omega\Lambda^{*\alpha\beta}_{\nu\nu'}(\mathbf{r})\partial u_\beta/\partial r_\alpha] = I_{\nu\nu'}[g]. \qquad (2.21)$$

Here g^e is connected with g by the relation

$$g^e_{\nu\nu'} = g_{\nu\nu'} - \frac{f_{\nu'} - f_\nu}{E_{\nu'} - E_\nu} \sum_{\nu_1\nu_2} F^{\nu_1\nu_2}_{\nu\nu'} g_{\nu_1\nu_2}, \tag{2.22}$$

$$\mathbf{E}'_\omega(\mathbf{r}) = \mathbf{E}_\omega(\mathbf{r}) - \frac{1}{e} \operatorname{grad} \zeta'(\mathbf{r}) - \frac{i\omega}{c}[\mathbf{u}(\mathbf{r}) \times \mathbf{B}], \tag{2.23}$$

$$\Lambda'^*(\mathbf{r}) = \Lambda^*(\mathbf{r}) - \Pi^*(\mathbf{r}). \tag{2.24}$$

The tensor operator $\Pi^*(\mathbf{r})$ is defined by means of an equation of motion of the operator of a momentum density

$$\frac{1}{i\hbar}(E_{\nu'} - E_\nu)\mathbf{P}_{\nu\nu'}(\mathbf{r}) = \frac{1}{c}[\mathbf{j}^*_{\nu\nu'}(\mathbf{r}) \times \mathbf{B}] - \frac{\partial}{\partial r_\beta}\Pi^{*\alpha\beta}_{\nu\nu'}(\mathbf{r}). \tag{2.25}$$

This operator $\Pi^*(\mathbf{r})$ has the meaning of the operator of the effective flux of the momentum of electrons and is connected with the nonrenormalized operator of the momentum flux $\Pi(\mathbf{r})$ as follows:

$$\Pi^*_{\nu\nu'}(\mathbf{r}) = \Pi_{\nu\nu'}(\mathbf{r}) + \sum_{\nu_1\nu_2} \frac{f_{\nu_1} - f_{\nu_2}}{E_{\nu_1} - E_{\nu_2}} F^{\nu_1\nu_2}_{\nu\nu'} \Pi^*_{\nu_1\nu_2}(\mathbf{r}). \tag{2.26}$$

In the low-frequency range we can neglect the term $-i\omega g_{\nu\nu'}$ in (2.21). Thus (2.21) becomes an equation in g^e ($I_{\nu\nu'}[g] = I_{\nu\nu'}[g^e]$), and it does not explicitly include Fermi–liquid terms. By virtue of a symmetry of the Fermi-liquid kernel $F^{\nu_1\nu_2}_{\nu\nu'} = F^{\nu'\nu}_{\nu_2\nu_1}$ we can write for any single-particle operator $D(\mathbf{r})$ the following equality, which is derived in [27]:

$$\sum_{\nu\nu'} g_{\nu\nu'}(\mathbf{r})D_{\nu'\nu}(\mathbf{r}) = \sum_{\nu\nu'} g^e_{\nu\nu'}(\mathbf{r})D^*_{\nu'\nu}(\mathbf{r}), \tag{2.27}$$

where $D^*(\mathbf{r})$ is the "bare" operator $D(\mathbf{r})$ renormalized due to the Fermi-liquid interaction. Hence the only effect of the electron–electron interaction in the low-frequency range is the renormalization of the electronic characteristics. This is a well-known result of the theory of a Fermi liquid.

Supposing $\mathbf{u}_\omega(\mathbf{r}) = \mathbf{u}_{q\omega} \exp(i\mathbf{qr})$, $\mathbf{E}_\omega(\mathbf{r}) = \mathbf{E}_{q\omega}\exp(i\mathbf{qr})$, and passing to Fourier trasforms in space variables, we obtain a simple expression for the Fourier component of the correction to a chemical potential of electrons ζ'_q and eliminate it from the kinetic equation. As a result we arrive at the equation

$$-i\omega g_{\nu\nu'} + \frac{1}{i\hbar}(E_{\nu'} - E_\nu)g^e_{\nu\nu'}$$

$$+ \frac{f_\nu - f_{\nu'}}{E_\nu - E_{\nu'}}\left[\tilde{\mathbf{j}}_{\nu\nu'}(-\mathbf{q})\mathbf{E}'_{q\omega} + i\omega\mathbf{G}_{\nu\nu'}(-\mathbf{q})\mathbf{u}_{q\omega}\right] = I_{\nu\nu'}[g], \tag{2.28}$$

where

$$G^\alpha_{\nu\nu'}(-\mathbf{q}) = -iq_\beta\tilde{\Lambda}^{\alpha\beta}_{\nu\nu'}(-\mathbf{q}) + \frac{1}{c}[\tilde{\mathbf{j}}_{\nu\nu'}(-\mathbf{q}) \times \mathbf{B}]_\alpha - i\omega\tilde{P}^\alpha_{\nu\nu'}(-\mathbf{q}),$$

$$\tilde{\Lambda}_{\nu\nu'}(-\mathbf{q}) = \tilde{\Lambda}'^*_{\nu\nu}(-\mathbf{q}) - n^*_{\nu\nu'}(-\mathbf{q})\frac{(n, \Lambda'^*)}{(n, n^*)},$$

$$\tilde{\mathbf{j}}_{\nu\nu'}(-\mathbf{q}) = \mathbf{j}^*_{\nu\nu}(-\mathbf{q}) - n^*_{\nu\nu'}(-\mathbf{q})\frac{(n, \mathbf{j}^*)}{(n, n^*)}.$$

The quantity $\tilde{\mathbf{P}}$ is defined similarly to $\tilde{\mathbf{j}}$, $\mathbf{j}^*_{\nu\nu'}(-\mathbf{q})$, $n^*_{\nu\nu'}(-\mathbf{q})$, $\Lambda'^*_{\nu\nu'}(-\mathbf{q})$, $\mathbf{P}^*_{\nu\nu'}(-\mathbf{q})$ are the Fourier-transforms of matrix elements of the appropriate operators; the notation (C, D) for arbitrary single-particle operators C and D is defined as follows:

$$(C, D) = -\sum_{\nu\nu'}\frac{f_\nu - f_{\nu'}}{E_\nu - E_{\nu'}}C_{\nu\nu'}(\mathbf{q})D_{\nu'\nu}(\mathbf{q}). \tag{2.29}$$

In semiclassical limit (2.28) coincides with the kinetic equation used in [28], where the deformation potential was defined as a semiclassical analog of a tensor Λ'. Neglecting the Fermi-liquid correlation we see that in the semiclassical limit (2.28) passes into the kinetic equation formulated earlier (see [98], [99]).

According to the structure of a nonequilibrium part of the electronic density matrix we can separate from the force $\mathbf{F}(\mathbf{r})$, which is included in the equation of motion of the lattice (2.7), a dynamical part $F^d(\mathbf{r})$. This dynamical part of the force determines the rate of absorption of the energy of ultrasonic waves by the electron system. Its Fourier component equals

$$\mathbf{F}^d_q = \sum_{\nu\nu'} g^e_{\nu\nu'}\mathbf{G}_{\nu'\nu}(\mathbf{q}). \tag{2.30}$$

The remaining part of the force includes the contribution from a "bare" elasticity of a lattice, and the term which originates from the averaging over the local equilibrium. In the low-frequency range the Fourier component of this force equals

$$F^L_{q\alpha} = -q_\beta q_\delta \lambda_{\alpha\beta\gamma\delta}u^\gamma_{q\omega} + q_\beta q_\delta(\tilde{\Lambda}_{\gamma\delta}\Lambda'_{\alpha\beta})$$
$$+ \frac{1}{c}[\mathbf{J}^L \times \mathbf{B}]_\alpha + iq_\beta\frac{1}{c}(\tilde{j}_\gamma, \Lambda'_{\alpha\beta})A'_{q\gamma} + \frac{1}{\hbar}[P_\gamma(-\mathbf{q}), \Pi_{\alpha\beta}(\mathbf{q})]q_\beta u^\gamma_q. \tag{2.31}$$

Here $\mathbf{J}^L_q$ represents the sum of the locally equilibrium contribution to the electron current and the lattice current, $-i\omega Q\mathbf{u}_{q\omega}$, $\mathbf{A}'_q = \mathbf{A}_q - [\mathbf{u}_{q\omega} \times \mathbf{B}]$; the last terms, as well as the first, are averaged over the equilibrium state of the electron liquid.

To apply the Fermi-liquid theory to treat concrete problems we have to define a form of the Fermi-liquid kernel $F^{\nu_1\nu_2}_{\nu\nu'}$ more specifically. Neglecting small contributions from a spin-orbit coupling we can write the matrix elements of the Fermi-liquid kernel as follows:

$$F^{\nu_1\nu_2}_{\nu\nu'} = \varphi^{\alpha_1\alpha_2}_{\alpha\alpha'}\delta_{\sigma\sigma'}\delta_{\sigma_1\sigma_2} + 4(\mathbf{s}_{\sigma\sigma'}\mathbf{s}_{\sigma_1\sigma_2})\psi^{\alpha_1\alpha_2}_{\alpha\alpha'}. \tag{2.32}$$

Here α is the set of orbital quantum numbers, σ is the spin number, and $\mathbf{s}$ is the operator of the electron spin.

In the semiclassical limit the set of orbital quantum numbers includes a band index n and quasi momentum $\mathbf{p}$. Therefore, ignoring interband transitions, we can assume that the kernel of the Fermi-liquid interaction depends only on quasi momenta and spins of interacting quasi particles:

$$F(\mathbf{p}, \sigma; \mathbf{p}', \sigma')_{\sigma\sigma'} = \varphi(\mathbf{p}, \mathbf{p}')I + 4(\mathbf{s}, \mathbf{s}')\psi(\mathbf{p}, \mathbf{p}'). \tag{2.33}$$

For an isotropic metal the functions $\varphi(\mathbf{p}, \mathbf{p}')$ and $\psi(\mathbf{p}, \mathbf{p}')$ depend only on the angle between the vectors $\mathbf{p}$ and $\mathbf{p}'$ laying on the Fermi sphere and we can expand them in series in spherical harmonics:

$$\varphi(\mathbf{p}, \mathbf{p}') = \frac{4\pi^3 \hbar^3}{p_F^2} v_F \sum_{j=0} \sum_{|m| \leq j} A_j Y_{jm}(\theta, \Phi) Y_{j-m}(\theta', \Phi'),$$

$$\psi(\mathbf{p}, \mathbf{p}') = \frac{4\pi^3 \hbar^3}{p_F^2} v_F \sum_{j=0} \sum_{|m| \leq j} B_j Y_{jm}(\theta, \Phi) Y_{j-m}(\theta', \Phi'),$$

$$(2.34)$$

where

$$Y_{jm} = \sqrt{\frac{(2j+1)}{4\pi} \frac{(j - |m|)!}{(j + |m|)!}} P_j^{|m|}(\cos\theta) \exp(im\Phi).$$

Here $P_j^{|m|}(x)$ are the associated Legendre polynomials; θ, Φ are the angles determining a position of the vector $\mathbf{p}$ on the Fermi surface in a spherical coordinate system; and p_F, v_F are the Fermi momentum and Fermi velocity of electrons.

The expansions (2.34) enable us to analyze Fermi-liquid effects systematically and completely in the framework of the isotropic model of metal. As a rule real metals have complicated in shape anisotropic Fermi surfaces which prevent the application of these expansions in concrete calculations. Therefore in studying the effects caused by the geometry of Fermi surfaces it is impossible to apply the expansions (2.34) even to obtain estimates of a qualitative character.

The simplest approach to the treatment of the Fermi-liquid effects in real metals is to suppose that the functions $\varphi(\mathbf{p}, \mathbf{p}')$ and $\psi(\mathbf{p}, \mathbf{p}')$ can be replaced by constants. This corresponds to the assumption that the radius of the the correlation interaction between electrons in the coordinate space is negligible. Using such approximation we can satisfactorily describe some effects but, nevertheless, it is too simplified to analyze adequately manifestations of the Fermi-liquid interaction in most of the concrete cases. Therefore it is necessary to develop more detailed approximations of the Fermi-liquid kernel for anisotropic Fermi liquids.

There exist two ways to take into account the anisotropy of the Fermi surface when we define the concrete form of the Fermi-liquid functions. The first is to make a coordinate transformation in the momentum space which allows us to transform the Fermi surface into a sphere. After such a transformation we can apply the usual expansions of the Fermi-liquid functions φ and ψ (depending on new variables) in spherical harmonics. This method was used in [100] to study the propagation of electromagnetic waves in metals with an ellipsoidal Fermi surface.

The second way is to define the character of dependence of the Fermi-liquid functions $\varphi(\mathbf{p}, \mathbf{p}')$ and $\varphi(\mathbf{p}, \mathbf{p}')$ on the quasi momenta $\mathbf{p}$, $\mathbf{p}'$ starting from a character of symmetry of the Fermi surface. Such approximation for $\varphi(\mathbf{p}, \mathbf{p}')$ and $\psi(\mathbf{p}, \mathbf{p}')$ for metals with a cubic symmetry of the crystalline lattice was used in [101]–[103].

Each of the above mentioned methods has both strong and feeble aspects, which appear to be essential in solving concrete problems. The advantage of the first way, based on conversion of the Fermi surface into a sphere, is that it allows us to analyze

Fermi-liquid effects in the electromagnetic response of electrons for an arbitrary direction of an external magnetic field relative to symmetry axes of the crystalline lattice. However, this method can be applied exclusively for Fermi surfaces which can be converted to spheres. To make such transformations possible the Fermi surface should satisfy rather severe restrictions.

The second way is also applicable to a broader class of Fermi surfaces, but detailed approximations for the Fermi-liquid functions are valid only when the external magnetic field is directed along a symmetry axis of a high order. To be solved, problems considered in the present work need, in approximations, both the first and second types. Accordingly, several alternatives of such approximations are offered below.

2.2 Approximation of the Fermi-Liquid Kernel for a Metal with a Simply Connected and Everywhere Convex Fermi Surface

Let us assume that the Fermi surface of a metal is everywhere convex and has a piecewise smooth surface, bounding a singly connected domain of a finite diameter in the quasi-momenta space. Under these assumptions we can carry out mapping of the Fermi surface on a sphere of the unit radius. An arbitrary point of the initial surface of the radius vector $\mathbf{p}$ passes into a point of the sphere of radius vector $\mathbf{n}_v (\mathbf{n}_v = \mathbf{v}_p/|\mathbf{v}_p|)$ if there is a one-to-one correspondence between points of the initial surface and its spherical mapping.

In "space of velocities" to which we transfer our Fermi surface, the Fermi–liquid functions $\varphi(\mathbf{v}, \mathbf{v}')$, $\psi(\mathbf{v}, \mathbf{v}')$ depend only on the angle between the vectors $\mathbf{v}$ and $\mathbf{v}'$. Their angular dependence can be presented as the expansions in spherical harmonics of the form (2.34), where p_F, v_F are maxima in magnitude of the quasi momentum and velocity of quasi particles on the Fermi surface. The angles θ, Φ now determine a position of a vector $\mathbf{n}_v$ on the spherical mapping of the Fermi surface.

The kinetic semiclassical equation (1.48) can be transformed to the form

$$\chi_p^e = \int_{-\infty}^{t} \left[e\mathbf{E}_{q\omega}\mathbf{v}(t') + \left(i\omega - \frac{1}{\tau}\right) \sum_{p''} \varphi(\mathbf{p}', \mathbf{p}'') \frac{\partial f_{p''}}{\partial E_{p''}} \chi_{p''} \right]$$

$$\times \exp \int_{t}^{t'} (-i\omega + 1/\tau + i\mathbf{q}\mathbf{v})\, dt''dt'. \tag{2.35}$$

Here $g_p = \chi_p\, \partial f_p/\partial E_p$; $g_p^e = \chi_p^e\, \partial f_p/\partial E_p$; the collision integral describes scattering with a relaxation time τ; the integration over "t" is performed over the time of the electron motion along the cyclotron orbit.

Introducing the averages

$$I_{jm} = \int_0^{2\pi} d\Phi \int_0^{\pi} \frac{\sin\theta d\theta}{w(\theta, \Phi)\kappa(\theta, \Phi)} Y_{j-m}(\theta, \Phi)\chi(\theta, \Phi), \tag{2.36}$$

where $w(\theta, \Phi) = |\mathbf{v}|/v_F, \kappa(\theta, \Phi) = p_F^2|K(\theta, \Phi)|$, $K(\theta, \Phi)$ is the Gaussian curvature of the Fermi surface, we can transform (2.35) to a set of equations

$$I_{jm}^e = W_{jm}^{\alpha} E_{q\omega}^{\alpha} - \sum_{j'=0} \sum_{|m'|\leq j'} A_{j'} g_{jj'}^{mm'} I_{j'm'}. \tag{2.37}$$

Here

$$g_{jj'}^{mm'} = \frac{eB(i\omega - 1/\tau)}{c} \frac{p_F^2}{v_F} \int_0^T dt \int dp_z$$

$$\times \int_{-\infty}^t \exp\left[\int_t^{t'} (-i\omega + 1/\tau + i\mathbf{q}\mathbf{v}) dt''\right] Y_{j-m}(\theta, \Phi) Y_{j'm'}(\theta', \Phi') dt',$$

$$\mathbf{W}_{jm} = \frac{ie^2 p_F^2 B}{v_F c} \int_0^T dt \int dp_z \int_{-\infty}^t \exp\left[\int_t^{t'} (-i\omega + 1/\tau + i\mathbf{q}\mathbf{v}) dt''\right]$$

$$\times \mathbf{v}(t') Y_{j-m}(\theta', \Phi') dt', \tag{2.38}$$

T is a period of the electron motion along the cyclotron orbit, averages I_{jm}^e are defined by relations (2.36), where χ is replaced by χ^e. These new averages are connected with I_{jm} by the relations

$$I_{jm}^e = I_{jm} + \sum_{j'=0} \sum_{|m'|\leq j'} A_{j'} I_{j'm'} \Gamma_{jj'}^{mm'}, \tag{2.39}$$

$$\Gamma_{jj'}^{mm'} = \int_0^{2\pi} d\Phi \int_0^{\pi} \frac{\sin\theta \, d\theta}{w(\theta, \Phi)\kappa(\theta, \Phi)} Y_{j-m}(\theta, \Phi) Y_{j'm'}(\theta, \Phi). \tag{2.40}$$

The expressions for $g_{jj'}^{mm'}$ and $\mathbf{W}_{jm}$ can be presented in more compact form when the integrands in (2.38) are expanded in Fourier series in variable $\psi(\psi = \Omega t; \Omega$ is the cyclotron frequency):

$$g_{jj'}^{mm'} = \left(\omega + \frac{i}{\tau}\right) \frac{p_F^2}{v_F} \sum_l \int dp_z m_\perp(p_z) \frac{Q_{jm}^l(\mathbf{q}) Q_{j'm'}^l(-\mathbf{q})}{l\Omega + \mathbf{q}\langle\mathbf{v}\rangle - \omega - i/\tau},$$

$$\mathbf{W}_{jm} = i\frac{ep_F^2}{v_F} \int dp_z m_\perp(p_z) \sum_l \frac{\mathbf{v}_l(\mathbf{q}) Q_{jm}^l(-\mathbf{q})}{l\Omega + \mathbf{q}\langle\mathbf{v}\rangle - \omega - i/\tau}. \tag{2.41}$$

Here $m_\perp(p_z)$ is the cyclotron mass and $\langle\mathbf{v}\rangle$ is the average of the velocity of electrons over the cyclotron period,

$$\mathbf{v}_l(-\mathbf{q}) = \frac{\Omega}{2\pi} \int_0^T dt\mathbf{v}(t) \exp\left[i\mathbf{q}\int_0^T (\mathbf{v}(t') - \langle\mathbf{v}\rangle) dt' - il\Omega t\right],$$

$$Q_{jm}^l = \frac{\Omega}{2\pi} \int_0^T dt Y_{jm}(\theta, \Phi) \exp\left[i\mathbf{q}\int_0^T (\mathbf{v}(t') - \langle\mathbf{v}\rangle) dt' - il\Omega t\right].$$

Using definition (2.36) we can rewrite the kinetic equation (2.35) as follows:

$$\chi_p + \sum_{j'=0} \sum_{|m'|\leq j'} A_{j'} Y_{j'm'}(\theta, \Phi) I_{j'm'}$$

$$= \exp\left[i\mathbf{q}\int_0^T (\mathbf{v}(t') - \langle\mathbf{v}\rangle)\,dt'\right]\sum_l \frac{\exp(il\Omega t)}{l\Omega + \mathbf{q}\langle\mathbf{v}\rangle - \omega - i/\tau}$$

$$\times \left\{-ie\mathbf{E}_{q\omega}v_l(t) + \omega\sum_{j'=0}\sum_{|m'|\le j'} A_{j'}I_{j'm'}Q^l_{j'm'}(t)\right\}. \tag{2.42}$$

This gives us the following expression for the current density:

$$J^\alpha_{q\omega} = \sigma^0_{\alpha\beta}E^\beta_{q\omega} + \frac{2i\omega}{(2\pi\hbar)^3}\frac{p_F^2}{v_F}\sum_{j'=0}\sum_{|m'|\le j'} A_{j'}Y_{j'm'}W^\alpha_{j'm'}. \tag{2.43}$$

Here the conductivity of the electron gas is

$$\sigma^0_{\alpha\beta} = \frac{2ie^2}{(2\pi\hbar)^3}\int m_\perp(p_z)\sum_l \frac{v_l^\alpha(q)v_l^\beta(-\mathbf{q})}{l\Omega + \mathbf{q}\langle\mathbf{v}\rangle - \omega - i/\tau}\,dp_z. \tag{2.44}$$

Expression (2.43) combined with the equations for the averages (2.37), (2.39) and Maxwell's equations makes a basis for the analysis of the high-frequency properties of metals, when the Fermi surface belongs to the considered class.

For the spherical Fermi surface the obtained results can be simplified. Following the usual method we can assume that the vector $\mathbf{q}$ is placed at the "yz"-plane of the coordinate system. Expanding the function $\exp[icq_y p_\perp/|e|B]$, which is included in the expressions for the Fourier components $v_e(-\mathbf{q})$ and $Q^l_{jm}(-\mathbf{q})$ is series in Bessel functions, we arrive at the formulas

$$Q^l_{jm}(-\mathbf{q}) = \sqrt{\frac{(2j+1)}{4\pi}\frac{(j-|m|!)}{(j+|m|!)}}P_j^{|m|}\left(\frac{p_z}{p_F}\right)J_{l-m}\left(\frac{cq_y p_\perp}{|e|B}\right),$$

$$v_l^x(-\mathbf{q}) = \frac{ip_\perp}{2m}\left[J_{l+1}\left(\frac{cq_y p_\perp}{|e|B}\right) - J_{l-1}\left(\frac{cq_y p_\perp}{|e|B}\right)\right],$$

$$v_l^y(-\mathbf{q}) = \frac{p_\perp}{2m}\left[J_{l+1}\left(\frac{cq_y p_\perp}{|e|B}\right) + J_{l-1}\left(\frac{cq_y p_\perp}{|e|B}\right)\right], \tag{2.45}$$

$$v_l^z(-\mathbf{q}) = \frac{p_z}{m}J_l\left(\frac{cq_y p_\perp}{|e|B}\right).$$

Here $p_\perp$ is the projection of a quasi momentum to a plane $p_z = 0$, $J_l(x)$ are the Bessel functions and $P_j^{|m|}(x)$ are the associated Legendre polynomials.

In this case, the quantities $g_{jj'}^{mm'}$ and $\mathbf{W}_{jm}$, defined by relations (2.41), are correspondingly equal:

$$g_{jj'}^{mm'} = \frac{(\omega + i/\tau)}{4\pi}\frac{mp_F^2}{v_F}\frac{(2j+1)(j-|m|)!}{(j+|m|)!}$$

$$\times \sum_l\int dp_z \frac{P_j^{|m|}(p_z/p_F)P_{j'}^{|m'|}(p_z/p_F)J_{l-m}(cq_y p_\perp/|e|B)J_{l-m'}(cq_y p_\perp/|e|B)}{l\Omega + q_z v_z - \omega - i/\tau}, \tag{2.46}$$

$$\mathbf{W}_{jm} = \frac{ie^2 p_F^2 m}{v_F} \int dp_z \sum_l \frac{(2j+1)(j-|m|)!}{(j+|m|)!}$$

$$\times \frac{v_l^\alpha(\mathbf{q}) P_j^{|m|}(p_z/p_F) J_{l-m}(cq_y p_\perp/|e|B)}{l\Omega + q_z v_z - \omega - i/\tau}. \tag{2.47}$$

Using the orthogonality of spherical harmonics we obtain

$$\Gamma_{jj'}^{mm'} = \delta_{jj'}\delta_{mm'}. \tag{2.48}$$

Starting from (2.45)–(2.48) we can transform the set of equations (2.37), (2.39) to the form

$$\tilde{I}_{jm} + \sum_{j'=0} \sum_{|m'|\leq j'} c_{j'm'} A_{j'} \tilde{I}_{j'm'} \beta_{jj'}^{mm'} = \mu_{jm}^\alpha E_{q\omega}^\alpha, \tag{2.49}$$

Where

$$c_{jm} = \frac{(2j+1)(j-|m|)!}{(j+|m|)!},$$

$$\beta_{jj'}^{mm'} = \frac{1}{2} \sum_l \int_{-1}^1 dx \frac{l\Omega + q_z v_z}{l\Omega + q_z v_z - \omega - i/\tau} P_j^{|m|}(x) P_{j'}^{|m'|}(x)$$

$$\times J_{l-m}\left(\frac{cq_y p_F\sqrt{1-x^2}}{|e|B}\right) J_{l-m'}\left(\frac{cq_y p_F\sqrt{1-x^2}}{|e|B}\right),$$

$$\mu_{jm}^\alpha = -\frac{iemp_F}{2\pi^2\hbar^3} \sum_l \int_{-1}^1 dx \frac{v_l^\alpha(\mathbf{q},x) P_j^m(x) J_{l-m}(cq_y p_F\sqrt{1-x^2}/|e|B)}{l\Omega + q_z v_z - \omega - i/\tau}. \tag{2.50}$$

The averages $\tilde{I}_{jm}$ are expressed in terms of the averages I_{jm} by the relations

$$\tilde{I}_{jm} = \frac{1}{\sqrt{c_{jm}}} \frac{mp_F}{2\pi^2\hbar^3} I_{jm}. \tag{2.51}$$

System (2.49) completely coincides with the appropriate result contained in the review article [27]. It can form the basis for the analysis of the electromagnetic response of an isotropic electron liquid of the metal in the framework of the semiclassical approach.

2.3 Approximation of the Fermi-Liquid Functions in Metals with a Cubic Symmetry of a Crystal Lattice

Let us consider the approximation of the Fermi-liquid functions based on the symmetry properties of the Fermi surface. Apparently, the functions $\varphi(\mathbf{p}, \mathbf{p}')$ and $\psi(\mathbf{p}, \mathbf{p}')$ should not vary when the vectors $\mathbf{p}$ and $\mathbf{p}'$ are replaced by the vectors $\tilde{\mathbf{p}}$ and $\tilde{\mathbf{p}}'$ connected to $\mathbf{p}$ and $\mathbf{p}'$ with any symmetry element included in a point group of crystalline structure of the considered metal. Therefore we can present the

Fermi-liquid functions in a form of expansions in basis functions of the irreducible representations of the corresponding symmetry group

$$\varphi(\mathbf{p}, \mathbf{p}') = \sum_{s=1}^{\infty} \sum_{j=1}^{d} \sum_{m=1}^{d_j} \varphi_{js}(p, p') R_{jm}^s(\theta, \Phi) R_{jm}^{*s}(\theta', \Phi'),$$

$$\psi(\mathbf{p}, \mathbf{p}') = \sum_{s=1}^{\infty} \sum_{j=1}^{d} \sum_{m=1}^{d_j} \psi_{js}(p, p') R_{jm}^s(\theta, \Phi) R_{jm}^{*s}(\theta', \Phi').$$

(2.52)

Here d is the order of the point group, index j labels irreducible representations of the group, d_j is the dimension of the jth irreducible representation, and $\{R_{jm}^s(\theta, \Phi)\}$ is the basis of the jth irreducible representation including the d_j functions. The index s labels sets of basis functions belonging to the same representation. Thus, $R_{jm}^s(\theta, \Phi)$ is the mth function belonging to the sth basis of the jth irreducible representation of the point symmetry group.

The functions $R_{jm}^s(\theta, \Phi)$ can be presented as linear combinations of the spherical harmonics $Y_{jm}(\theta, \Phi)$ transforming in the required way by symmetry operations (see [104]). Each set at a given s includes the harmonics with an identical value of the index l. We have to take into account an assemblage of the sets of functions, transforming in a similar way under the symmetry operations included in the point symmetry group, because these sets include finite numbers of elements. Therefore taking into consideration only one basis for each irreducible representation, we can write the Fermi-liquid functions $\varphi(\mathbf{p}, \mathbf{p}')$ and $\psi(\mathbf{p}, \mathbf{p}')$ as polynomials depending on the coordinates of the vectors $\mathbf{p}$ and $\mathbf{p}'$. To obtain the expansions of $\varphi(\mathbf{p}, \mathbf{p}')$ and $\psi(\mathbf{p}, \mathbf{p}')$ in series, it is necessary to take into account a set of bases for each irreducible representation of the symmetry group. For each representation the sets with larger values of the index "s" include spherical harmonics with larger values of the index l.

For isotropic metal the point group is a complete rotation group and it includes an infinite number of elements. In this case, we have no necessity to take into account a lot of sets of basis functions belonging to the same representation to expand the Fermi-liquid functions in series. Because of the infinite number of irreducible representations of the rotation group, the expansions of $\varphi(\mathbf{p}, \mathbf{p}')$ and $\psi(\mathbf{p}, \mathbf{p}')$ automatically contain an infinite number of terms. The orthonormalized basic set of the complete rotation group consists of spherical harmonics, so the expansions (2.52) in this case coincide with the well-known expansions (2.34) which are used in the theory of an isotropic electronic liquid.

For metals whose crystalline lattice is of bcc or fcc type the point group coincides with the complete cubic group. The orthonormalized basis functions for this group (cubic harmonics), composed from the spherical harmonics for $l < 4$ (see [104], [105]), are presented in Table 2.1. Including in this expansion (2.52) only these harmonics, we can write the functions $\varphi(\mathbf{p}, \mathbf{p}')$ and $\psi(\mathbf{p}, \mathbf{p}')$ as polynomials of the fourth degree of the coordinates of vectors $\mathbf{p}$ and $\mathbf{p}'$. We obtain a simpler approximation, taking into account only those basic functions in expansion (2.52), which include spherical harmonics with $l \leq 2$ (it is equivalent to keeping only the three first Fermi-liquid coefficients A_j or B_j in the expansions for $\varphi(\mathbf{p}, \mathbf{p}')$ and $\psi(\mathbf{p}, \mathbf{p}')$ for an isotropic metal). The approximation is

$$
\begin{pmatrix} \varphi(\mathbf{p}, \mathbf{p}') \\ \psi(\mathbf{p}, \mathbf{p}') \end{pmatrix} = \begin{pmatrix} \varphi_0 \\ \psi_0 \end{pmatrix} + \begin{pmatrix} \varphi_1 \\ \psi_1 \end{pmatrix} (p_x p_x' + p_y p_y' + p_z p_z')
$$

$$
+ \begin{pmatrix} \varphi_{21} \\ \psi_{21} \end{pmatrix} (p_z p_z' p_x p_x' + p_z p_z' p_y p_y' + p_x p_x' p_y p_y')
$$

$$
+ \begin{pmatrix} \varphi_{22} \\ \psi_{22} \end{pmatrix} (p_x^2 - p_y^2)(p_x'^2 - p_y'^2)
$$

$$
+ \frac{1}{3} \begin{pmatrix} \varphi_{22} \\ \psi_{22} \end{pmatrix} (2p_z^2 - p_x^2 - p_y^2)(2p_z'^2 - p_x'^2 - p_y'^2). \quad (2.53)
$$

Representation	s	l	Cubic harmonics
Γ_1	1	0	1
	2	4	$\sqrt{\frac{7}{12}}Y_{4,0} + \sqrt{\frac{5}{24}}(Y_{4,1} + Y_{4;-1})$
Γ_4	1	1	$Y_{1,0}; \frac{i}{\sqrt{2}}(Y_{1,-1} + Y_{1,1}); \frac{1}{\sqrt{2}}(Y_{1,-1} - Y_{1,1})$
	2	3	$Y_{3,0}; \sqrt{\frac{3}{8}}Y_{3,1} + \sqrt{\frac{5}{8}}Y_{3,-3}; \sqrt{\frac{3}{8}}Y_{3;-1} + \sqrt{\frac{5}{8}}Y_{3,3}$
Γ_5	1	2	$\frac{1}{2\sqrt{2}}(Y_{2,-1} - Y_{2,1}); \frac{i}{2\sqrt{2}}(Y_{2,-1} + Y_{2,1}); -\frac{i}{\sqrt{2}}(Y_{2,-2} - Y_{2,2})$
	2	4	$\frac{1}{\sqrt{2}}(Y_{4,2} - Y_{4,-2}); \frac{1}{2\sqrt{2}}(Y_{4,-1} - \sqrt{7}Y_{4,3}); -\frac{1}{2\sqrt{2}}(Y_{4,1} - \sqrt{7}Y_{4,-3})$
Γ_3	1	2	$Y_{2,0}; \frac{1}{\sqrt{2}}(Y_{2,2} + Y_{2,-2})$
	2	4	$\frac{1}{\sqrt{2}}(Y_{4,2} + Y_{4,-2}); \sqrt{\frac{5}{12}}Y_{4,0} - \sqrt{\frac{7}{24}}(Y_{4,4} + Y_{4,-4})$
Γ_2	1	3	$\frac{1}{\sqrt{2}}(Y_{3,2} + Y_{3,-2})$

We use a coordinate system whose axes x, y, z are directed along symmetry axes of the fourth order.

The coefficients φ_0, ψ_0, φ_1, ψ_1, φ_{21}, ψ_{21}, φ_{22}, ψ_{22} depend on p and p' on the Fermi surface. For a singly connected Fermi surface we can assume as a first approximation that these coefficients are constants. For a multiply connected Fermi surface such an approximation is unjustified. In this case it is necessary to take into account that the coefficients will accept different values for various sheets. Let us suppose that an electron with the quasi momentum $\mathbf{p}$ concerns the ith sheet of the Fermi surface and an electron with the quasi momentum $\mathbf{p}'$ concerns its kth sheet. Under this condition we can write the functions $\varphi(\mathbf{p}, \mathbf{p}')$ and $\psi(\mathbf{p}, \mathbf{p}')$ as follows:

$$
\begin{pmatrix} \varphi(\mathbf{p}, \mathbf{p}') \\ \psi(\mathbf{p}, \mathbf{p}') \end{pmatrix} = \begin{pmatrix} \varphi_0^{ik} \\ \psi_0^{ik} \end{pmatrix} + \begin{pmatrix} \varphi_1^{ik} \\ \psi_1^{ik} \end{pmatrix} (p_x p_x' + p_y p_y' + p_z p_z')
$$

$$
+ \begin{pmatrix} \varphi_{21}^{ik} \\ \psi_{21}^{ik} \end{pmatrix} (p_z p_z' p_x p_x' + p_z p_z' p_y p_y' + p_x p_x' p_y p_y')
$$

$$
+ \begin{pmatrix} \varphi_{22}^{ik} \\ \psi_{22}^{ik} \end{pmatrix} (p_x^2 - p_y^2)(p_x'^2 - p_y'^2)
$$

$$
+ \frac{1}{3} \begin{pmatrix} \varphi_{22}^{ik} \\ \psi_{22}^{ik} \end{pmatrix} (2p_z^2 - p_x^2 - p_y^2)(2p_z'^2 - p_x'^2 - p_y'^2). \quad (2.54)
$$

Assume that the Fermi surface has a hexagonal symmetry and that the axis z is directed along a symmetry axis of the sixth order. Keeping in the expansion (2.52) only those basis functions which are composed from spherical harmonics with $l \leq 2$, we will have the following expression instead of (2.54):

$$
\begin{pmatrix} \varphi(\mathbf{p}, \mathbf{p}') \\ \psi(\mathbf{p}, \mathbf{p}') \end{pmatrix} = \begin{pmatrix} \varphi_0^{ik} \\ \psi_0^{ik} \end{pmatrix} + \begin{pmatrix} \varphi_{01}^{ik} \\ \psi_{01}^{ik} \end{pmatrix} p_z p_z' + \begin{pmatrix} \varphi_{10}^{ik} \\ \psi_{10}^{ik} \end{pmatrix} (p_x p_x' + p_y p_y')
$$

$$
+ \begin{pmatrix} \varphi_{11}^{ik} \\ \psi_{11}^{ik} \end{pmatrix} p_z p_z' (p_x p_x' + p_y p_y') + \begin{pmatrix} \varphi_{21}^{ik} \\ \psi_{21}^{ik} \end{pmatrix} p_x p_x' p_y p_y'
$$

$$
+ \begin{pmatrix} \varphi_{22}^{ik} \\ \psi_{22}^{ik} \end{pmatrix} (p_x^2 - p_y^2)(p_x'^2 - p_y'^2)
$$

$$
+ \begin{pmatrix} \varphi_{23}^{ik} \\ \psi_{23}^{ik} \end{pmatrix} (2p_z^2 - p_x^2 - p_y^2)(2p_z'^2 - p_x'^2 - p_y'^2). \quad (2.55)
$$

These representations (2.55) for the Fermi-liquid functions can also be used for an axisymmetric Fermi surface. The coefficients φ^{ik} and ψ^{ik} with a fixed set of upper indices can be arranged as a square matrix, whose dimension coincides with the number of sheets of the Fermi surface. This matrix should be symmetric ($\varphi^{ik} = \varphi^{ki}$). Besides the matrix elements describing the Fermi-liquid interaction of quasi particles belonging to the same sheet will be the matrix elements for all sheets which pass each other as a result of symmetry operations. Also this coincidence should take place for matrix elements describing the interaction between electrons of two different couples of sheets, if the appropriate couples of sheets pass each other under a symmetry operation.

Moreover, we can assume that the coefficients φ^{ik} accept the same values when corresponding ith and kth sheets belong to that part of the Fermi surface which transforms to itself under any symmetry operator. The number of such parts of the Fermi surface is significantly less than the total number of sheets and each of them has a complete symmetry of the Fermi surface.

The nonequilibrium component $g_s(\mathbf{p})$ of the distribution function for the quasi particles of a sort "s" (i.e., electrons or holes associated with the sth part of the Fermi surface which transforms to itself under any symmetry operator) can be determined from the equation

$$g_s^e(\mathbf{p}) = \int_{-\infty}^{t} dt' \left\{ e_s E_\alpha v_s^\alpha(t' + i\omega - 1/\tau_s) \right.$$

$$\left. \times \sum_{p''s} \varphi(\mathbf{p}_s', \mathbf{p}_s'')g_{s'}(\mathbf{p}'')\exp\left[\int_t^{t'}(-i\omega+1/\tau_s+i\mathbf{q}\mathbf{v}_s(t''))dt''\right]\right\}. \quad (2.56)$$

Here the functions $g_s^e(\mathbf{p})$ are connected with $g_s(\mathbf{p})$ by the relations

$$g_s^e(\mathbf{p}) = g_s(\mathbf{p}) - \frac{\partial f_s(\mathbf{p})}{\partial E_s(\mathbf{p})} \sum_{p's'} \varphi(\mathbf{p}_s, \mathbf{p}_s')g_{s'}(\mathbf{p}'). \quad (2.57)$$

Introducing the quantities $\pi_s^{\alpha\beta} = p_s^\alpha p_s'^\beta$, where p_s^α are coordinates of the quasi momentum $\mathbf{p}_s$, we can define following averages:

$$I_{0s} = \sum_p g_s(\mathbf{p}),$$

$$I_{1s}^\alpha = \sum_p g_s(\mathbf{p})p_s^\alpha, \quad (2.58)$$

$$I_{2s}^{\alpha\beta} = \sum_p g_s(\mathbf{p})\pi_s^{\alpha\beta}.$$

Starting from the kinetic equation (2.56) we arrive at the system of linear equations for the averages (2.58), which gives the response of an electronic liquid to electromagnetic disturbances. Let us consider, for example, transverse oscillations of the electron liquid of a bcc or fcc metal arising due to an electromagnetic disturbance. Assume that the disturbance propagates along a symmetry axis of the fourth order parallel to the external magnetic field ($\mathbf{q}\|\mathbf{B}$). Choosing a coordinate system whose axes are directed along symmetry axes of the fourth order we can obtain the following system of equations for circular components of averages defined by the relations (2.58):

$$I_{1s}^\pm + \left[1 - i(\omega + i/\tau_s)G_{0s}^\pm\right]\sum_{s'} A_1^{ss'} I_{1s'}^\pm - i(\omega + i/\tau_s)G_{1s}^\pm \sum_{s'} A_2^{ss'} I_{2s'}^\pm$$

$$= ie_s W_{0s}^\pm E_{q\omega}^\pm, \quad (2.59)$$

$$I_{2s}^\pm + \left[1 - i(\omega + i/\tau_s)G_{2s}^\pm\right]\sum_{s'} A_2^{ss'} I_{2s'}^\pm - i(\omega + i/\tau_s)\tilde{G}_{1s}^\pm \sum_{s'} A_1^{ss'} I_{1s'}^\pm$$

$$= ie_s W_{1s}^\pm E_{q\omega}^\pm.$$

Here

$$I_{1s}^{\pm} = \sum_p g_s(\mathbf{p})(p_s^x \pm i p_s^y) \equiv \sum_p g_s(\mathbf{p}) p_s^{\pm},$$

$$I_{2s}^{\pm} = \sum_p g_s(\mathbf{p})(\pi_s^{xz} \pm i \pi_s^{yz}) \equiv \sum_p g_s(\mathbf{p}) \pi_s^{\pm}, \qquad (2.60)$$

$$E_{q\omega}^{\pm} = E_{q\omega}^x \pm i E_{q\omega}^y,$$

and

$$G_{0s}^{\pm} = \frac{1}{Q_{0s}} \sum_l \int dp_z \frac{p_s^{\pm}(\mathbf{q}) p_s^x(-\mathbf{q}) m_{\perp s}(p_z)}{l\Omega_s + \mathbf{q}\langle\mathbf{v}\rangle_s - \omega - i/\tau_s},$$

$$G_{1s}^{\pm} = \frac{1}{Q_{2s}} \sum_l \int dp_z \frac{p_s^{\pm}(\mathbf{q}) \pi_s^x(-\mathbf{q}) m_{\perp s}(p_z)}{l\Omega_s + \mathbf{q}\langle\mathbf{v}\rangle_s - \omega - i/\tau_s},$$

$$\tilde{G}_{1s}^{\pm} = \frac{Q_{2s}}{Q_{1s}} G_{1s}^{\pm},$$

$$G_{2s}^{\pm} = \frac{1}{Q_{2s}} \sum_l \int dp_z \frac{\pi_s^{\pm}(\mathbf{q}) \pi_s^x(-\mathbf{q}) m_{\perp s}(p_z)}{l\Omega_s + \mathbf{q}\langle\mathbf{v}\rangle_s - \omega - i/\tau_s},$$

$$W_{0s}^{\pm} = \frac{1}{4\pi^2\hbar^3} \sum_l \int dp_z \frac{p_s^{\pm}(\mathbf{q}) v_s^x(-\mathbf{q}) m_{\perp s}(p_z)}{l\Omega_s + \mathbf{q}\langle\mathbf{v}\rangle_s - \omega - i/\tau_s},$$

$$W_{1s}^{\pm} = \frac{1}{4\pi^2\hbar^3} \sum_l \int dp_z \frac{\pi_s^{\pm}(\mathbf{q}) v_s^x(-\mathbf{q}) m_{\perp s}(p_z)}{l\Omega_s + \mathbf{q}\langle\mathbf{v}\rangle_s - \omega - i/\tau_s}, \qquad (2.61)$$

$$A_1^{ss'} = \frac{1}{4\pi^2\hbar^3} \int m_{\perp s}(p_z) \langle p_\perp^2 \rangle_s \varphi_1^{ss'} dp_z,$$

$$A_2^{ss'} = \frac{1}{4\pi^2\hbar^3} \int m_{\perp s}(p_z) p_{zs}^2 \langle p_\perp^2 \rangle_s \varphi_{21}^{ss'} dp_z,$$

$$Q_{0s} = \frac{1}{4\pi^2\hbar^3} \int m_{\perp s}(p_z) \langle p_\perp^2 \rangle_s dp_z,$$

$$Q_{2s} = \frac{1}{4\pi^2\hbar^3} \int m_{\perp s}(p_z) p_{zs}^2 \langle p_\perp^2 \rangle_s dp_z.$$

The integration over p_z has to be accompanied by summation over the sheets included, to the sth part of the Fermi surface. The quantity $\langle\mathbf{v}\rangle_s$ represents the average of the longitudinal component of the velocity of the quasi-particles of the sort s over the cyclotron orbit

$$\langle v \rangle_s = \frac{2\pi}{\Omega} \int_0^T v_{zs}(t)\, dt. \qquad (2.62)$$

The average over the cyclotron orbit $\langle p_\perp^2 \rangle_s$ can be defined in a similar way.

The obtained equations allow us to find components of the conductivity tensor. In particular, for a singly connected Fermi surface we can write the circular components of the conductivity $\sigma_{\pm} = \sigma_{xx} \pm i\sigma_{yx}$ in the form

$$\sigma_{\pm} = ie^2 \Delta_1^{\pm}/\Delta^{\pm}. \qquad (2.63)$$

Here

$$
\begin{aligned}
\Delta_1^\pm = \Gamma^\pm(v, v) & \left[1 + \frac{\omega + i/\tau}{Q_2}\alpha_2\Gamma^\pm(p, p)\right]\left[1 - \frac{\omega + i/\tau}{Q_0}\alpha_1 D^\pm(p, v, v)\right] \\
& - \left(\omega + \frac{i}{\tau}\right)\alpha_2\Gamma^\pm(v, \pi)\frac{\Gamma^\pm(\pi, v)}{Q_2} - \frac{(\omega + i/\tau)^2}{Q_0 Q_2} \\
& \times \alpha_1\alpha_2\big[\Gamma^\pm(v, \pi)\Gamma^\pm(\pi, v)D^\pm(p, \pi, v) - \Gamma^\pm(p, \pi)\Gamma^\pm(\pi, p)D^\pm(v, \pi, p)\big],
\end{aligned}
\tag{2.64}
$$

$$
\begin{aligned}
\Delta^\pm = & \left[1 + \frac{\omega + i/\tau}{Q_0}\alpha_1\Gamma^\pm(p, p)\right]\left[1 + \frac{\omega + i/\tau}{Q_2}\alpha_2\Gamma^\pm(\pi, \pi)\right] \\
& - \left(\omega + \frac{i}{\tau}\right)^2 \alpha_1\alpha_2\frac{\Gamma^\pm(p, \pi)\Gamma^\pm(\pi, p)}{Q_0, Q_2}.
\end{aligned}
\tag{2.65}
$$

The quantities $\Gamma^\pm(a, b)$ and $D^\pm(a, b, c)$ are defined by the relations

$$
\Gamma^\pm(a, b) = \frac{1}{4\pi^2\hbar^3}\sum_l \int dp_z \frac{a_l^\pm(\mathbf{q})b_l^x(-\mathbf{q})m_\perp(p_z)}{l\Omega + \mathbf{q}\langle\mathbf{v}\rangle - \omega - i/\tau},
\tag{2.66}
$$

$$
D^\pm(a, b, c) = \Gamma^\pm(a, a) - \frac{\Gamma^\pm(a, b)\Gamma^\pm(c, a)}{\Gamma^\pm(c, b)}.
\tag{2.67}
$$

The dimensionless coefficients α_1 and α_2 are expressed in terms of the Fermi-liquid parameters A_1 and A_2 (see (2.61)) by the relations $\alpha_{1,2} = A_{1,2}/(1 + A_{1,2})$.

When we neglect the Fermi-liquid interaction, this expression (2.63) passes the well-known result for circular components of the transverse conductivity of a gas of electrons

$$
\sigma^\pm = \frac{ie^2}{4\pi^2\hbar^3}\sum_l \int dp_z \frac{v_l^\pm(\mathbf{q})v_l^x(-\mathbf{q})m_\perp(p_z)}{l\Omega + \mathbf{q}\langle\mathbf{v}\rangle - \omega - i/\tau},
\tag{2.68}
$$

which agrees with (2.44).

Expression (2.63) also describes the conductivity when a metal has an ionic lattice of the hexagonal symmetry and the magnetic field is directed along the symmetry axis of the sixth order. In this case the Fermi–liquid parameters equal

$$
A_1 = \frac{1}{4\pi^2\hbar^3}\int dp_z m_\perp(p_z)\langle p_\perp^2\rangle\varphi_{10},
$$

$$
A_2 = \frac{1}{4\pi^2\hbar^3}\int dp_z m_\perp(p_z)p_z^2\langle p_\perp^2\rangle\varphi_{11}.
\tag{2.69}
$$

Circular components of the transverse conductivity, for a metal with the axisymmetric singly connected Fermi surface in the geometry when the external magnetic field is directed along the symmetry axis, can be written similarly.

2.4 Approximation of the Fermi-Liquid Functions for a Metal with an Axial-Symmetric Fermi Surface

In the previous section we touched upon the problem of the approximation of the Fermi-liquid functions in metals with axisymmetric Fermi surfaces. However, this model of metal with an axial-symmetric Fermi surface and the same value of cyclotron mass for all cyclotron orbits (provided that the magnetic field is directed along the symmetry axis) should be considered in more detail.

Again we will proceed from the symmetry properties of the Fermi surface. Let us assume that the symmetry axis of the Fermi surface coincides with the z-axis of the coordinate system. Such a model saves some useful features of the simple isotropic model which permits us to avoid excessive complexities in calculations. At the same time, this model enables us to analyze the influence of the Fermi surface shape on the observables rather thoroughly. For the axisymmetric Fermi surface the functions $\varphi(\mathbf{p}, \mathbf{p}')$ and $\psi(\vec{\mathbf{p}}\vec{\mathbf{p}}')$ do not vary under an identical change of direction of the projections $\mathbf{p}_\perp$ and $\mathbf{p}'_\perp$ of the quasi momenta $\mathbf{p}$ and $\mathbf{p}'$ on a plane $p_z = 0$. On the Fermi surface these functions actually depend only on the cosine of an angle Φ between the vectors $\mathbf{p}_\perp$ and $\mathbf{p}'_\perp$, and on the longitudinal components of the quasi momenta p_z and p'_z.

We can separate out even and odd in $\cos\Phi$ parts of the Fermi-liquid functions and write them in the form

$$\varphi(\mathbf{p}, \mathbf{p}') = \varphi_0(p_z, p'_z, \cos\Phi) + (\mathbf{p}_\perp \mathbf{p}'_\perp)\varphi_1(p_z, p'_z, \cos\Phi),$$
$$\psi(\mathbf{p}, \mathbf{p}') = \psi_0(p_z, p'_z, \cos\Phi) + (\mathbf{p}_\perp \mathbf{p}'_\perp)\psi_1(p_z, p'_z, \cos\Phi),$$
(2.70)

where φ_0, φ_1, ψ_0, ψ_1 are even functions of $\cos\Phi$. By virtue of the invariancy of the Fermi surface under replacement both $\mathbf{p} \to -\mathbf{p}$ and $\mathbf{p}' \to -\mathbf{p}'$, the functions φ_0 and φ_1 (as well as ψ_0 and ψ_1), should not vary under a simultaneous change of signs of p_z and p'_z. Using it, we can separate the functions φ_0, φ_1, ψ_0, ψ_1 into the parts which are even and odd in p_z, p'_z and rewrite (2.70) as

$$\varphi(p_z, p'_z, \cos\Phi) = \varphi_{00} + p_z p'_z \varphi_{01} + (\mathbf{p}_\perp \mathbf{p}'_\perp)(\varphi_{10} + p_z p'_z \varphi_{11}),$$
$$\psi(p_z, p'_z, \cos\Phi) = \psi_{00} + p_z p'_z \psi_{01} + (\mathbf{p}_\perp \mathbf{p}'_\perp)(\psi_{10} + p_z p'_z \psi_{11}).$$
(2.71)

Here the functions $\varphi_{00}, \varphi_{01}, \varphi_{10}, \varphi_{11}, \psi_{00}, \psi_{01}, \psi_{10}, \psi_{11}$ are even in all their arguments.

Each term in (2.71) corresponds to a particular part in the expansions (2.33) of the Fermi-liquid functions in spherical harmonics. First, terms correspond to the terms in the expressions (2.34) with even values of indices j and m; second, to the terms with odd j and even m; third, to the terms with both j and m odd; fourth, to that part of the expansions (2.34) which contains the harmonics with even j and odd m. When we keep in the expansion (2.33) terms with $j \le 2$, it complies with the assumption that the functions $\varphi_{00}, \varphi_{01}, \varphi_{10}, \varphi_{11}, \psi_{00}, \psi_{01}, \psi_{10}, \psi_{11}$ are constants. These constants can be expressed through the Fermi-liquid parameters

A_j, B_j. For example we have

$$\varphi_{00} = \pi^2\hbar^3 A_0 \frac{v_F}{p_F^2}, \qquad \varphi_{01} = \varphi_{10} = 3\pi^2\hbar^3 A_1 \frac{v_F}{p_F^4}, \qquad \varphi_{11} = 15\pi^2\hbar^3 A_2 \frac{v_F}{p_F^5}.$$

We can write a simple approximation for the Fermi-liquid function $\varphi(\mathbf{p}, \mathbf{p}')$ assuming that φ_{00}, φ_{01}, φ_{10}, φ_{11} are the constants for an arbitrary axisymmetric Fermi surface. Similar approximation can be used for the function $\psi(\mathbf{p}, \mathbf{p}')$. In these approximations we keep the first three terms in expressions (2.55). Using them we can analyze the main Fermi-liquid effects when the perturbation propagates along the magnetic field.

In the case considered, the system of equations for the averages $I_1^\pm$, $I_2^\pm$, which are defined by the relations similar to (2.59), (2.60), takes the form

$$I_1^\pm \left\{ 1 + A_1 - \frac{A_1}{Q_0} \Phi_0^\pm \right\} - I_2^\pm \frac{A_2}{Q_2} \Phi_1^\pm = \frac{ie}{m} \Phi_0^\pm E_{q\omega}^\pm, \qquad (2.72)$$

$$I_2^\pm \left\{ 1 + A_2 - \frac{A_2}{Q_2} \Phi_2^\pm \right\} - I_1^\pm \frac{A_1}{Q_0} \Phi_1^\pm = \frac{ie}{m} \Phi_1^\pm E_{q\omega}^\pm. \qquad (2.73)$$

Here

$$\Phi_m^\pm(q, \omega) = \left(i\omega - \frac{1}{\tau} \right) \frac{1}{4\pi^2\hbar^3} \int dp_z \frac{m_\perp p_\perp^2 p_z^m}{\mp\Omega + qv_z - \omega - i/\tau},$$

$$Q_0 = \frac{1}{4\pi^2\hbar^3} \int dp_z m_\perp p_\perp^2, \qquad (2.74)$$

$$Q_2 = \frac{1}{4\pi^2\hbar^3} \int dp_z m_\perp p_\perp^2 p_z^2.$$

These equations give expressions for the components of the conductivity tensor which contains complete information about the electromagnetic response of the electron system. In particular, circular components of the transverse conductivity $\sigma_\pm = \sigma_{xx} \pm \sigma_{yx}$ equal

$$\sigma_\pm = \frac{ie^2}{4\pi^2\hbar^3 m} \frac{\left[\Phi_0^\pm \left(1 - \frac{\alpha_2}{Q_2} \Phi_2^\pm \right) - \frac{\alpha_2}{Q_2}(\Phi_1^\pm)^2 \right]}{\left[\left(1 - \frac{\alpha_1}{Q_0} \Phi_0^\pm \right) \left(1 - \frac{\alpha_2}{Q_2} \Phi_2^\pm \right) - \frac{\alpha_1\alpha_2}{Q_0 Q_2}(\Phi_1^\pm)^2 \right]}. \qquad (2.75)$$

Here we introduce new parameters instead of φ_{10} and φ_{11}:

$$A_1 = \frac{2m_\perp}{(2\pi\hbar)^3} \int p_\perp^2 \varphi_{10} \, dp_z,$$

$$A_2 = \frac{2m_\perp}{(2\pi\hbar)^3} \int p_\perp^2 p_z^2 \varphi_{11} \, dp_z.$$

For a spherical Fermi surface these parameters coincide with the Fermi-liquid coefficients A_1 and A_2, $\alpha_{1,2} = A_{1,2}/(1 + A_{1,2})$.

The semiclassical expressions (2.71) for the Fermi-liquid functions φ and ψ can easily be extended to analyze the quantum effects in a strong magnetic field. For a metal with an axisymmetric Fermi surface the quasi-particle motion is described by a set of orbital quantum numbers, which includes (supposing that we neglect interband transitions) a projection of the quasi momentum in a direction of the magnetic field p_z, a discrete number of the Landau level n, and a coordinate of a center of the cyclotron orbit x_0. Assuming that the wave vector of the perturbation is placed in a yz-plane, we can present the matrix elements $\varphi_{\alpha\alpha'}^{\alpha_1\alpha_2}$ and $\varphi_{\alpha\alpha'}^{\alpha_1\alpha_2}$ as

$$\varphi_{\alpha\alpha'}^{\alpha_1\alpha_2} = \varphi(\mathbf{q}, Nlp_z, N_1l_1p_{z_1})\delta_{p_z',p_z-\hbar q_z}\delta_{p_{z_2},p_{z_1}-\hbar q_z}\delta_{x_0',x_0+\lambda^2 q_y}\delta_{x_{02},x_{01}+\lambda^2 q_y},$$
$$\psi_{\alpha\alpha'}^{\alpha_1\alpha_2} = \psi(\mathbf{q}, Nlp_z, N_1l_1p_{z_1})\delta_{p_z',p_z-\hbar q_z}\delta_{p_{z_2},p_{z_1}-\hbar q_z}\delta_{x_0',x_0+\lambda^2 q_y}\delta_{x_{02},x_{01}+\lambda^2 q_y},$$

(2.76)

where $N = \frac{1}{2}(n + n')$, $l = n - n'$, and $\lambda^2 = \hbar c/|e|B$ is the magnetic length.

Under conditions when we can use the semiclassical approach (the distances between adjacent energy levels is small in comparison with the Fermi energy) the matrix elements (2.76), as well as matrix elements of other operators, can be replaced by their semiclassical analogs. The set of quantum numbers Nlp_z corresponds to cylindrical coordinates of a quasi particle in the momenta space $p_\perp, p_z, \Phi, p_\perp = (\hbar/\lambda)\sqrt{2N+1}$, and l represents the number of harmonics in the expansions in Fourier series in angles Φ. The semiclassical analog of the matrix element $\varphi(\mathbf{q}, Nlp_z, N_1, l_1 p_{z_1})$ is the Fourier transform of the function

$$\varphi(\mathbf{q}; \mathbf{p}, \mathbf{p}') \exp\left[-\frac{icq_y p_\perp}{|e|B}\sin\Phi\right]\varphi(\mathbf{p}, \mathbf{p}')\exp\left[-\frac{icq_y p_\perp'}{|e|B}\sin\Phi'\right] \quad (2.77)$$

in its expansion in azimuthal angles.

With the help of a well-known identity

$$\exp(\pm iz\sin\Phi) = \sum_{m=-\infty}^{\infty} J_m(z)\exp(\pm im\Phi), \quad (2.78)$$

where $J_m(z)$ are Bessel functions, we obtain

$$\varphi_{ll'}(\mathbf{q}; \mathbf{p}, \mathbf{p}') = \frac{1}{4\pi^2}\sum_{m=-\infty}^{\infty} J_m\left(\frac{cq_y p_\perp}{|e|B}\right)\sum_{m'=-\infty}^{\infty} J_{m'}\left(\frac{cq_y p_\perp'}{|e|B}\right) \quad (2.79)$$

$$\times \int_0^{2\pi} d\Phi\exp\left[i(m-l)\Phi\right]\int_0^{2\pi} d\Phi'\exp\left[i(m'-l')\Phi'\right]\varphi(\mathbf{p}, \mathbf{p}').$$

The Fermi-liquid functions φ and ψ for any axisymmetric Fermi surface actually depend only on p_z, p_z' and $\cos(\Phi - \Phi')$ and we can write them in the form (2.71). Each of the four functions on the right-hand side of (2.71) is even in all arguments, therefore, their expansions in Fourier series in a variable $\Phi - \Phi'$ are

$$\varphi_{ik}(p_z, p_z', \cos\theta) = \sum_{s=-\infty}^{\infty} \varphi_{ik}^{|2s|}(p_z, p_z')\exp[2is(\theta)]. \quad (2.80)$$

Here $i, k = 0, 1$, $\theta = \Phi - \Phi'$.

Substituting expression (2.71) into (2.79) and using the expansion (2.80) we arrive at the result

$$\varphi_{ll'}(\mathbf{q}, p_\perp, p_z, p_\perp', p_z') = \sum_{s=-\infty}^{\infty}\left\{\left[\varphi_{00}^{|2s|}(p_z, p_z') + p_z p_z'\varphi_{01}^{|2s|}(p_z, p_z')\right]\right. \quad (2.81)$$

$$\times J_{l-2s}\left(\frac{cq_y p_\perp}{|e|B}\right) J_{l'-2s}\left(\frac{cq_y p'_\perp}{|e|B}\right) + \left[\varphi_{10}^{|2s|}(p_z, p'_z) + p_z p'_z \varphi_{11}^{|2s|}(p_z, p'_z)\right]$$

$$\times \left[J_{l-2s-1}\left(\frac{cq_y p_\perp}{|e|B}\right) J_{l'-2s-1}\left(\frac{cq_y p'_\perp}{|e|B}\right) + J_{l-2s+1}\left(\frac{cq_y p_\perp}{|e|B}\right) J_{l'-2s+1}\left(\frac{cq_y p'_\perp}{|e|B}\right) \right] \Biggr\} .$$

Correspondingly, the matrix elements $\varphi(\mathbf{q}, Nlp_z; N'l'p'_z)$ are described by the expansions of the form

$$\varphi(\mathbf{q}, Nlp_z; N'l'p'_z) = \sum_{s=-\infty}^{\infty} \Biggl\{ \left[\varphi_{00}^{|2s|}(p_z, p'_z) + p_z p'_z \varphi_{01}^{|2s|}(p_z, p'_z)\right]$$

$$\times J_{l-2s}\left(\frac{\hbar c q_y}{\lambda |e|B}\sqrt{2N+1}\right) J_{l'-2s}\left(\frac{\hbar c q_y}{\lambda |e|B}\sqrt{2N'+1}\right)$$

$$+ \left[\varphi_{10}^{|2s|}(p_z, p'_z) + p_z p'_z \varphi_{11}^{|2s|}(p_z, p'_z)\right]$$

$$\times \left[J_{l-2s-1}\left(\frac{\hbar c q_y}{\lambda |e|B}\sqrt{2N+1}\right) J_{l'-2s-1}\left(\frac{\hbar c q_y}{\lambda |e|B}\sqrt{2N'+1}\right) \right.$$

$$\left. + J_{l-2s+1}\left(\frac{\hbar c q_y}{\lambda |e|B}\sqrt{2N+1}\right) J_{l'-2s+1}\left(\frac{\hbar c q_y}{\lambda |e|B}\sqrt{2N'+1}\right) \right] \Biggr\} . \qquad (2.82)$$

The functions $\varphi_{ik}^{|2s|}(p_z, p'_z)$ are even in p_z, p'_z. Besides, they should remain invariable when their arguments change places. The simplest functions possessing these properties are functions of the form

$$\varphi_{ik}^{|2s|}(p_z, p'_z) = P_{ik}^{|2s|}(p_z) P_{ik}^{|2s|}(p'_z), \qquad (2.83)$$

where both cofactors are even functions in their arguments. Further we will use this approximation.

The obtained results create a body of mathematical techniques permitting us to study the response of an electronic liquid of real metals to external disturbances. In subsequent sections these techniques will be applied to study concrete problems.

CHAPTER 3

Local Anomalies of the Fermi Surface Curvature and the Anomalous Skin Effect in Metals

3.1 The Effect of Local Geometry of the Fermi Surface on the Conductivity of a Metal

When the Fermi surface of a metal has points or lines of zero or anomalously large Gaussian curvature it can essentially influence the conductivity of metal under conditions of a strong spatial dispersion. We begin by analyzing this influence. First we consider a metal with an axisymmetric Fermi surface. Let us choose a coordinate system whose z-axis is directed along the symmetry axis of the Fermi surface. We shall write the energy momentum relation for electrons as

$$E(\mathbf{p}) = \frac{\mathbf{p}_\perp^2}{2m_\perp} + E_{\parallel}(p_z), \qquad (3.1)$$

This energy-momentum relation (3.1) describes a fairly wide class of axisymmetric one-sheet Fermi surfaces.

Provided that the wave vector $\mathbf{q}$ is directed along the z-axis we can write the following expressions for transverse components of the conductivity σ_{xx} and σ_{yy}:

$$\sigma_{xx} = \sigma_{yy} = \frac{ie^2 p_m S(0)}{4\pi^3\hbar^3 m_\perp v_m q} \frac{\Phi_0(u)(1 - \bar{\alpha}_2 u \Phi_2(u)) + \bar{\alpha}_2 u \Phi_1^2(u)}{(1 - \bar{\alpha}_1 u \Phi_0(u))(1 - \bar{\alpha}_2 u \Phi_2(u)) - \bar{\alpha}_1 \bar{\alpha}_2 u^2 \Phi_1^2(u)},$$
$$(3.2)$$

where the functions $\Phi_m(u)$ are given by

$$\Phi_m(u) = \int_{-1}^{1} \frac{\bar{s}(x) x^m}{u - \bar{v}(x) + i/q v_m \tau} dx \qquad (3.3)$$

and

$$\bar{\alpha}_{1,2} = \frac{A_{1,2}}{(1 + A_{1,2})\bar{Q}_{0,2}}, \qquad \bar{Q}_m = \int_{-1}^{1} \bar{s}(x)x^m dx,$$

$$\bar{v}(x) = v_z/v_m, \qquad \bar{s}(x) = S(p_z)/S(0),$$

$$u = \omega/qv_m, \qquad x = p_z/p_m,$$

where p_m, v_m are the maximum values of the longitudinal components of the quasi momemtum and the velocity of electrons, and $S(p_z)$ is the area of the cross-section of the Fermi surface by a plane $p_z = \text{const}$.

Calculating the integrals $\Phi_m(u)$ we have to take into account that, due to the mirror symmetry of the Fermi surface defined by (3.1), the longitudinal component of the velocity $\bar{v}(x)$ is an odd function, and the sectional area $\bar{s}(x)$ is an even function of x. Combining contributions from the symmetric segments of the Fermi surface we can carry out integration over that half of the Fermi surface which corresponds to positive x. To proceed we go over, in the expressions (3.3), to integration in the velocity space. As a result we can derive the following expressions:

$$\Phi_0(u) = 2u \left(1 + \frac{i}{\omega\tau}\right) \sum_i \int \frac{\bar{s}_i(\bar{v})(dx_i/d\bar{v})}{u^2(1 + i/\omega\tau)^2 - \bar{v}^2} d\bar{v}; \qquad (3.4)$$

$$\Phi_1(u) = 2 \sum_i \int \frac{\bar{s}_i(\bar{v})x_i(\bar{v})\bar{v}(dx_i/d\bar{v})}{u^2(1 + i/\omega\tau)^2 - \bar{v}^2} d\bar{v}; \qquad (3.5)$$

$$\Phi_2(u) = 2u \left(1 + \frac{i}{\omega\tau}\right) \sum_i \int \frac{\bar{s}_i(\bar{v})x_i^2(\bar{v})(dx_i/d\bar{v})}{u^2(1 + i/\omega\tau)^2 - \bar{v}^2} d\bar{v}. \qquad (3.6)$$

It is implied in these expressions that the Fermi surface is divided into segments with numbers i, so that the dependence of the longitudinal component of quasi momentum p_z on v_z is unique on each segment. Correspondingly, summation over j has to be performed in (3.4)–(3.6). For large q, when $u \ll 1$, the main contributions to the integrals (3.3) are from the vicinities of those cross-sections of the Fermi surface where the longitudinal component of the velocity turns to zero, i.e., from the effective strips. Therefore we select from the sums over i included in (3.4)–(3.6) those terms which correspond to segments of the Fermi surface containing effective strips, and we shall consider their contribution to the integrals $\Phi_m(u)$.

The value of the contribution from the effective strips to the integrals (3.4)–(3.6) depends on the behavior of the derivative of the longitudinal velocity with respect to p_z, which is in turn determined by the Gaussian curvature of the appropriate segments of the Fermi surface. Actually, the Gaussian curvature of the axisymmetric Fermi surface can be represented as

$$K(p_z) = \frac{m_\perp \left(v_z^2 + \frac{1}{m_\perp}p_\perp^2 \frac{\partial v_z}{\partial p_z}\right)}{(p_\perp^2 + m_\perp^2 v_z^2)^2}. \qquad (3.7)$$

We see that the curvature of the effective strips is proportional to $\partial v_z/\partial p_z$. Thus, when the Fermi surface has finite and nonzero curvature everywhere, we can assume that the functions $dx/d\bar{v}$ in the integrands of (3.4)–(3.6) are nonzero constants. For the Fermi surface of finite nonzero curvature the main term in the expansion of σ in powers of the small parameter u equals

$$\sigma_0 = \frac{e^2 p_0^2}{4\pi\hbar^3 q}, \qquad p_0^2 = 2\sum_j \left(1 - \tfrac{1}{2}\delta_{j0}\right)\frac{1}{K_j}. \tag{3.8}$$

Here p_0 has dimensions of momentum, summation over j is performed over the effective cross-sections of the Fermi surface, K_j is the Gaussian curvature of the jth effective cross-section, and $j = 0$ corresponds to the central cross-section of the Fermi surface where the condition $v_z = 0$ should be satisfied according to the symmetry of the surface.

The expansion of the conductivity in powers of the small parameter u can be written in the form

$$\sigma = \sigma_0(1 + \Lambda_1 u + \Lambda_2 u^2 + \cdots) \equiv \sigma_0\bar{\sigma}. \tag{3.9}$$

The dimensionless coefficients $\Lambda_1, \Lambda_2, \ldots$ depend on the frequency and Fermi-liquid parameters. For an axisymmetric Fermi surface, which corresponds to the energy-momentum relation of the form (3.1), the first coefficients of expansion (3.9) are accordingly equal:

$$\Lambda_1 = -\frac{ig}{\pi}\left[a\left(1 + \frac{i}{\omega\tau}\right) + \frac{\pi^2}{g^2}\bar{\alpha}_1 - b^2\bar{\alpha}_2\right], \tag{3.10}$$

$$\Lambda_2 = d\left(1 + \frac{i}{\omega\tau}\right)^2 - 2\bar{\alpha}_1 a\left(1 + \frac{i}{\omega\tau}\right) - 2\bar{\alpha}_2\frac{b}{c}g\left(1 + \frac{i}{\omega\tau}\right) - 2\bar{\alpha}_1\bar{\alpha}_2 b^2 - \bar{\alpha}_1^2\frac{\pi^2}{g^2}, \tag{3.11}$$

where a, b, c, d, g are dimensionless constants of the order of unity whose values depend on the shape of the Fermi surface, (expressions for these constants are derived in Appendix 1). For a spherical Fermi surface $c = g = 1, d = -1, a = -4, b = \frac{4}{3}$.

It is known that the Fermi surfaces to some metals have points and lines where their Gaussian curvature turns to zero or tends to infinity. The lines with a zero curvature are characteristic, for example, for surfaces having dents or bulges. The curvature can be anomalously large at the points on the edges of narrow lenses and needle-shaped sheets, and also at the conic points.

Let us assume that the effective cross-sections of the Fermi surface include cross-sections of zero or anomalously large curvature. Under such conditions the expansion (3.9) will not correctly describe the transverse components of the conductivity. The anomalous curvature affects the conductivity because it is closely related to the width of the effective strips on the Fermi surface, that is, to the number of effective electrons associated with these strips. When the curvature of an effective cross-section turns to zero, the width of the corresponding effective strip becomes anomalously large and its contribution to the conductivity grows.

In contrast, when the curvature of an effective cross-section tends to infinity, then an anomalously small number of effective electrons is associated with this cross-section, and its contribution to the conductivity is smaller than those from the effective cross-sections with finite curvature.

In the vicinity of the effective cross-section with an anomalous curvature the function $K(\bar{v})$ can be conveniently described by the approximation

$$K(\bar{v}) = w(\bar{v})|\bar{v}|^{-\beta}, \tag{3.12}$$

where $w(\bar{v})$ is a function which has a finite nonzero value at $\bar{v} = 0$. Negative values of the exponent β correspond to lines of zero curvature, and the closer β is to -1, the closer will be the effective strip on the Fermi surface to the cylindrical shape. When the exponent β is positive, the curvature of the Fermi surface on the effective cross-section tends to infinity. The approximation (3.12) is also applicable in neighborhoods of "conventional" effective cross-sections of finite nonzero curvatures. For such cross-sections we have $\beta = 0$.

When the set of effective cross-sections of the Fermi surface includes one with an anomalous curvature, the expression for the transverse conductivity in the region of small u contains an additional term σ_a arising due to the "anomalous" cross-section:

$$
\sigma_a(q) = \sigma_0 \rho \left(1 - i \tan \left(\frac{\pi \beta}{2} \right) \right) u^\beta \left(1 + \frac{i}{\omega \tau} \right)^\beta
$$
$$
\times \left[1 - \frac{i\pi}{g} \bar{\alpha}_1 \rho \left(1 - i \tan \left(\frac{\pi \beta}{2} \right) \right) u^{\beta+1} \right.
$$
$$
\left. + \frac{\rho}{g} \theta u^{\beta+2} + (2\pi b)^4 \bar{\alpha}_1 \bar{\alpha}_2 u^2 + \frac{\pi^2}{g^2} \bar{\alpha}_1^2 u^2 + \cdots \right]. \tag{3.13}
$$

Here $\rho = 2 \left(1 - \frac{1}{2}\delta_{l0} \right) / (p_0^2 w_l(0))$, $\theta = \eta_l \pi^2 m_\perp v_m p_m / S_l(\beta + 2)$, l labels an effective cross-section with abnormal curvature, η_l equals 1 when the effective cross-section corresponds to the minimum of the function $S(p_z)$ and -1 when it corresponds to its maximum, and S_l is the area of the given cross-section. Expression (3.13) for the abnormal part of the conductivity includes only the main terms, whose order of smallness does not exceed u^2 for negative values of the parameter β. Summation over the effective cross-sections, which has to be performed in the calculation of p_0^2 in (3.8) and also of the constants a, b, d, is carried out only over "conventional" effective strips of the Fermi surface.

It follows from the expressions for the coefficients in the expansion of the conductivity in powers of u (3.10), (3.11), and also for the contribution from the "anomalous" effective cross-section (3.13), that the Fermi-liquid interaction does not qualitatively change the dependence of the transverse components of the conductivity of u. Taking account of it leads to rather small changes in the coefficients Λ_1, Λ_2. We come to a similar conclusion analyzing the "abnormal" contribution to the conductivity. Terms including the Fermi-liquid parameters $\bar{\alpha}_1, \bar{\alpha}_2$ in expression (3.13) are of the order of u^2, therefore the main part of σ_a does not depend on the Fermi-liquid interaction. According to these conclusions we will omit all respective terms in the following consideration. Instead of (3.10), (3.11), (3.13) we will use the simplified expressions

$$\Lambda_1 = -\frac{ig}{\pi}a\left(1 + \frac{i}{\omega\tau}\right), \qquad \Lambda_2 = d\left(1 + \frac{i}{\omega\tau}\right)^2,$$

$$\sigma_a = \sigma_0\rho\left(1 - i\tan\left(\frac{\pi\beta}{2}\right)\right)\left(1 + \frac{\rho}{g}\theta u^{\beta+2}\right)u^\beta\left(1 + \frac{i}{\omega\tau}\right)^\beta. \quad (3.14)$$

For the effective cross-section of zero curvature ($\beta < 0$), the "anomalous" contribution σ_a to the conductivity is proportional to a dimensionless parameter ρ whose
value is determined by the relative number of effective electrons associated with
this cross-section. When $\rho \sim 1$ (the number of effective electrons in the anomalous cross-section is comparable with the total number of effective electrons on
the Fermi surface) the first term in (3.13) is the principal term in the expansion of
the conductivity in powers of u. When $\rho \ll 1$ (the relative number of effective
electrons associated with the anomalous cross-section is small) the term σ_a is considerably smaller in magnitude than the principal term of the "normal" part σ_0 of
the conductivity, but it can be larger by an order of magnitude greater than all the
other terms in the expansion (3.9). The "anomalous" contribution to the conductivity for the effective cross-section with infinite curvature is, in any case, smaller
than σ_0 by an order of magnitude. But the first term σ_0 is still a first-order correction
for the principal conductivity value even in this case, when the parameter β is not
too large ($0 < \beta < 1$) and ρ is not too small.

We can generalize these results to include Fermi surfaces which are not axisymmetric. Assume that the Fermi surface has a mirror symmetry in a momentum
space relative to a plane $p_z = 0$ and that the conductivity tensor is diagonalized
in a chosen coordinate system. Then we can write the following expression for
$\sigma_{xx}(\omega, q)$ (see [106]):

$$\sigma_{xx}(\omega, q) = \frac{ie^2}{4\pi^3\hbar^3 q}\sum_j\int d\varphi\int d\theta\frac{\cos^2\varphi\sin^3\theta}{|K_j(\theta, \varphi)|[(\omega + i/\tau)/qv_j - \cos\theta]}. \quad (3.15)$$

Here the integration is carried out with respect to spherical coordinates in the
velocity space. It is assumed that the Fermi surface can be divided into parts with
numbers j and the dependence of the momentum $\mathbf{p}$ on the magnitude of the velocity
v and the angles θ, φ is unique for each part. The magnitude of the velocity depends
on the angles θ, φ on each segment of the Fermi surface. Integration with respect
to the angles θ, φ is performed over each segment and the area of the surface
element is written in the form $dS_j = \sin\theta d\theta d\varphi/|K_j(\theta, \varphi)|$, where $K_j(\theta, \varphi)$ is
the Gaussian curvature of the surface. In the following formulas we include the
term $1/\tau$ corresponding to the collision frequency into ω.

Let us consider the asymptotic expression for the conductivity under $\omega/qv \ll 1$.
We can write the well-known result for the principal term of the conductivity
component $\sigma_{xx}(\omega, q)$ (see [107], [108]) in a form similar to (3.8):

$$\sigma_0(q) = \frac{e^2}{4\pi^3\hbar^3 q}\sum_l\int d\varphi\frac{\cos^2\varphi}{|K_l(\pi/2, \varphi)|} \equiv \frac{e^2}{4\pi^3\hbar^3 q}p_0^2. \quad (3.16)$$

The indices xx are omitted for simplicity here and in the following expressions. Summation over l is carried out over all segments of the Fermi surface containing effective strips which correspond to $\theta = \pi/2(v_z = 0)$; it is assumed that the curvature $K_l(\pi/2, \varphi)$ takes finite and nonzero values at any point of any effective strip. For a spherical Fermi surface, p_0 equals the Fermi momentum p_F.

Proceeding in a way similar to that described in Appendix 1, we can calculate the next term of the expansion of the conductivity in powers of the small parameter ω/qv. For the Fermi surface whose curvature everywhere is finite and nonzero we arrive at the result

$$\sigma_1(\omega, q) = \frac{e^2 p_0^2}{4\pi\hbar^3 q} \frac{i\omega}{qv_0}, \tag{3.17}$$

$$\frac{1}{v_0} = \frac{2}{\pi^2 p_0^2} \sum_l \int d\varphi \int_{\theta_l}^{\pi/2} d\theta \frac{\cos^2\varphi \sin\theta}{\cos^2\theta}$$

$$\times \left[\frac{1 + \cos^2\theta/\cos^2\theta_l}{|K_l(\pi/2; \varphi)| v_l(\pi/2; \varphi)} - \frac{\sin^2\theta}{|K_l(\theta, \varphi)| v_l(\theta, \varphi)} \right]$$

$$- \frac{2}{\pi^2 p_0^2} \sum_{j\neq l} \int d\varphi \int_{\theta \leq \pi/2} d\theta \frac{\cos^2\varphi \sin^3\theta}{|K_j(\theta, \varphi)| v_j(\theta, \varphi) \cos^2\theta}, \tag{3.18}$$

where θ_l is a minimum value of the angle θ for given φ. For the spherical Fermi surface we have $v_0 = \pi v_F/4$, where v_F is the Fermi velocity of the electrons. When the curvature at any effective line tends to zero, it changes the asymptotics of the conductivity. We can also expect some changes in the asymptotics of the conductivity arising due to the curvature tending to infinity.

To proceed we assume that the curvature tends to zero or has a singularity at some points of the effective line passing through one of the segments of the Fermi surface. We suppose that the anomaly is attributed to that radius of curvature which corresponds to the direction perpendicular to a tangent to the effective line. When the anomaly exhibits itself along all effective lines, we can use the following approximation for the curvature $K(\theta, \varphi)$ at $\theta \leq \pi/2$:

$$K(\theta, \varphi) = W(\theta, \varphi)(\cos\theta)^{-\beta}, \tag{3.19}$$

where the function $W(\theta, \varphi)$ assumes finite and nonzero values everywhere.

As well as in the case of an axisymmetric Fermi surface, negative values of the exponent β correspond to the line of zero curvature (line of parabolic points) at $\theta = \pi/2$. The closer β to -1, the closer to a cylinder is the effective strip of the Fermi surface. For positive values of β the curvature tends to infinity when $\theta \to \pi/2$. Thus the approximation (3.19) can be applied to describe a rather wide class of the effective strips which are different in shape.

The contribution to the conductivity from the "anomalous" segment of the Fermi surface is given by

$$\sigma_a(\omega, q) = \frac{ie^2\omega}{2\pi^3\hbar^3 q} \left[\int d\varphi \int_\alpha^{\pi/2} d\theta \left(\frac{\sin^2\theta}{|W(\theta, \varphi)| v(\theta, \varphi)} - \frac{\sin^2\theta_0}{|W(\theta_0, \varphi)| v(\theta, \varphi)} \right) \right.$$

$$\times \frac{\cos^2 \varphi \sin \theta (\cos \theta)^\beta}{(\omega/qv(\theta, \varphi))^2 - \cos^2 \theta}$$

$$+ \int d\varphi \frac{\cos^2 \varphi \sin^2 \theta_0}{|W(\theta_0, \varphi)| v(\theta_0, \varphi)} \int_\alpha^{\pi/2} d\theta \frac{\sin \theta (\cos \theta)^\beta}{(\omega/qv(\theta, \varphi))^2 - \cos^2 \theta}.$$

$$(3.20)$$

To proceed, we can use the approximation ($\beta < 1$):

$$\int_\alpha^{\pi/2} d\theta \frac{\sin \theta (\cos \theta)^\beta}{u^2 - \cos^2 \theta} \approx -\frac{\pi}{2}\left(i + \tan\left(\frac{\pi\beta}{2}\right)\right) u^{\beta-1} - \frac{(\cos \alpha)^{\beta-1}}{\beta - 1}. \quad (3.21)$$

This approximation is also valid for $1 < \beta < 2$, and for $\beta \geq 2$ we have to omit the term of the order of $u^{\beta-1}$ in (3.21). When $\beta = 1$ this integral (3.21) is equal to $-i\pi/2 - \ln(\cos \alpha/u)$.

Using this asymptotic expression we can calculate the "abnormal" contribution to the conductivity $\sigma_a(\omega, q)$ for small ω/qv. Introducing the largest magnitude of the velocity on the effective line v_a we have

$$\sigma_a(\omega, q) = \frac{e^2}{4\pi\hbar^3 q} p_a^2 \left[\left(\frac{\omega}{qv_a}\right)^\beta (1 - i\mu_\beta) + i\left(\frac{\omega}{qv_a}\right)\eta_\beta\right], \quad (3.22)$$

$$p_a^2 = \frac{1}{\pi} \int d\varphi \frac{\cos^2 \varphi}{|W(\pi/2, \varphi)|} \left(\frac{v_a}{v(\pi/2, \varphi)}\right)^\beta, \quad (3.23)$$

$$\mu_\beta = \tan\left(\frac{\pi\beta}{2}\right), \quad \beta \neq 1, \quad \mu_1 = \left(\frac{2}{\pi}\right)\ln\left(\frac{qv_a}{\omega}\right), \quad (3.24)$$

$$\eta_\beta = \frac{2}{\pi^2 p_a^2} \int d\varphi \cos^2 \varphi \int_\alpha^{\pi/2} d\theta \frac{\sin \theta}{(\cos \theta)^{2-\beta}}$$

$$\times \left[\frac{v_a}{v(\pi/2, \varphi)|W(\pi/2, \varphi)|} - \frac{v_a \sin^2 \theta}{v(\theta, \varphi)|W(\theta, \varphi)|}\right] + \theta_\beta, \quad (3.25)$$

$$\theta_\beta = -\frac{v_a(\cos \alpha)^{\beta-1}}{(\beta - 1)v(\pi/2, \varphi)|W(\pi/2, \varphi)|}, \quad \beta \neq 1,$$

$$\theta_1 = \frac{v_a}{v(\pi/2, \varphi)|W(\pi/2, \varphi)|} \ln \frac{v_a}{v(\pi/2, \varphi)\cos \alpha}. \quad (3.26)$$

We remark that for $\beta < 0$ the formula (3.25) is valid when the expression in square brackets in the integrand for $\theta \to \pi/2$ tends to zero as $\cos^2 \theta$. When this expression is linear in $\theta - \pi/2$, the expression for η_β will change. The corresponding contribution to the conductivity $(\omega/qv_a)^{\beta+1}$ under this condition is very small.

As well as in the case of an axisymmetric Fermi surface which was analyzed before, the value of the contribution to the conductivity from the "abnormal" effective strip depends at first on the character of the curvature anomaly at a given strip and, secondly, on the relative number of effective electrons concentrated there. If the curvature of the Fermi surface in the vicinity of the considered effective cross-section tends to infinity ($\beta > 0$), the main term in the asymptotic expression for conductivity is the term given by (3.16), circumscribing the contribution from conventional effective strips with finite nonzero curvatures. The abnormal contribution to the conductivity (3.22) represents the first correction to the principal term when $0 < \beta < 1$ and the ratio p_a^2/p_0^2 is not too small (the effective strip on the Fermi surface is not too narrow). For $\beta > 1$ the contribution from the "anomalous" effective strip is smaller in magnitude than the main approximation of the conductivity (3.13), and also the correction (3.17). Thus for $\beta > 1$ the effective line of infinite curvature loosely influences the conductivity, besides the case when the Fermi surface has only effective lines of infinite curvature. In this case the first two terms in the expression for conductivity drop out. The conductivity coincides with σ_a and is completely determined by the contribution from the "anomalous" effective strips.

For negative $\beta(-1 < \beta < 0)$ the contributions from abnormal effective lines can predominate in the conductivity when $p_a^2/p_0^2 \sim 1$, i.e., the number of electrons associated with the nearly cylindrical effective strips on the Fermi surface is comparable with the total number of effective electrons. When $p_a^2/p_0^2 \ll 1$ the contribution from the lines of zero curvature determines the first correction to the principal term of the conductivity (3.16).

The coincidence of an effective line on the Fermi surface with a line of anomalous curvature is not necessary for the appearance of the "abnormal" contribution to the asymptotic expression for the conductivity. Under certain conditions it is sufficient that the curvature tends to zero at isolated points on the effective line. Let us assume that the curvature vanishes at the point whose position on the Fermi surface is determined by the angles $\theta = \pi/2$ and $\varphi = \varphi_0$. We can represent the curvature $K(\theta, \varphi)$ in the vicinity of this point as follows:

$$K(\theta, \varphi) = W(\theta, \varphi) \sin(\varphi - \varphi_0)^{-\nu} (\cos^{-\beta}\theta + a^2 \sin^{-\gamma}(\varphi - \varphi_0)), \qquad (3.27)$$

where a^2 is a dimensionless positive constant, the function $W(\theta, \varphi)$ (as well as in approximation (3.19)) has a finite nonzero value, and the exponent β accepts values from the interval $(-1, 0)$. The exponents ν, γ in (3.27) are also negative and these parameters can be larger than unity in absolute values. When the absolute values of ν and γ are rather large, i.e., the Fermi surface is strongly flattened near the point of zero curvature, this point gives the contribution to the conductivity similar to the first term in (3.22):

$$\sigma_\alpha = \frac{e^2}{4\pi\hbar^3 q} p_1^2 \left(1 - i \tan\left(\frac{\pi s}{2}\right)\right) \left(\frac{\omega}{q v_\alpha}\right)^s, \qquad (3.28)$$

where

$$s = \beta(1 + \nu + \gamma)/\gamma,$$

and

$$p_1^2 = \frac{\operatorname{cosec}[\pi(1 + 1/\gamma)]}{\gamma a^{2(1+1/\gamma)}} \frac{\cos^2\varphi_0}{W(\pi/2, \varphi_0)} \left[\frac{v_a}{v(\pi/2, \varphi_0)}\right]^s. \qquad (3.29)$$

For $(\nu + \gamma) > -1$ the contribution to the conductivity originating from this point of zero curvature is small and can be neglected.

Our analysis shows that for large q the conductivity essentially varies when the Fermi surface contains effective lines of anomalous curvature. Therefore we can expect that the local geometry of the Fermi surface will influence observables, in particular, the surface impedance of a metal, under conditions of an anomalous skin effect.

3.2 Frequency Dependence of the Surface Impedance of a Metal with an Anomalous Local Geometry of the Fermi Surface

At abnormal skin effect in metals the value of the surface impedance is determined by the curvature of the effective segments of the Fermi surface which correspond to the effective electrons moving at small angles to the boundary of the metal [107]. The surface impedance can be expressed in terms of the inversed Gaussian curvature averaged over the effective part of the Fermi surface, supposing that the curvature is finite and nonzero everywhere [107], [108]. However, on the Fermi surfaces of the majority of metals exist some points or lines of zero or anomalously large curvature. The surface impedance of a semi-infinite metal, whose Fermi surface has points of flattening, was calculated in [83]. The obtained results show that the anomalies of curvature lead to some noticeable effects of qualitative character.

The contribution to the surface impedance from those segments of the Fermi surface, whose curvature is anomalously large or close to zero, can change its magnitude and frequency dependence. It is obviously important to analyze the character of these changes for probable types of anomalies of the curvature. Such a problem was solved in [54]. The calculation of the surface impedance of a massive metal was carried out for a particular type of axisymmetric Fermi surface. For these surfaces the angular dependence of their Gaussian curvature can be expressed in terms of their unique parameters. In the present section the results of [54], and also their generalization on the wider class of Fermi surfaces, are expounded.

Assume that a metal occupies a half-space $z < 0$. Concerning the Fermi surface, we shall assume that it is mirror symmetric in the momenta space about the plane $p_z = 0$. The conductivity tensor is assumed to be diagonalized in the chosen coordinate system. In the case of the specular reflection of electrons from the metal boundary we can write

$$Z_{xx} = 8i\omega \int_0^\infty \frac{dq}{4\pi i\omega\sigma_{xx}(\omega, q) - c^2 q^2}. \tag{3.30}$$

The results expounded below concern asymptotics of the surface impedance under conditions of the anomalous skin effect. The surface scattering of electrons is not taken into account by formula (3.30), because of its small influence on sliding electrons which are effective under the considered conditions. We also omit the additive, rather small, contributions to the surface impedance arising due to the Fermi-liquid interaction of electrons.

Under abnormal skin effect conditions the impedance can be represented as an expansion in inverse powers of the anomaly parameter ξ which under these conditions is large compared to unity. The main contribution to the integral (3.30) is from large q, therefore we can expand the conductivity component $\sigma_{xx}(\omega, q)$ in the integrand in powers of the small parameter ω/qv. Performing integration in (3.30) over a new variable $t(t = (q\delta)^{-1}, \delta$ is the skin depth in the abnormal skin effect), and representing the conductivity as the sum of terms (3.16) and (3.17), we can calculate the first two terms in the expansion of the surface impedance in inverse powers of the anomaly parameter

$$Z = -\frac{8i\omega}{c^2}\delta \int_0^\infty dt \frac{1}{1 - it^3(1 + it/\xi)} \approx \frac{\omega}{c^2}\delta \frac{8\pi}{3\sqrt{3}}\left[1 - i\sqrt{3} - \frac{2}{3\xi}(1 + i\sqrt{3})\right],$$
(3.31)

where $\delta = (c^2\hbar^3/e^2 p_0^2\omega)^{1/3}$, and $\xi = v_0/\omega\delta$ is the anomaly parameter.

We can obtain a similar asymptotic for the surface impedance of a metal with an axisymmetric Fermi surface. In this case, the parameter p_0^2 in the expression for σ_0 is defined by the relation (3.8), and the factor $2/(3\xi)$ in the expression for Z should be replaced by $(2\Lambda_1/3\xi)$, where Λ_1 is defined by (3.14). To consider the surface impedance for a metal with the axisymmetric Fermi surface it is more preferable to define the anomaly parameter by a relation $\xi = v_m/(\omega\delta)$. Then we arrive at the following expression for the surface impedance containing the two first terms of its expansion in powers of ξ^{-1}:

$$Z \approx \frac{\omega}{c^2}\delta \frac{8\pi}{3\sqrt{3}}\left[1 - i\sqrt{3} - \frac{2\Lambda_1}{3\xi}(1 + i\sqrt{3})\right].$$
(3.32)

Formulas (3.31) and (3.32) are valid, in particular for a spherical Fermi surface (see [109]) and they coincide in this case. The expression of the form (3.31) for an anisotropic Fermi surface was obtained in [110].

Our results (3.31), (3.32) feature the surface impedance of a metal whose Fermi surface has its Gaussian curvature finite and nonzero everywhere. For Fermi surfaces of the considered type the frequency dependence of the surface impedance is of the same character, as for a spherical surface. The main approximation of the surface impedance is proportional to $\omega^{2/3}$ and the first correction to it is proportional to $\omega^{4/3}$.

When any of the effective strips of the Fermi surface has locally flattened or nearly cylindrical segments or some points where one of the main radii of curvature vanishes (the curvature has a singularity), the asymptotic expression for the conductivity for large q and, correspondingly, asymptotic for the surface impedance have to be changed. The effect of this anomalous local geometry of the Fermi surface on the impedance is very strong for $\beta < 0$, $p_a^2/p_0^2 \sim 1$. Under these conditions the "abnormal" contribution dominates over the other terms in the expression for conductivity and determines the principal term of the surface impedance. As a result we have

$$Z = \frac{\omega}{c^2} \delta_a \xi_a^{\beta/(3+\beta)} \frac{8\pi}{3+\beta} \left(\cos \frac{\pi\beta}{2} \right)^{1/(3+\beta)} \cot \left(\frac{\pi}{3+\beta} \right) \left[1 - i \tan \left(\frac{\pi}{3+\beta} \right) \right].$$

(3.33)

Here $\xi_a = v_a/\omega\delta_a$, $\delta_a = (c^2\hbar^3/e^2 p_a^2\omega)^{1/3}$, and the value of p_a^2 is defined by relation (3.23).

For a metal with an axisymmetric Fermi surface we can also describe the main term of the expansion of the surface impedance by formula (3.33), provided that the main body of effective electrons are associated with a nearly cylindrical effective strip ($\rho \sim 1$). In this case, we have to replace the factor $\delta_a \xi_a^{\beta/(3+\beta)}$ by $\rho^{1/(3+\beta)} \xi^{\beta/(3+\beta)}$. The same formula is valid when the Fermi surface curvature tends to infinity at every point of every effective line and $0 < \beta < 1$. In the cases mentioned above, the anomalies of the curvature essentially change both the value and frequency dependence of the surface impedance. For the Fermi surface with finite and nonzero curvature everywhere the main approximation of the surface impedance is proportional to ξ^{-1} but in the considered case it appears to be proportional to $\xi^{(-3/(\beta+3))}$. The parameter $3/(\beta + 3)$ accepts values from an interval $(1, \frac{3}{2})$, when the main contribution to the impedance is brought in by electrons associated with the nearly cylindrical effective strip and from an interval $(\frac{3}{4}, 1)$ when the effective electrons are concentrated in a neighborhood of a line whose curvature tends to infinity.

Thus for $\beta < 0$ the surface impedance is essentially larger, and for $\beta > 0$ smaller, in magnitude than the surface impedance of a conventional metal. Frequency dependence of the surface impedance also changes. Now it is proportional to $\omega^{2/(\beta+3)}$. For a nearly cylindrical effective strip the exponent $2/(\beta + 3)$ varies in the interval $(\frac{2}{3}, 1)$ where the value 1 corresponds to the precisely cylindrical form of the strip. For the effective strip with a singularity of curvature ($0 < \beta < 1$) the given exponent accepts values from the interval $(\frac{1}{2}, \frac{2}{3})$.

Therefore, when the main body of effective electrons is associated with a nearly cylindrical effective strip, the impedance should increase faster with an increase in frequency than in the "conventional" case. If all the effective strips are characterized by a moderate singularity of the curvature ($0 < \beta < 1$) the surface impedance should rise slower than for a metal with finite and nonzero curvature everywhere of the Fermi surface.

Note that the Fermi surfaces of most of the metals do not satisfy the conditions which are necessary for the asymptotic (3.33) to be valid. We can rather expect that the contributions from the "abnormal" segments of the Fermi surface do not dominate over the remaining terms in the expression for the surface impedance. However, this contribution from the "abnormal" effective strip can determine the first-order correction to the main term in the expansion of the conductivity in powers of the small parameter ω/qv_a. It gives the following expression for the surface impedance:

$$Z = \frac{\omega}{c^2}\frac{8\pi}{3\sqrt{3}}\delta \times \left\{1 - i\sqrt{3} - \frac{1}{\xi^\beta}\left(\frac{p_a}{p_0}\right)^2\frac{\beta+1}{\sqrt{3}}\frac{\cot[(\beta+1)\pi/3]}{\cos(\pi\beta/2)}\right.$$

$$\left. \times \left[1 - i\tan\left(\frac{\pi}{3}(\beta+1)\right)\right]\right\}. \tag{3.34}$$

We can apply this formula to describe the contribution to the surface impedance from an effective line of zero curvature, when the relative number of effective electrons associated with this effective line is rather small $(p_a^2/p_0^2 \ll 1)$ and also from a point of zero curvature disposed on the effective line. In this last case we have to replace p_a^2 by the parameter p_1^2 defined by (3.29). The same asymptotic expression describes the surface impedance when the curvature is anomalously large and $0 < \beta < 1$. The correction to the main approximation of the surface impedance is now proportional to $\xi^{-(\beta+1)}$. By comparing (3.34) and (3.31) we conclude that in the considered case this correction appears to be larger than in a "usual" case when $\beta = 0$. Besides, the correction is now proportional to $\omega^{2(\beta+1)/3}$, i.e., its frequency dependence also changes.

At last, when the last term in expression (3.22) for $\sigma_a(\omega, q)$ is the principal term, we obtain the following asymptotic for the surface impedance:

$$Z = -2\sqrt{2}i\pi\frac{\omega}{c^2}\delta_a\left(\frac{\xi_a}{\eta_\beta}\right)^{1/4}, \qquad \beta > 1. \tag{3.35}$$

This asymptotic expression corresponds to the absence of any nonscattering damping of the electromagnetic field which occurs due to the negligible contribution from the effective electrons.

Formula (3.35) is immediately applicable for a positive value of the parameter η_β. However, negative values of η_β are also possible. It follows from definition (3.25) under the assumption that v_z has a very rapid change (a jump) in the vicinity of the effective line, so that $W(\pi/2, \varphi) \to \infty$. We can obtain negative values for this parameter η_β using models of the Fermi surface which are characterized by a jump in the longitudinal velocity v_z on the unique effective line. Such models were used in [88], [89] and also here (see Chapter 6). For $\eta_\beta < 0$ the integrand in (3.30) has a pole which is disposed near the real axis when $\omega\tau \gg 1$. It means that a new weakly damping wave can propagate in the metal. For this wave we have

$$q^4 = \frac{\omega + i/\tau}{v_0}\frac{|\eta_\beta|}{\delta_a^3} = \eta_\beta\frac{e^2 p_a^2\omega}{\hbar^3 v_a}\frac{(\omega + i/\tau)}{c^2}. \tag{3.36}$$

Such a wave was predicted in [88], [89]. It arises due to the collective motion of the electrons which (in the given special case) have nonzero v_z in an immediate neighborhood of the effective line. For negative η_β we can arrive at the result for the surface impedance multiplying the right-hand side of (3.35) by the factor $(1 + i)/\sqrt{2}$ and replacing η_β by $|\eta_\beta|$. The real part of the surface impedance at $\omega\tau \gg 1$ appears to be due to the excitation of this weakly damping electromagnetic wave mentioned above.

Thus in the absence of an external magnetic field the surface impedance of a semi-infinite metal, whose Fermi surface has points or lines of zero or anomalously large curvature, can be described by formulas (3.33)–(3.35). The obtained asymptotic formulas show that an anomaly of curvature on an effective line changes the frequency dependence of the surface impedance, and under certain conditions it can essentially change its magnitude. This follows from the relation discussed above between the curvature of the Fermi surface and the number of effective electrons.

CHAPTER 4

Cyclotron Resonance in Metals in a Normal Magnetic Field

4.1   On the Nature of Cyclotron Resonance in Metals in a Normal Magnetic Field

No satisfactory theoretical explanation has yet been given for the cyclotron resonance observed in cadmium [45]–[47], zinc [48], and potassium [49], in a magnetic
field applied perpendicularly to the metal surface. The resonance peaks in the real
part of the surface impedance in potassium have been suggested as being associated with the contribution of the Fermi-liquid cyclotron wave propagating along
the magnetic field. The detailed calculation of the surface impedance performed
for the spherical Fermi surface in [111] demonstrates, however, only a very weak
singular resonance feature due to the wave in the case of the specular reflection of
electrons from the metal surface. The contribution from this wave does not account
either for the relatively large peak height or for its shape. A similar disagreement
between theoretical predictions and experimental results is found in [47], where
the resonance peak in the derivative of the real part of the surface impedance
for cadmium was attributed to the contribution from the vicinities of the limiting
points of the electron lens. The theoretical estimate made for the case of the specular reflection of electrons in this study was significantly underestimated, too. The
disagreement between theoretical and experimental results can be considerably
reduced, if the diffuse reflection of electrons from the metal surface is assumed in
the calculation of the impedance. This assumption seems to be unrealistic since it
hardly agrees with the anomalous character of the skin effect in all experiments
with metals exhibiting cyclotron resonance in normal magnetic fields.

   A different approach to the interpretation of this phenomenon has been put
forward in [54], where the analysis is centered on the effect produced by the
shape of the Fermi surface on the behavior of the metal surface impedance. It has

been shown that when the Fermi surface of a metal has lines of anomalous (zero or infinite)curvature they affect the anomalous skin effect characteristics and the cyclotron resonance in the normal magnetic field can be significantly enhanced. The height and shape of the impedance peak at the cyclotron frequency in potassium has been correctly estimated and described under an assumption that the Fermi surface of this metal had lines of zero curvature (this assumption agrees with the results concerning the Fermi surface shape in potassium which follows from the charge density wave theory [112]–[120]).

The treatment of [54] was not sufficiently complete. Only the first two terms were retained in the expansion of the surface impedance in the inverse powers of the anomaly parameter. Therefore the analysis failed to appreciate the full extent of the contributions made by the anomalies of the curvature of the Fermi surface to the origination and development of the resonance. The analysis also ignored the Fermi-liquid interaction, and thus the resonance frequency was not investigated (in [111], [122] it was assumed that the frequency of the resonance peak in the real part of the surface impedance found for potassium was shifted with respect to the cyclotron frequency Ω and coincided with the boundary frequency of the Fermi-liquid wave ω_0). The analysis of these problems was performed in [59]–[61]. The main results of these works are set out below.

At first we consider a metal with an axisymmetric Fermi surface. This model enables us to calculate and analyze resonance contributions to the surface impedance arising due to the local geometry of the Fermi surface and also to those arising due to the Fermi-liquid cyclotron wave. We can compare the results of the analysis with the corresponding results obtained for the spherical Fermi surface in [111]. Furthermore, this model of an axisymmetric Fermi surface can be used to describe the cyclotron resonance in a normal magnetic field observed in cadmium and zinc.

Assume that a metal occupies the half-space $z < 0$. The external magnetic field will be assumed to be normal to the metal surface and parallel to the symmetry axis of the axially symmetric Fermi surface. In the case of the specular reflection of electrons from the metal boundary, the circular components of the surface impedance tensor are

$$Z_\pm = \frac{8i\omega}{c^2} \int_0^\infty \frac{dq}{4\pi i\omega\sigma_\pm/c^2 - q^2}, \tag{4.1}$$

where q and ω are the wave vector and frequency of the alternating electromagnetic field, and $\sigma_\pm$ are the circular components of the transverse conductivity.

When we calculate the surface impedance (in the case of polarization corresponding to the propagation of the Fermi-liquid cyclotron wave) it is a good point in (4.1) to turn to the integration over a new variable $t = (q\delta)^{-1}$ where δ is the skin depth in the anomalous skin effect. The integration range can then be divided into two segments with different asymptotic behaviors of conductivity

$$Z = Z_1 + Z_2,$$
$$Z_1 = -\frac{8i\omega}{c^2}\delta \int_0^{t_0} \frac{dt}{1 - it^3\overline{\sigma}(t/\xi)}, \tag{4.2}$$

$$Z_2 = -\frac{8i\omega}{c^2}\delta \int_{t_0}^{\infty} \frac{dt}{1 - it^3\overline{\sigma}(t/\xi)}.$$ (4.3)

Here $\xi = v_m/(\omega\delta)$ is the anomaly parameter. The parameter t_0 is chosen so that $u = t/\xi$ is smaller than unity for all t up to $t = t_0$. Under these conditions we can calculate Z_1 using the expansion (3.9) for $\overline{\sigma}$. The coefficients Λ_n in this expansion now depend on the external magnetic field.

Choosing the polarization which corresponds to the propagation of Fermi-liquid cyclotron waves, helicons, and dopplerons we can write the following expressions for the first coefficients in the expansion of $\overline{\sigma}$ in powers of the small parameter "u":

$$\Lambda_1 = -\frac{ig}{\pi}\left(a\chi + \frac{\pi^2}{g^2}\overline{\alpha}_1 - b^2\overline{\alpha}_2\right),$$ (4.4)

$$\Lambda_2 = d\chi^2 - 2\overline{\alpha}_1 a\chi - 2\alpha_2\frac{b}{c}g\chi - 2\overline{\alpha}_1\overline{\alpha}_2 b^2 - \overline{\alpha}_1^2\frac{\pi^2}{g^2},$$ (4.5)

here $\chi = 1 - \Omega/\omega + i/\omega\tau$.

Under the conditions of the anomalous skin effect the surface impedance can be presented in the form of expansion in the inverse powers of the anomaly parameter. The main terms of this expansion originate from the addend Z_1. The remaining addend Z_2 does not contribute either to the principal term of the impedance or to the first correction to it. However, we cannot neglect this term because it contains the resonance contribution from the Fermi-liquid cyclotron wave.

The first terms of the expansion of Z_1 in powers of $(\xi)^{-1}$ can readily be found by expanding the integrand in (4.1) in powers of the parameter ρ given by

$$\overline{\sigma} = 1 + \rho.$$ (4.6)

This parameter is small for $\xi \gg 1$. We obtain

$$Z_1 = \frac{8\omega}{c^2}\delta\left[\frac{\pi}{3\sqrt{3}}(1 - i\sqrt{3}) + \frac{2\pi}{9\sqrt{3}}\frac{\Lambda_1}{\xi}(1 + i\sqrt{3})\right.$$
$$\left. + \frac{\Lambda_1^2 - \Lambda_2}{\xi^2}\left(\ln t_0 + \frac{i\pi}{3}\right) - \frac{1}{2t_0^2} + \frac{\Lambda_2}{3\xi^2} - \frac{\Lambda_1}{t_0\xi}\right] + \delta Z_1.$$ (4.7)

To arrive at the result (4.7) we used three first terms of the expansion of $\overline{\sigma}$ in powers of the parameter ρ and we kept the terms of the order of $(t/\xi)^2$ in the expression for ρ. Taking into account next the terms in the expansion of $\overline{\sigma}$ in powers of ρ or keeping next terms in the expansion of ρ itself we obtain that the corresponding integrals are divergent. This means that we cannot expand the correction δZ_1 in the inverse powers of the anomaly parameter. However, the correction term δZ_1 can be written as

$$\delta Z_1 = \frac{8\omega}{c^2}\delta \int_0^{t_0}\left(\frac{1}{t^3\overline{\sigma}(t/\xi)} - \frac{1}{t^3} + \frac{\Lambda_1}{\xi t^2} - \frac{\Lambda_1^2 - \Lambda_2}{\xi^2 t}\right)dt.$$ (4.8)

This expression is correct to within the terms of the order of $(\xi)^{-3}$.

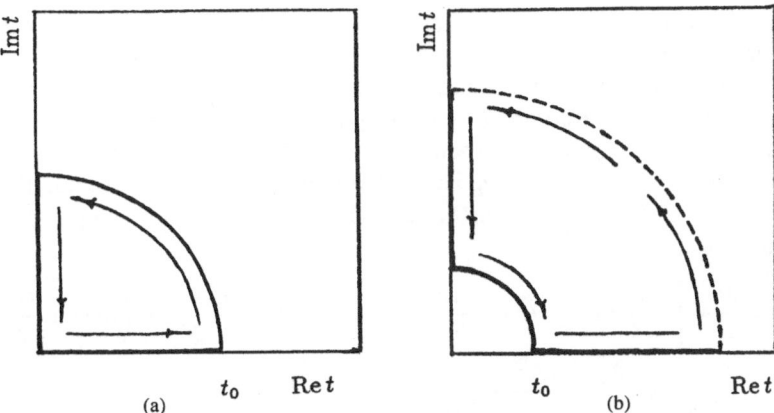

FIGURE 4.1. Contours used for calculation of the quantities (a) δZ_1 and (b) δZ_2.

When we calculate the second term in the expression for the impedance, we can ignore the unity in the denominator of the integrand in (4.3):

$$Z_2 = \frac{8\omega}{c^2}\delta \int_{t_0}^{\infty} \frac{dt}{t^3\overline{\sigma}(t/\xi)}. \tag{4.9}$$

Equations (4.7)–(4.9) depend on the boundary parameter t_0 which has been introduced arbitrarily. As should be expected, however, we have $dZ/dt_0 = 0$, that is, the impedance is independent of t_0. Therefore, we can select a specific value of the parameter t_0 (from the range of $t_0 \sim \xi$) and then derive various equivalent forms of (4.7)–(4.9) which will be more or less suitable for analysis of the behavior of the surface impedance as a function of the external magnetic field and also for comparison with the available results.

To proceed we can perform analytic continuation of the integrands in (4.8) and (4.9) to the first quadrant of the complex plane. To calculate the term δZ_1 we will use a contour shown in Figure 4.1a. The contour consists of segments of real and imaginary axes and a circular arc of the radius t_0. Inside of this contour the integrand can be presented as a sum of the convergent Taylor expansion. Therefore the integrand is an analytic function in this region. To calculate Z_2 we choose the contour shown in Figure 4.1b. In this case the integrand has a pole inside the contour of integration. The contribution from the residue concerning this pole to Z_2 equals

$$\delta Z_2 = \frac{4\pi^2}{c^2}\omega\delta\frac{1}{\xi^2}\left(\frac{A_2}{1+A_2}\right)^2 W\sqrt{\frac{\chi(1+A_2)-A_2}{a^*A_2}}. \tag{4.10}$$

Here

$$W = \left(\int_{-1}^{1}\overline{s}(x)\overline{v}(x)x\,dx\right)^2 \Big/ \overline{Q}_0\overline{Q}_2 a^*.$$

For a spherical Fermi surface we have $a^* = \frac{8}{35}$ and $W = \frac{21}{35}$, and we obtain the same result as in [111].

Under the conditions of the anomalous skin effect, the term δZ_2 which is of the order of $1/\xi^2$ is small in comparison with Z_1. It cannot be neglected, however, since it is precisely this term that is associated with the contribution made by the Fermi-liquid cyclotron wave. When we have identified the cyclotron wave contribution δZ_2, we can perform some transformations and obtain the following expression for the surface impedance Z:

$$
Z = \frac{8\omega}{c^2}\delta \left[\frac{\pi}{3\sqrt{3}}(1 - i\sqrt{3}) + \frac{2\pi}{9\sqrt{3}}\frac{\Lambda_1}{\xi}(i - \sqrt{3}) + \frac{\Lambda_1^2 - \Lambda_2}{\xi^2}\left(\ln t_0 + \frac{i\pi}{3}\right) \right.
$$
$$
\left. + \frac{1}{2t_0} + \frac{\Lambda_2}{3\xi^2} - \frac{\Lambda_1^2}{2\xi^2} - \frac{i\Lambda_1}{t_0\xi} - Y_1 - Y_2 \right] + \delta Z_2. \tag{4.11}
$$

Here we have

$$
Y_1 = \frac{1}{\xi^2}\int_{\xi/t_0}^\infty \left(\frac{y}{\overline{\sigma}(i/y)} - y + i\Lambda_1 + \frac{\Lambda_1^2 - \Lambda_2}{y}\right)dy, \tag{4.12}
$$

$$
Y_2 = \frac{1}{\xi^2}\int_0^{\xi/t_0} \frac{y}{\overline{\sigma}(i/y)}dy. \tag{4.13}
$$

Using (4.4), (4.5) for $\Lambda_{1,2}$ we can readily show that, for the spherical Fermi surface, expression (4.13) fully coincides with a similar result derived in [111] by a different method for $t_0 = \xi/\chi$.

It is better to select another boundary value t_0 to analyze the dependence of the surface impedance on the external magnetic field, $t_0 = \xi$. Then (4.11)–(4.13) will be rewritten as

$$
Z = \frac{8\omega}{c^2}\delta \left\{ \frac{\pi}{3\sqrt{3}}(1 - i\sqrt{3}) + \frac{2g}{9\sqrt{3}\xi}\left(a\chi + \frac{\pi^2}{g^2}\overline{\alpha}_1 - b^2\overline{\alpha}_2\right)(1 + i\sqrt{3}) \right.
$$
$$
+ \frac{1}{\xi^2}\left[-\left(\frac{g^2a^2}{\pi^2} + d\right)\chi^2 + 2\overline{\alpha}_2 bg\left(\frac{gab}{\pi^2} + c\right)\chi - \frac{g^2}{\pi^2}b^4\overline{\alpha}_2^2\right]\left(\ln\xi + \frac{i\pi}{3}\right)
$$
$$
+ \frac{1}{\xi^2}\left(\frac{1}{2} + \frac{d\chi^2}{3} + \frac{g^2}{2\pi^2}a^2\chi^2 - \frac{g}{\pi}a\chi\right) - \frac{\overline{\alpha}_1}{\xi^2}\left(\frac{a\chi}{6} + \frac{\pi}{g}\right)
$$
$$
- \frac{\overline{\alpha}_2 bg}{\xi^2}\left(-\frac{2\chi}{3c} + \frac{b}{\pi} - \frac{g}{\pi^2}ab\chi\right) - \frac{\pi^2}{6g^2\xi^2}\overline{\alpha}_1^2
$$
$$
\left. - \frac{b^2}{6\xi^2}\overline{\alpha}_1\overline{\alpha}_2 + \frac{b^4g^2}{2\pi^2\xi^2}\overline{\alpha}_2^2 - Y_1 - Y_2 \right\} + \delta Z_2, \tag{4.14}
$$

$$
Y_1 = \frac{1}{\xi^2}\int_1^\infty dy \left\{ \frac{y}{\overline{\sigma}(i/y)} - y + \frac{g}{\pi}\left(d\chi + \frac{\pi^2}{g^2}\overline{\alpha}_1 + b^2\overline{\alpha}_2\right) \right.
$$
$$
\left. + \frac{1}{y}\left[-\left(\frac{g^2a^2}{\pi^2} + d\right)\chi^2 + 2\overline{\alpha}_2 bg\left(\frac{abg}{\pi^2} + c\right)\chi - \frac{g^2}{\pi^2}b^4\overline{\alpha}_2^2\right] \right\}, \tag{4.15}
$$

$$
Y_2 = \frac{1}{\xi^2}\int_0^1 \frac{ydy}{\overline{\sigma}(i/y)}. \tag{4.16}
$$

The result indicates the lack of resonance at the cyclotron frequency, since not a single term in (4.14) exhibits a resonance behavior. The lack of resonance at the cyclotron frequency for $t_0 = (\chi/\xi)^{-1}$ is proved in the Appendix 2. We see that in the case of the anomalous skin effect in a metal with an axially symmetric Fermi surface, which everywhere has a finite nonzero curvature, the behavior of the surface impedance does not exhibit a qualitative difference from that of the surface impedance of the metal with a spherical Fermi surface. In both cases the only resonance contribution to the surface impedance is the term δZ_2 describing a root singularity in the impedance derivative at the boundary frequency of the Fermi-liquid cyclotron wave $\omega_0 = \Omega(1 + A_2)$. Regarding the behavior of the surface impedance itself, the term δZ_2 gives rise to a break at the boundary frequency of the Fermi-liquid cyclotron wave against the background of its steady dependence on the magnetic field.

The amplitude of the impedance singularity due to the Fermi-liquid cyclotron wave has been estimated under the conditions typical for experimental observations of the cyclotron resonance in the normal magnetic field ($\omega\tau \sim 10$, $\xi^3 \sim 10^3 - 10^4$). Under the assumption that the magnitude of the Fermi-liquid parameter A_2 is not greater than 0.1, the estimate is not more than 10^{-4}–10^{-5} of the real part of the impedance. This estimate is smaller, by at least two orders of magnitude, than the resonance peak in potassium discovered in the experiments of [49]. The root singularity in the impedance derivative caused by the cyclotron wave is also too weak to be useful for describing the resonance in cadmium and zinc. It thus may be suggested that no satisfactory theoretical description of the cyclotron resonance in the normal magnetic field can be obtained for a metal whose Fermi surface has a nonzero curvature everywhere.

These results emphasize the part played by singularities of the Fermi surface curvature in the generation of the cyclotron resonance in a normal magnetic field. When the effective cross-sections of the Fermi surface include some of an anomalous curvature, the expression for the transverse conductivity in the range of small u contains the following additional term σ_a:

$$\sigma_a(q) = \sigma_0\rho\left[1 - i\tan\left(\frac{\pi\beta}{2}\right)\right](u\chi)^\beta\left\{1 - \frac{i\pi}{g}\overline{\alpha}_1\rho\left[1 - i\tan\left(\frac{\pi\beta}{2}\right)\right]\chi^\beta u^{\beta+1}\right.$$
$$\left. + \frac{\rho}{g}\Theta(u\chi)^{\beta+2} + (2\pi b)^4\overline{\alpha}_1\overline{\alpha}_2(u\chi)^2 + \frac{\pi^2}{g^2}\overline{\alpha}_1^2 u^2\chi^{2-\beta} + \cdots\right\}. \quad (4.17)$$

Here we use the same notation as in expression (3.13) describing the "anomalous" contribution to the conductivity in the absence of the external magnetic field.

We first consider the case when the effective cross-sections of the Fermi surface include one of a zero curvature. It can be the only cross-section of its kind if it is the central cross-section of the Fermi surface. Otherwise, the Fermi surface must have two effective cross-sections, with an anomalous curvature, which are symmetrically arranged with respect to a plane $p_z = 0$. The relative contribution of these cross-sections to the surface impedance depends on the relative number of the effective electrons associated with them. If the effective cross-sections of zero curvature contain the same number (or more) of electrons as the remaining effective cross-sections ($\rho \sim 1$), the first term in (4.17) is the principal term in the expansion of the conductivity in powers of u and it contributes strongly to the surface impedance. As a result the principal term of the impedance in the case of the anomalous skin effect is given by

$$Z = \frac{8\pi\omega}{c^2}\delta\left(\frac{\xi\beta}{\rho\cos(\pi\beta/2)}\right)^{1/(\beta+3)}\frac{\cot(\pi/(\beta+3))}{\beta+3}\left[1 - i\tan\left(\frac{\pi}{\beta+3}\right)\right]\chi^{-\beta/(\beta+3)}.$$

$$(4.18)$$

It can be seen from the approximation (4.18) that when the Fermi surface has an effective cross-section of zero curvature, which is associated with a considerable part of the effective electrons, the surface impedance of the metal is significanty smaller than in the conventional case. The impedance then substantially depends on the external magnetic field. Near the cyclotron frequency the real and imaginary parts of the surface impedance increase sharply and then level off, and the increase is of the same order of magnitude as the principal term of the surface impedance (see Figure 4.2). The derivatives of the real and imaginary parts of the impedance must accordingly exhibit sharp and high peaks at the cyclotron frequency.

No experimental observations of such impedance behavior have been made in conventional metals. The explanation is that it is unrealistic to assume a large (of the order of unity) value of the parameter ρ. We must rather expect that in real metals the relative number of electrons associated with the cross-section of the anomalous curvature is small, that is, $\rho \ll 1$. Under these conditions the principal term in the

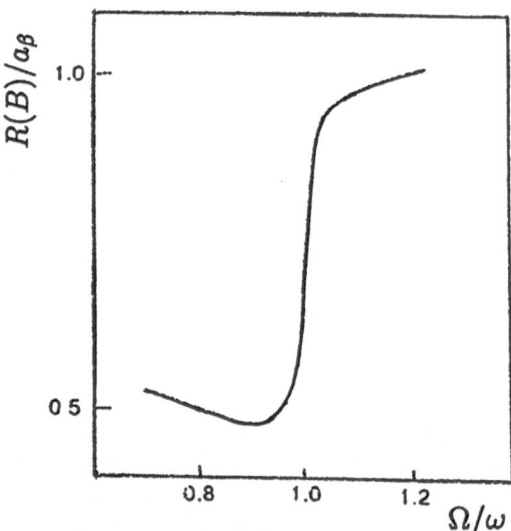

FIGURE 4.2. Dependence of the real part of the surface impedance of a metal on the magnetic field near the cyclotron resonance, under the assumption that the relative number of effective electrons corresponding to the nearly cylindrical effective cross-section of the Fermi surface is of the order of unity ($\rho \sim 1$). The curve is plotted for $\beta = -0.3$, $\omega\tau = 10$.

surface impedance expansion in inverse powers of the anomaly parameter remains the same, as in a metal whose Fermi surface has a finite and nonzero curvature everywhere. However, we obtain additional higher-order terms corresponding to the contribution of the anomalous effective cross-section. If we keep only the terms of the second order in the small parameters ξ^{-1} and ρ, we obtain the following expression for the additional contribution Z_a to the surface impedance:

$$
\begin{aligned}
Z_a = & -\frac{8\pi v_m}{9c^2\xi}\left(\frac{\chi}{\xi}\right)^\beta \rho\left(\frac{\beta+1}{\cos(\pi\beta/2)}\right)\cot\left(\frac{\pi(\beta+1)}{3}\right)\left[1-i\tan\left(\frac{\pi(\beta+1)}{3}\right)\right] \\
& +\frac{8\pi v_m}{27c^2\xi}\frac{(\beta+2)(2\beta+1)}{(\cos(\pi\beta/2))^2}\rho^2\left(\frac{\chi}{\xi}\right)^{2\beta}\cot\left(\frac{\pi(2\beta+1)}{3}\right)\left[1-i\tan\left(\frac{\pi(2\beta+1)}{3}\right)\right] \\
& -\frac{16(\beta+4)}{3\xi^2c^2}g\left(b\chi+\frac{\pi^2}{g^2}\overline{\alpha}_1-b^2\overline{\alpha}_2\right)\rho\left(\frac{\chi}{\xi}\right)^\beta \\
& \times \frac{\left[1+i\tan(\pi\beta/2)\right]\left[1-i\tan(2\pi(\beta+1)/3)\right]}{\tan(2\pi(\beta+1)/3)} \\
& +\frac{16\pi v_m}{9c^2\xi^2}(1+\beta)\rho^2\left(\frac{\chi}{\xi}\right)^{2\beta}\overline{\alpha}_1\left[1+i\tan\left(\frac{\pi\beta}{2}\right)\right] \\
& \times\left[1-i\tan\left(\frac{2\pi(\beta+1)}{3}\right)\right]\cot\left(\frac{2\pi(\beta+1)}{3}\right).
\end{aligned}
\tag{4.19}
$$

All the terms in (4.19) exhibit resonance features at the cyclotron frequency at $\beta < 0$. It can thus be seen that the Fermi-liquid interaction does not produce a shift in the resonance but only a change in the resonance peak shape. Moreover, this change is insignificant since the first term, which does not include the Fermi-liquid parameters, is greater than the other terms in the order of magnitude.

A comparison of (4.19) and (4.14) demonstrates that the term Z_a is the largest of the correction terms added to the principal term in the surface impedance approximation. The asymptotic expression for this term has the following form:

$$
\begin{aligned}
Z = & \frac{8\pi v_m}{3\sqrt{3}c^2\xi}\left\{1-i\sqrt{3}-\frac{\rho}{\sqrt{3}}\frac{\beta+1}{\cos(\pi\beta/2)}\cot\left(\frac{\pi(\beta+1)}{3}\right)\right. \\
& \left.\times\left[1-i\tan\left(\frac{\pi(\beta+1)}{3}\right)\right]\left(\frac{\chi}{\xi}\right)^\beta\right\}.
\end{aligned}
\tag{4.20}
$$

The first correction term added to the principal term in the surface impedance approximation for $\beta < 0$ describes a resonance peak at the cyclotron frequency (Figure 4.3). The shape of the peak in the real part of the impedance is close to that recorded for potassium (see [49]). The peak height depends on the parameter ρ and can be rather large when ρ is not too small. For instance, for $\rho \sim 10^{-2}$, $\omega\tau \sim 10$, and $\xi^3 \sim 10^4$, the resonance amplitude in Re Z is approximately 10^{-2} of the principal term of the surface impedance. This is considerably larger than the amplitude of the resonance feature (4.10) which arises due to the Fermi-liquid wave, and has the same order of magnitude as the recorded resonance amplitude.

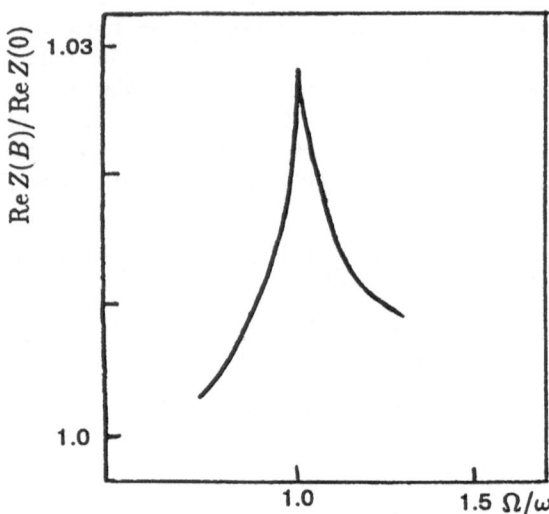

FIGURE 4.3. Dependence of the real part of the surface impedance of a metal on the magnetic field near the cyclotron resonance in the case when an extremal section of the Fermi surface coincides with a line of parabolic points and $\rho \ll 1$. The curve is plotted for $\rho = 0.01$, $\beta = -0.5$, $\omega\tau = 50$, $\xi^3 = 10^4$.

When ρ is not too small the asymptotic expression (4.20) for the impedance also remains applicable when the effective cross-sections of the Fermi surface include one of an infinite curvature and $0 < \beta < 1$. Under these conditions the first correction term of the surface impedance in the vicinity of the cyclotron frequency decreases monotonically upon an increase in the magnetic field. However, the resonance at the cyclotron frequency should take place in the derivatives of the real and imaginary parts of impedance. Resonance features of this type have been observed for cadmium and zinc. The Fermi surface of these metals exhibits a narrow electron lens. We can expect the curvature at the lens edges to be very large. Under this assumption the resonance occurring in the surface impedance derivative may be attributed to the contribution of the effective cross-section of an anomalously large curvature passing through the lens edge.

Finally we consider the Fermi surface which has a single effective cross-section of an infinitive curvature and $\beta > 1$. In this case the first term in expansion (3.9) disappears. The principal term of the conductivity is given by the second term of this expansion (3.9), which equals

$$- i e^2 p_m S(0) u \chi\, Q^* / (4\pi\hbar^3 m_\perp v_m q), \qquad (4.21)$$

where $Q^* = \int_0^1 (\bar{s}(x)\, dx / \bar{v}^2(x))$, and the integral is converging.

For $\beta > 1$ the remaining terms of expansion (4.17) for the transverse conductivity are small compared to this term (4.21) and they do not change the asymptotic expression for the surface impedance which is given by

$$Z = \frac{2\pi\omega}{c^2}\bar{\delta}\left(\frac{\bar{\xi}}{\chi Q^*}\right)^{1/4}(1-i).\tag{4.22}$$

Here $\bar{\delta} = (c^2\hbar^3/\omega e^2\bar{p}^2)^{1/3}$, $\bar{\xi} = v_m/\omega\bar{\delta}$, $\bar{p}^2 = p_m S(0)/m_\perp v_m \pi^2$ have the dimensions of momentum squared. The obtained result describes the dependence of the surface impedance on the magnetic field which is very distinct from all those analyzed above. Now the principal term of the impedance exhibits a resonance peak at the cyclotron frequency. It is caused by the additional pole of the integrand (4.1) which appears when $\beta > 1$. For $\omega\tau \gg 1$, this pole is arranged near the real axis and corresponds to the weakly damping electromagnetic wave

$$\omega - \Omega + i/\tau = \frac{(q\bar{\delta})^4 v_m}{Q^*\bar{\delta}}.\tag{4.23}$$

Such a wave can also propagate in metal in the absence of the external magnetic field (see (3.36)). In the case under consideration the resonance arises due to excitation of these weakly damping waves.

4.2 Cyclotron Resonance in Potassium in a Normal Magnetic Field

The cyclotron resonance in potassium was associated for the first time with the presence of zero-curvature regions on the Fermi surface in [121]. The authors of this work proposed that the resonance is due to the cylindrical region on the Fermi surface, which has been split off by a gap from its main part. However, the assumption concerning the existence of a finite strictly cylindrical region on the Fermi surface, which has the same cyclotron frequency as the main sphere, is hardly justified. The resonance should rather be associated with a local vanishing of the curvature on the main part of the Fermi surface, under the assumption that deviations from the spherical shape of the Fermi surface in potassium are on the whole insufficiently large for changing the cyclotron frequency. Under these assumptions, it is possible to estimate and describe the shape of the cyclotron resonance peak correctly.

Considering the Fermi surface of potassium within the framework of a nearly free electrons model we can show that the curvature of the Fermi surface vanishes at some points and lines. Actually, using the above mentioned approximation, we arrive at the following dispersion relation:

$$E = \frac{k^2}{2m} + \frac{g^2}{2m} - \frac{1}{m}\sqrt{(\mathbf{k}\cdot\mathbf{g})^2 + m^2 V^2},\tag{4.24}$$

where m is the mass of a free electron, $\mathbf{k} = \mathbf{p} - \mathbf{g}$, $\mathbf{g} = \hbar\mathbf{G}/2$, $\mathbf{G}$ is a reciprocal lattice wave vector, and V is the Fourier component of the potential energy of an electron in the lattice field which corresponds to the vector **G**. The velocity of electrons is given by

$$\mathbf{v} = \frac{\partial E}{\partial \mathbf{k}} = \frac{\mathbf{k}}{m} - \frac{\mathbf{g}}{m} \frac{(\mathbf{k} \cdot \mathbf{g})}{\sqrt{(\mathbf{k} \cdot \mathbf{g})^2 + m^2 V^2}}. \tag{4.25}$$

At those points of the Fermi surface where the longitudinal component of the velocity v_z vanishes, the Gaussian curvature of the surface is

$$K = \frac{1}{v^4} \left\{ \frac{\partial v_z}{\partial p_z} \left(\frac{\partial v_y}{\partial p_y} v_x^2 + \frac{\partial v_x}{\partial p_x} v_y^2 - 2 \frac{\partial v_x}{\partial p_y} v_x v_y \right) - \left(\frac{\partial v_z}{\partial p_y} v_x - \frac{\partial v_z}{\partial p_x} v_y \right)^2 \right\}. \tag{4.26}$$

Correspondingly, the principal radii of the curvature R_1 and R_2 are

$$R_{1,2}^{-1} = \frac{1}{2v^3} \left\{ (v_y^2 + v_x^2) \frac{\partial v_z}{\partial p_z} + v_y^2 \frac{\partial v_x}{\partial p_x} + v_x^2 \frac{\partial v_y}{\partial p_y} - 2 v_x v_y \frac{\partial v_y}{\partial p_y} \right\}$$

$$\pm \left\{ \left[(v_y^2 + v_x^2) \frac{\partial v_z}{\partial p_z} - v_y^2 \frac{\partial v_x}{\partial p_x} - v_x^2 \frac{\partial v_y}{\partial p_y} + 2 v_x v_y \frac{\partial v_x}{\partial p_y} \right]^2 \right.$$

$$\left. + 4(v_y^2 + v_x^2) \left(v_x \frac{\partial v_z}{\partial p_y} - v_y \frac{\partial v_z}{\partial p_x} \right)^2 \right\}^{1/2} \frac{1}{2v^3}. \tag{4.27}$$

When we choose the coordinate system so that the reciprocal lattice vector **G** is arranged on a plane $p_z = 0$, the longitudinal velocity of electrons v_z does not depend on the transverse components of the quasi momentum p_x, p_y. In this case, we can simplify the expressions for the radii of curvature

$$R_1^{-1} = -\frac{1}{v} \frac{\partial v_z}{\partial p_z}, \tag{4.28}$$

$$R_2^{-1} = -\frac{1}{v^3} \left\{ v_y^2 \frac{\partial v_x}{\partial p_x} + v_x^2 \frac{\partial v_y}{\partial p_y} - 2 v_x v_y \frac{\partial v_x}{\partial p_y} \right\}. \tag{4.29}$$

Here R_2 is the radius of curvature at the points of the cross-section of the Fermi surface with the plane $p_z = 0$. In our case, this cross-section is the unique line on the surface where the longitudinal velocity v_z vanishes. The quantity R_1 represents the radii of curvature of the cross-sections of the Fermi surface with planes perpendicular to the plane $p_z = 0$ at the points of their intersection with the cross-section $p_z = 0$. In other words, R_2^{-1} describes the curvature of the "equator" of the Fermi surface and R_1^{-1} gives the curvature of its "meridians" at the points of their intersection with the "equator."

It follows from formulas (4.25), (4.28), (4.29) that the curvature of the "meridians" at the points of their intersection with the "equator" is finite and nonzero. On the contrary, the curvature of the "equator" vanishes at some points. To be convinced, we can arrange the coordinate system on the xy-plane so that the angle between the vector **G** and the "x"-axis equals $\pi/4$. Near the boundary of the Brillouin zone the dependence of R_2^{-1} on an azimuthal angle Φ can be written as follows:

$$R_2^{-1} \approx -\frac{1}{mv}\left[1 - \frac{p_F^2}{8mV}(1 - \sin 2\Phi)^2\right], \tag{4.30}$$

Here p_F is the Fermi momentum. When $\Phi \approx \pi/4 \pm 4\sqrt{2}\, mV/p_F^2$, this corresponds to the parabolic points at the central cross-section.

In the framework of the considered model the Fermi surface represents, roughly speaking, a sphere with "knobs" located at those segments which are close to the boundaries of the Brillouin zone. The line of zero curvature passes along the boundary between a knob and the main body of the Fermi surface. The radius of curvature R_2 tends to infinity at the points of intersection of these lines of zero curvature and the "equator." However, the contributions from the vicinities of these points to the transverse conductivity are too small to cause the appearance of an "anomalous" term of noticeable value which stipulates the cyclotron resonance. Thus it appears that (in the framework of the nearly free electron approximation) the cyclotron resonance in a normal magnetic field cannot exhibit itself when the magnetic field is directed along the symmetry axis of high order of the monocrystalline sample.

Assume that the magnetic field is slightly deviated from the symmetry axis so that the reciprocal lattice vector $\mathbf{G}$ has a nonzero component G_z($|G_z| \ll |G_x|, |G_y|$). The effective line ($v_z = 0$) passes through the distorted (nonspherical in shape) segments of the Fermi surface when the angle θ between the magnetic field ("z"-axis) and the symmetry axis is of the order of $\sqrt{V/\zeta}$ or smaller. Under these conditions we can find on the effective line some points at which the curvature radius R_1, corresponding to the "meridional" cross-sections of the Fermi surface, tends to infinity. Vicinities of these points can contribute strongly to the transverse conductivity.

The coordinates of these points can be found as a result of solving the following set of equations:

$$\begin{cases} p_F^2 = \mathbf{k}^2 + \mathbf{g}^2 - 2\sqrt{(\mathbf{k}\cdot\mathbf{g})^2 + m^2 V^2}, \\[2mm] k_z = \dfrac{(\mathbf{k}\cdot\mathbf{g})g_z}{\sqrt{(\mathbf{k}\cdot\mathbf{g})^2 + m^2 V^2}}, \\[2mm] 1 - \dfrac{m^2 V^2 g_z^2}{\sqrt{\left[(\mathbf{k}\cdot\mathbf{g})^2 + m^2 V^2\right]^3}} = 0, \\[2mm] k_x g_y = k_y g_x. \end{cases} \tag{4.31}$$

The first equation of the system (4.31) means that the desired point is on the Fermi surface, and the second that it belongs to the effective line where $v_z = 0$. The remaining equations are satisfied when the partial derivatives of v_z with respect to k_z and ψ are equal to zero. The system (4.31) has no solutions far away from the boundaries of the Brillouin zone when $mV \ll |(\mathbf{k} \cdot \mathbf{g})|$. However, near the boundaries, when $mV \geq |(\mathbf{k} \cdot \mathbf{g})|$, we can find the solution of the system and write it in the form

$$k_z^2 \approx \tfrac{2}{3} g_z^2 \left(1 - \frac{mV}{g_z^2} \right),$$

$$k_x^2 \approx \frac{2}{3} \frac{g_x^2}{g_\perp^2} mV \left\{ 1 + \sqrt{15} - \frac{g_z^2}{mV} - \sqrt{\frac{3}{5} \frac{mV}{g_z^2}} - \frac{3}{2} \frac{g^2 - p_F^2}{mV} \right\}, \qquad (4.32)$$

$$k_x^2 \approx \frac{2}{3} \frac{g_y^2}{g_\perp^2} mV \left\{ 1 + \sqrt{15} - \frac{g_z^2}{mV} - \sqrt{\frac{3}{5} \frac{mV}{g_z^2}} - \frac{3}{2} \frac{g^2 - p_F^2}{mV} \right\}.$$

The solution exists for $g_z^2 > mV$. Therefore we have one more constraint on the range of possible values of the angle θ. As a result we have

$$\sqrt{\frac{V}{\varsigma}} > \theta > \sqrt{\frac{V}{2\varsigma}}. \qquad (4.33)$$

Thus for any angle θ belonging to the interval (4.33) we can find some points on the effective line $v_z = 0$ whose coordinates satisfy (4.31). The dependence of the longitudinal velocity v_z on the variables p_z and ψ (the latter describes a position of the electron on its cyclotron orbit) near these points is given by the relation

$$v_z(p_z, \psi) = a(p_z - p_0)^2 + b(\psi - \psi_0)^2 + 2c(p_z - p_0)(\psi - \psi_0). \qquad (4.34)$$

The following equalities have to be satisfied in these points:

$$\frac{\partial v_z}{\partial p_z} = 0, \qquad \frac{\partial v_z}{\partial p_x} - \frac{v_x}{v_y} \frac{\partial v_z}{\partial p_y} = 0.$$

Therefore, as follows from (4.29), $R_1^{-1} = 0$, $R_2^{-1} = -1/mv$.

As a result of this analysis we arrive at the conclusion that when the magnetic field deviates from the symmetry axis, the effective line $v_z = 0$ passes through flattened segments of the Fermi surface. The electron density of the states is enormously large on these flattened segments. The electrons associated with these flattened parts of the Fermi surface can form the "anomalous" contribution to the surface impedance which corresponds to the cyclotron resonance in potassium in a normal magnetic field.

The experimental results obtained in the 1980s refuted the concept of the Fermi surfaces of alkali metals as closed, nearly spherical, surfaces. It was shown in works by A.W. Overhauser (see [112]–[114]) that inhomogeneity in the distribution of density of electrons in space (charge-density wave) can arise in these metals. These charge-density waves lead to discontinuities of the Fermi surface. When the wave vector of the charge-density wave $\mathbf{Q}$ is close to $2k_F$, open orbits in the direction of $\mathbf{Q}$ can appear. Experimental data (see [115], [116]) imply the existence of small energy gaps and open orbits in potassium and sodium.

The observed distortions of the Fermi sphere must be accompanied by transitions from positive to negative curvature, i. e. , zero-curvature points and lines. Therefore we can assume that zero-curvature points are spread widely over the Fermi surface and fall on cyclotron orbits of effective electrons over a wide range of magnetic field orientations. However, the distortions of the Fermi sphere as a whole may be not large enough to noticeably change the cyclotron frequency. Under these assumptions we can readily describe the cyclotron resonance observed in a polycrystalline sample of potassium in a normal magnetic field.

In view of the smallness of the distortion of the Fermi sphere, we can assume in the following calculations that the conductivity is diagonalized in circular components for an arbitrary direction of the magnetic field. In addition, we assume that departures of the Fermi surface from the spherical shape are significant only for the longitudinal component v_z of the electron velocity, while the transverse velocity components $v_{x,y}$ remain the same as for the undistorted Fermi sphere. Under these assumptions, and taking into account the Fermi-liquid coefficients A_j with $j \leq 2$, we can obtain the following expression for the transverse conductivity (by choosing the circular polarization corresponding to the resonance):

$$\sigma = \frac{ie^2}{4\pi\hbar^3 q} \frac{\Phi_0 \left[1 - \frac{2}{15}\alpha_1 u \Phi_2\right] + \frac{2}{15}\alpha_2 u \Phi_1^2}{\left[1 - \frac{4}{3}\alpha_1 u \Phi_0\right]\left[1 - \frac{2}{15}\alpha_2 u \Phi_2\right] - \frac{8}{45}\alpha_1 \alpha_2 \Phi_1^2 u^2}, \tag{4.35}$$

$$\Phi_m = \int_{-p_m}^{p_m} dp_z \frac{v_\perp^2(p_z)p_z^m}{w - \langle v_z \rangle}, \tag{4.36}$$

where $w = \omega - \Omega + i/\tau$. This formula coincides with the corresponding result obtained for the Fermi sphere besides the only discrepancy. The variable v_z in the denominators of the integrands of the integrals Φ_m is replaced by its average over the cyclotron orbit $\langle v_z \rangle$. Suppose that the longitudinal velocity is weakly dependent on ψ so that the following inequality is satisfied:

$$\frac{1}{2\pi} \left| \int_0^{2\pi} [v_z(p_z, \psi') - v_z(p_z, \psi)]d\psi' \right| \ll |v_z(p_z, \psi)|. \tag{4.37}$$

In this case, we can write the integrals (4.36) in the form

$$\Phi_m = \frac{1}{2\pi} \int_0^{\pi} d\psi \int dp_z \frac{v_\perp^2(p_z)p_z^m}{w - v_z(p_z, \psi)}, \tag{4.38}$$

which is more convenient for the following calculations.

Let us consider the influence of zero-curvature points on the surface impedance of a metal. If we disregard the Fermi-liquid correlation, formula (4.35) for the conductivity can be considerably simplified:

$$\sigma(\omega, q) = \frac{ie^2 m_\perp}{8\pi^3 \hbar^3 q} \int_0^{\pi} d\psi \int dp_z \frac{v_\perp^2(p_z)}{w - v_z(p_z, \psi)} \approx \frac{e^2 p_F^2}{4\pi\hbar^3 q}(1+\eta). \tag{4.39}$$

The quantity η here describes the contribution of a zero-curvature point and differs from zero if such a point exists on the effective cross-section of the Fermi surface. A possible type of such a point is a point of the minimum of the function $v_z(p_z, \psi)$:

$p_z = p_0$, $\psi = \psi_0$. In the vicinity of the point of minimum we can write the expression for $v_z(p_z, \psi)$ in the form (4.34) assuming that

$$a > 0, \qquad ac - b^2 > 0.$$

The corresponding expression for η is

$$\eta = \rho(\pi - i \ln(w_0/w)), \tag{4.40}$$

where ρ and w_0 are parameters characterizing local properties of the Fermi surface. In order of magnitude, ρ coincides with the relative number of effective electrons from the neighborhood of a zero-curvature point, in which expression (4.34) is applicable.

In view of the smallness of the Fermi sphere distortions, the parameter ρ assumes a value much smaller than unity. Consequently, the contribution of the neighborhood of the zero curvature point does not appear in the main approximation for the surface impedance. However, it determines the first correction to the main approximation. In the linear approximation in ρ, we obtain the following expression for the ratio of the real parts of the surface impedance in a magnetic field ($R(B)$) and zero field ($R(0)$):

$$\frac{R(B)}{R(0)} = 1 - \frac{\rho}{3} \left[\sqrt{3} \ln \sqrt{\delta^2 + (\omega\tau)^{-2}} + \text{sign } \delta \arctan[(\delta\omega\tau)^2 + 1]^{-1/2} \right], \tag{4.41}$$

where $\delta = 1 - B/B_r$, and B_r is the resonance value of the magnetic field. Formula (4.41) describes a positive asymmetric resonance peak with the vertex at the resonance frequency. The width and height of the peak in the resonance region ($|\delta| \leq 0.1$) are in good agreement with the observed values for reasonable values of ρ and $(\omega\tau)^{-1}$ (of the order of $10^{-1} - 10^{-2}$).

Figure 4.4 shows the result of a comparison of a curve 2 described by the expression (4.41) with a curve 1 obtained experimentally [49]. The theoretical curve was plotted for $\rho = 0.03$, $(\omega\tau)^{-2} = 2 \times 10^{-4}$. The reference point for the measured values of $R(B)/R(0)$, which cannot be determined experimentally, was chosen so that the resonance values for the theoretical and experimental curves are identical, and the cyclotron frequency Ω was determined from the cyclotron resonance in a field parallel to the boundary.

The resonance can be associated with contributions from zero curvature points of other types. For example, a narrower peak is formed if $ac - b^2 \to 0$ in (4.34). If the function $v_z(p_z, \psi)$ depends quadratically on $(p_z - p_0)$ in a finite range of ψ, i.e., $v_z(p_z, \psi) = a(p_z - p_0)^2$, we obtain $\eta = \rho(w_0/w)^{1/2}$. It gives the following expression instead of (4.41):

$$\frac{R(B)}{R(0)} = 1 - \rho \left[\text{Re} \left(\delta + \frac{i}{\omega\tau} \right)^{-1/2} + (2 + \sqrt{3}) \text{Im} \left(\delta + \frac{i}{\omega\tau} \right)^{-1/2} \right]. \tag{4.42}$$

If we take into account the Fermi-liquid interaction, the contribution to the surface impedance from the main part of the Fermi surface, which practically does

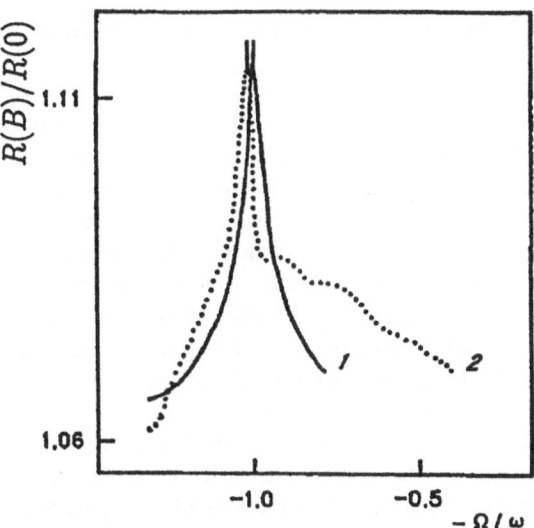

FIGURE 4.4. Relative magnitude of the real part of the surface impedance of potassium versus the magnetic field near the cyclotron resonance found in the experiments of [49] (dotted line) compared to the plot of expression (4.41) (solid line).

not differ from a sphere, is described by formula (4.14), where $a^* = \frac{8}{35}$, $W = \frac{21}{35}$, $a = -4$, $c = d = g = 1$. However, the impedance contains the contribution Z_a from the neighborhoods of zero-curvature points lying on the effective line. Taking for the parameter ρ a value 0.01 which provides good agreement with the experiment, we can calculate Z_a keeping terms proportional to the first power of ρ and no less than the square of the small parameter $1/\xi$ in the order of magnitude. In the case when the dependence of the longitudinal velocity on p_z and ψ in the vicinity of a zero-curvature point is described by relation (4.34), the resonance component Z_a can be written in the form

$$Z_a = -\frac{8\pi\omega}{9\sqrt{3}c^2}\delta\rho(i - \sqrt{3})\ln\left(\frac{w_0\omega\delta}{\chi}\right)$$
$$- \frac{320}{27\sqrt{3}}\omega\delta\frac{\rho}{\xi}\left(\chi - \frac{3\pi^2}{16}\alpha_1 + \frac{5}{3}\alpha_2\right)\ln\left(\frac{w_0\omega\delta}{\chi}\right). \qquad (4.43)$$

Thus, the resonance singularity occurs just at the cyclotron frequency. The introduction of corrections associated with the Fermi-liquid interaction leads only to an insignificant change in the shape of the resonance peak and does not shift its vertex. Therefore we can assume that the slight shift of the peak of the theoretical curve about the experimental peak in Figure 4.4 is due to the fact that effective electrons in the normal field have a shifted resonance frequency because of the departure of the Fermi surface from the sphere. Thus we can conclude that the resonance in potassium in a normal magnetic field occurs just at the cyclotron frequency and arises due to zero-curvature points on the Fermi surface.

4.3   Cyclotron Resonance in a Normal Magnetic Field in Cadmium

A rather intricate configuration is typical for Fermi surfaces for most of the metals. Therefore, one may suppose that along with zero-curvature points and lines, the Fermi surfaces also contain such points and lines where the Gaussian surface grows tending to infinity. The presence of anomalies of this type can also reveal itself in high-frequency properties of the metal. It has been shown in [88] and [89], that when the Fermi surface has a line of kinks, a weakly damped Reuter–Sondheimer wave can propagate in the metal. The line of kinks is not a unique curvature anomaly which can exist in metals. One can expect, for example, the curvature to tend to infinity at the edges of narrow lenses or needle-like cavities of the Fermi surfaces as well as at conic points.

It was shown before that under the conditions of the anomalous skin effect, the line of curvature singularities reveals itself in the most pronounced way in the metal surface impedance, provided that it coincides (in part or wholly)with one of the effective strips at the Fermi surface. In this case, the singularity of the curvature provides the cyclotron resonance to appear in the impedance derivative in a magnetic field perpendicular to the boundary. In this connection there are grounds for again turning to the analysis of the data of [47], on experimental observations of the resonance feature at the cyclotron frequency in the surface impedance of cadmium.

The observations were carried out on a single-crystal sample; the magnetic field **B** was directed along the [0001] axis at right angles to the surface of the metal. The peak found in the derivative of the real part of the impedance, dR/dB, was ascribed to a contribution from the vicinities of the reference points of the electron lens. However, this assumption enables us to describe the resonance feature in agreement with the experiments only under condition that the scattering of electrons by the surface is diffuse. This assumption disagrees with the anomalous character of the skin effect in the experiments of [47]. The present analysis is based on the assumption that the resonance arises due to the contribution of infinite curvature lines to the surface impedance of metal. On the basis of such ideas we can describe the observed resonance peak without recourse to the assumption (insufficiently grounded under anomalous skin effect conditions) of a diffuse character for the electron scattering.

In view of the fact that the magnetic field in the experiments of [47] is directed along the symmetry axis of a higher order we are free to assume that the conductivity tensor diagonalizes in circular components. According to the results of the experiments we have to calculate that component of the transverse conductivity which corresponds to the "minus" polarization for the electrons. It is convenient to separate the Fermi surface of cadmium into three parts: the hole part of the Fermi surface, the electron lens, and the remaining part of the electron Fermi surface. Assume that the hole Fermi surface everywhere has a finite and nonzero curvature and that the electron Fermi surface has an effective line of infinite curvature when the magnetic field is directed along the [0001] axis. This line passes along the edge of the lens.

Under these assumptions we can calculate circular components of the conductivity solving the system of equations (2.59) for a three-component Fermi liquid. To simplify the following calculations we assume the cyclotron mass to be constant for each group of quasi particles. To proceed, we can expand the resonance component of the conductivity in powers of a small parameter u ($u = \omega/qv_m$ and v_m is the maximum value of the longitudinal velocity of electrons of the lens). The principal term in this expansion is described by (3.16). When we calculate this principal term of the transverse conductivity, summation over all Fermi surface efficient strips, excluding that which passes along the edge of the lens, has to be performed.

The electron lens in cadmium is axially symmetric, hence the curvature at this portion of the Fermi surface depends solely on the angle θ defining the orientation of the electron velocity vector respective to the symmetry axis. In the vicinity of the anomalous effective cross-section (which corresponds to $\theta = \pi/2$), it is convenient to employ an approximation

$$|K(\theta)| = |\cos\theta|^{-\beta}/p_a^2, \tag{4.44}$$

where p_a^2 is the constant of dimensions of momentum squared.

The contribution to the conductivity from the vicinity of the effective line of infinite curvature running along the lens edge for the resonance circular polarization takes the form

$$\sigma_a(q) = \sigma_0(q)\rho(1 - i\mu_\beta)(u\chi)^\beta. \tag{4.45}$$

Here μ_β is defined by the formula (3.24);

$$R = \frac{p_a^2}{p_0^2}, \qquad \chi = 1 - \frac{\Omega}{\omega} + \frac{i}{\omega\tau}, \tag{4.46}$$

Ω is the cyclotron frequency of electrons associated with the lens edge and τ is the relaxation time.

For positive β the quantity $\sigma_a(q)$, within the range of small u ($u \ll 1$), is a small correction to the main approximation of the conductivity $\sigma_0(q)$. Its magnitude depends on the dimensionless parameter ρ which is determined by the relative number of the efficient electrons associated with the anomalous cross-section. It should be expected that under real conditions the electrons from the anomalous effective strip include a small part of the total number at the Fermi surface, hence $\rho \ll 1$.

Within terms of the order of u^2 the transverse conductivity at small u equals

$$\sigma = \sigma_0\big[1 + \Lambda_1 u + \Lambda_2 u^2 + \rho(1 - i\mu_\beta)(u\chi)^\beta\big]. \tag{4.47}$$

The complex coefficients $\Lambda_{1,2}$ depend on frequency, magnetic field, and Fermi-liquid interaction parameters.

For Λ_1 we obtain

$$\Lambda_1 = -\frac{i}{\pi}\Big(\sum_s g_s a_s \chi_s + \pi A_1 - A_2\Big). \tag{4.48}$$

Here $\chi_s = 1 \mp \Omega_s/\omega + i/\omega\tau_s$, Ω_s is the cyclotron frequency for the sth group of efficient electrons or holes, and the minus sign in the expression for χ_s corresponds to electrons and the plus sign to holes. The constants g_s, a_s are of the order of unity. We can calculate these constants using the formulas of Appendix 1. The two last terms in expression (4.48) arise due to the Fermi-liquid interaction. We can represent the term A_1 in the form

$$A_1 = (g_s Q_{0s})^{-1}(E + A_1)^{-1}_{ss'} A_{1s's''} a_{s''}. \tag{4.49}$$

Here, summation over s, s', s'' is performed over three groups of charge carriers and A_1 is the matrix whose matrix elements are the Fermi-liquid parameters $A_{1ss'}$. These parameters are defined by (2.61). The constants Q_{0s} are also defined by (2.61) and E is the unit matrix of third order. The last term in (4.49) equals

$$A_2 = g_s b_s Q_{2s}^{-1}(E + A_2)^{-1}_{ss'} A_{2s's''} b_{s''}. \tag{4.50}$$

Here, summation has also has to be performed over all groups of charge carriers, Fermi-liquid parameters $A_{2ss'}$ and constants Q_{2s} are defined by the relations (2.61), and the constants b_s can be calculated in the same way as described in Appendix 1.

The coefficient Λ_2 can be written as follows:

$$\Lambda_2 = \sum_s d_s(\chi_s)^2 - 2\overline{A}_1(\chi) - 2\overline{A}_2(\chi) - 2A_{12} - \pi^2 \tilde{A}_1^2, \tag{4.51}$$

where the Fermi-liquid terms are

$$\overline{A}_1(\chi) = \frac{a_s}{Q_{0s}}(E + A_1)^{-1}_{ss'} A_{1s's''} \chi_{s''}, \tag{4.52}$$

$$\overline{A}_2(\chi) = \frac{b_s}{Q_{2s}} g_s(E + A_1)^{-1}_{ss'} A_{2s's''} \chi_{s''}, \tag{4.53}$$

$$A_{12} = \left[\sum_{ss's''} b_s Q_{0s''}^{-1}(E + A_1)^{-1}_{ss'} A_{1s's''}\right]\left[\sum_{ss's''} b_s Q_{2s''}^{-1}(E + A_2)^{-1}_{ss'} A_{2s's''}\right], \tag{4.54}$$

$$\tilde{A}_1^2 = \left[\sum_{ss's''} g_s Q_{0s''}^{-1}(E + A_1)^{-1}_{ss'} A_{1s's''}\right]\left[\sum_{ss's''} g_{s''} Q_{0s''}^{-1}(E + A_1)^{-1}_{ss'} A_{1s's''}\right]. \tag{4.55}$$

Here, summation has to be done over two groups of charge carriers: electrons which are not associated with the lens and holes and the dimensionless constants d_s and b_s are of the order of unity.

As follows from (4.47) for $\beta < 1$ and not too small ρ the contribution from the lens edge is a first-order correction to the main term of the transverse conductivity; for $1 < \beta < 2$, it is the next (after the first correction) term in the expansion of $\sigma_\pm$ in powers of u. Finally, if $\beta > 2$ then the contribution of efficient electrons from the "anomalous" cross-section of the Fermi surface can be ignored when we calculate the conductivity asymptotics. This is due to the fact that an increase in β, which corresponds to enhancing singularities of the curvature at the efficient line, leads to a decrease in the number of effective electrons.

The main terms of the expansion of the resonance–polarization impedance component in the anomaly parameter ξ will be of the form

$$Z = \frac{8\omega\delta}{c^2}\left\{\frac{\pi}{3\sqrt{3}}(1 - i\sqrt{3}) + \frac{2\pi}{9\sqrt{3}}\frac{\Lambda_1}{\xi}(i - \sqrt{3}) + \frac{\Lambda_1^2 - \Lambda_2}{\xi^2}\ln\xi - \rho U_\beta\left(\frac{\chi}{\xi}\right)^\beta\right\},$$

(4.56)

where

$$U_\beta = (\beta + 1)\frac{\cot\left[\pi(\beta + 1)/3\right]}{\cos(\pi\beta/2)}\left[1 - i\tan\left(\frac{\pi(\beta + 1)}{3}\right)\right],$$

(4.57)

δ is the skin depth and $\xi = v_m/\omega\delta$.

When $0 < \beta < 1$, the contribution from the "anomalous" efficient cross-section of the Fermi surface is the first correction to the main approximation of the surface impedance. Owing to this term there should be a resonance in the derivatives of both the real and imaginary parts of the surface impedance at the cyclotron frequency of electrons associated with the electron lens edge. The remaining corrections to the main approximation of the impedance are not of special importance since they are small in magnitude and are not of resonant character.

Figure 4.5 shows the field dependence of the derivative dR/dB:

$$\frac{dR}{dB} = \frac{\rho|\chi|^{\beta-2}}{\xi^{\beta+1}}M(\chi'\cos\Phi - \chi''\sin\Phi),$$

(4.58)

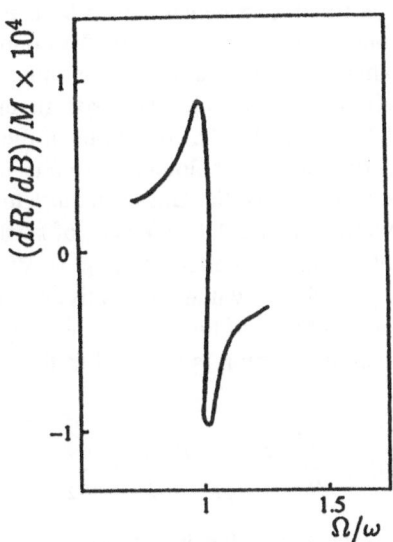

FIGURE 4.5. Cyclotron-resonance induced singularity in the magnetic field dependence of dR/dB in the case when the curvature of the Fermi surface of a metal has a singularity at one of the extremal sections. The curve is plotted for $\beta = 0.5$, $\omega\tau = 50$, $\rho = 0.01$, $\xi^3 = 10^4$. The value of M is determined by the expression (4.57).

where R is the real part of the impedance contribution from the edge of the electron, $\chi = \chi' + i\chi''$:

$$\Phi = \frac{\pi}{3}(1 - 2\beta) + \beta \arctan\left(\frac{\chi''}{\chi'}\right) + \pi\beta\theta(\omega - \Omega),$$

$$M = \frac{8\pi\omega(\delta)^{\beta/(\beta+1)}}{\cos(\pi\beta/2)\sin\left[\pi(\beta+1)/3\right]c^2}.$$

The shape of the resonance feature is consistent with that observed in experiments. Because of the axial symmetry of the electron lens in cadmium the resonance reveals itself as a solitary line at the cyclotron frequency of the electrons associated with the lens edge and does not have harmonics at multiple frequencies.

4.4   Local Flattening of the Fermi Surface and Cyclotron Resonance in a Normal Magnetic Field in Metals

It was shown in [79]–[87] that, under certain conditions, local flattening of the Fermi surface may cause considerable variations of the observed characteristics of metals. The obtained results give grounds for studying the role of local flattening of the Fermi surface in the emergence of cyclotron resonance observed in cadmium [47] and earlier in zinc [48] in a magnetic field normal to the metal boundary. Both these experiments revealed a resonance singularity in the field dependence of the derivative of the real part of the impedance of a bulk monocrystalline sample. In both cases, the resonance is associated with the contribution from electrons belonging to a lens arranged at the center of the Brillouin zone in such a way that it is axially symmetric relative to the axis [0001]. We will analyze the data obtained in [47], [48] for $\mathbf{B}||\mathbf{n}||[11\bar{2}0]$. Under the conditions of the anomalous skin effect, cyclotron orbits of effective electrons belonging to a lens pass in the vicinity of its vertices, and hence the vanishing of the Gaussian curvature at the vertices of the lens must considerably affect the field dependence of surface impedance.

Suppose that a metal occupies a half-space $z < 0$. We choose the coordinate system in the xy-plane in such a way that the x-axis coincides with the symmetry axis of the lens. For such a choice of the coordinate axis we can assume that the following energy-momentum relation is satisfied in the vicinity of the vertex of the lens:

$$E(\mathbf{p}) = \frac{p_1^2}{2m_1}\left(\frac{p_y^2 + p_z^2}{p_1^2}\right)^l + \frac{p_2^2}{m_2}\varepsilon\left(\frac{p_x}{p_2}\right). \tag{4.59}$$

Here, p_1 and p_2 are, respectively, the lens radius and thickness at the center, $\varepsilon(p_x/p_2)$ is an even function of p_x which increases monotonically for $p_x > 0$, and $\varepsilon(0) = 0$, $\varepsilon(1) = 1$. For $l = 1$ and $\varepsilon(p_x/p_2) = (p_x/p_2)^2$, the energy-momentum relation (4.59) has an ellipsoidal Fermi surface corresponding to it. In this case, m_1 and m_2 are the principal values of the effective mass tensor.

The Gaussian curvature at any point of the lens corresponding to (4.59) is described by the expression

$$K(\mathbf{p}) = \frac{l}{m_1 v^4} \left(\frac{p_y^2 + p_z^2}{p_1^2} \right)^{l-1} \left[(v_y^2 + v_z^2) \frac{\partial v_x}{\partial p_x} + v_x^2 \frac{l(2l-1)}{m_1} \left(\frac{p_y^2 + p_z^2}{p_1^2} \right)^{l-1} \right],$$
(4.60)

where v_a ($\alpha = x, y, z$) are the velocity components of the electrons belonging to the lens and $|v|$ is the magnitude of the velocity.

For $l > 1$, the Gaussian curvature $K(p_x, p_y, p_z)$ vanishes at the points $(\pm p_2, 0, 0)$ with the vertices of the lens coinciding in the selected coordinate system. In view of the axial symmetry of the lens, the curvatures of both principal cross-sections vanish at this point. In other words, the vertices are the points corresponding to flattening of the Fermi surface. The shape of the lens in the vicinity of its vertices will be the closer to a plane, the larger the value of the parameter l.

The electrical conductivity tensor components for a metal with a closed Fermi surface have the following form:

$$\sigma_{\alpha\beta} = \frac{2ie^2}{(2\pi\hbar)^3} \sum_n \int_0^{2\pi} d\psi \int dp_z$$

$$\times \int_{-\infty}^0 d\eta e^\eta \frac{m(p_z)v_a(p_z, \psi)v_{n\beta}(p_z)\exp(-in\psi)}{\omega + i/\tau - n\Omega(p_z) - qv_z(p_z, \psi + \delta\psi(\eta))}.$$
(4.61)

Here Ω is the cyclotron frequency which depends on p_z, ω and q are the frequency and wave vector, $\psi = \Omega t$, and τ is the effective relaxation time. For a metal with a complex many-sheet Fermi surface (e.g., cadmium or zinc), integration with respect to p_z in (4.61) is carried out within the limits determined by the form of each sheet, and summation is carried out over all sheets of the Fermi surface.

Under the conditions of the anomalous skin effect, the correction $\delta\psi(\eta)$ in the first approximation is proportional to η and is connected with it through the relation $\delta\psi(\eta) \approx -i\eta\Omega/qv_z(p_z, \psi)$. For the anomalous skin effect, $\omega/q|v| \ll 1$ and the correction $\delta\psi(\eta)$ is small in comparison with ψ. Hence we can expand the last term in the denominator of (4.61) in a series in $\delta\psi$, and confine the expansion to the first two terms

$$qv_z(p_z, \psi + \delta\psi(\eta)) \approx qv_z(p_z, \psi) - i\eta\Omega \frac{[\partial v_z(p_z, \psi)/\partial\psi]}{v_z(p_z, \psi)}$$

$$\equiv qv_z(p_z, \psi) - i\eta\Delta(p_z, \psi).$$
(4.62)

For convenience in obtaining the asymptotic form of the conductivity in the anomalous skin effect, we go over in (4.61) to integration in velocity space. In this case, the Fermi surface must be divided into segments with numbers j, the dependence of the momentum p on $|v|$, and the angles θ, φ determining the position of the velocity vector in the spherical system of coordinates being unique on each segment. Substitution of integration variables in integrals with respect to p_z and

ψ, gives the result

$$\sigma_{\alpha\beta} = \frac{2ie^2}{(2\pi\hbar)^3 q} \sum_n \int_{-\infty}^{0} d\eta \sum_j \int d\theta \int d\varphi \frac{\sin\theta}{|K_j(\theta,\varphi)|}$$

$$\times \frac{n_{\alpha j}(\theta,\varphi) V_{n\beta}^j(\theta,\varphi) e^\eta}{|\omega - n\Omega_j(\theta,\varphi) + i\Delta(\theta,\varphi)\eta + i/\tau|/q|v_j(\theta,\varphi)| - \cos(\theta)}. \quad (4.63)$$

where

$$V_{n\beta}^j(\theta,\varphi) = \frac{v_{n\beta j}(\theta,\varphi)\exp(-in\psi)}{|v_j(\theta,\varphi)|},$$

$$n_{\alpha j}(\theta,\varphi) = \frac{v_{\alpha j}(\theta,\varphi)}{|v_j(\theta,\varphi)|}.$$

The principal terms in the expansion of the electrical conductivity tensor components in powers of $\omega/q|v|$ are independent of the magnetic field. For σ_{xx}^0, we obtain the familiar relation (see [107], [108]):

$$\sigma_{xx}^0(q) = \frac{e^2}{4\pi^2\hbar^3 q} \sum_r \int \frac{\cos^2\varphi\, d\varphi}{|K_r(\pi/2,\varphi)|} \equiv \frac{e^2}{4\pi\hbar^3 q} p_0^2, \quad (4.64)$$

where summation over r is carried out over all effective strips on the electron and hole parts of the Fermi surface with the exception of the effective line passing through the vertices of the electronic lens, and $K_r(\pi/2,\varphi)$ is the Gaussian curvature at the corresponding points on the effective rth strip, which is assumed to be nonzero everywhere. The principal term in the asymptotic representation of the component σ_{yy} is described by a formula analogous to (4.64), where the numerator of the integrand in each term is replaced by $\sin^2\varphi$. The nondiagonal components of the electrical conductivity tensor must vanish in the experimental geometry of [47], [48] in which the magnetic field (and hence the z-axis of the chosen coordinate system) is directed along the second-order symmetry axis [11$\bar{2}$0].

For a metal with everywhere finite nonzero curvature of the Fermi surface the first correction to the principal term in the expansion for σ_{xx} has the form

$$\sigma_{xx}^{(1)}(\omega,q) = i\sigma_{xx}^0(q)\frac{\omega}{q}\sum_n \left(\frac{w_n}{v_0}\right). \quad (4.65)$$

Here

$$\left(\frac{w_n}{v_0}\right) = \frac{2}{\pi^2 p_0^2}\sum_r \int_{-\infty}^{0} e^\eta d\eta \int d\theta \int d\varphi \frac{\cos\varphi\sin\theta}{\cos^2\theta}$$

$$\times \left[\frac{V_{nx}^r(\pi/2,\varphi) w_{nr}(\pi/2,\varphi,\eta)}{|K_r(\pi/2,\varphi)||v_r(\pi/2,\varphi)|}\left(1 + \frac{\cos^2\theta}{\cos^2\theta_r}\right) - \frac{V_{nx}^r(\theta,\varphi) w_{nr}(\theta,\varphi,\eta)}{|K_r(\theta,\varphi)||v_r(\theta,\varphi)|}\sin\theta\right]$$

$$- \frac{2}{\pi^2 p_0^2}\sum_{j\neq r}\int_{-\infty}^{0} e^\eta d\eta \int d\theta \int d\varphi \frac{\cos\varphi\sin^2\theta\, V_{nx}^j(\theta,\varphi) w_{nj}(\theta,\varphi,\eta)}{|K_j(\theta,\varphi)||v_j(\theta,\varphi)|}, \quad (4.66)$$

where

$$w_{nr}(\theta, \varphi, \eta) = 1 - \frac{n\Omega_r(\theta, \varphi)}{\omega} + \frac{i\Delta_r(\theta, \varphi)\eta}{\omega} + \frac{i}{\omega\tau}.$$

Summation over r in (4.66) is carried out over all parts of the sheets of the Fermi surface through which the effective strips pass. The last term in (4.66) takes into account the contributions from those parts of the Fermi surface which do not contain the effective lines. Integration with respect to θ and φ in each term of the sums over r and j is carried out within the limits set by the shape and size of the corresponding segment of the Fermi surface. The mirror symmetry of the latter allows us to assume that θ varies from a certain minimum value θ_r to its maximum value $\pi/2$ in all terms, if we consider the Fermi surface region containing an effective strip, and less than $\pi/2$ for the remaining segments. It can be seen from (4.66) that the quantity v_0 in the formula (4.65) has the dimensions of velocity, while its value for a one-sheet Fermi surface is of the order of the Fermi velocity of electrons. The first correction to the main approximation of the electrical conductivity component σ_{yy} in this case is also described by an expression similar to (4.65).

However, if locally flat regions exist on one of the effective strips of the Fermi surface, the quantity (4.65) can no longer be treated as a first correction to the principal term in the asymptotic representation of σ_{xx}, since the contribution to the conductivity from the neighborhood of the flattening points is much larger than this quantity. The contribution to σ_{xx} from the effective strip passing through the vertex of the electron lens can be presented as the sum of two terms. The first term has the same order of magnitude as the correction (4.65) and is given by

$$\sigma_{axx}^{(1)}(\omega, q) = i\sigma_{xx}^0(q)\frac{\omega}{q}\sum_n \left\langle\!\!\left\langle \frac{w_n}{v_a} \right\rangle\!\!\right\rangle, \tag{4.67}$$

where

$$\left\langle\!\!\left\langle \frac{w_n}{v_a} \right\rangle\!\!\right\rangle = \frac{8}{\pi^2 p_0^2} \int_{-\infty}^0 e^\eta d\eta \int_0^{\pi/2} d\theta \int_0^{\pi/2} \frac{d\varphi \cos\varphi \sin\theta}{\cos^2\theta |K(\theta, \varphi)|}$$
$$\times \left[\frac{V_{nx}(\pi/2, \varphi)w_n(\pi/2, \varphi, \eta)}{|v(\pi/2, \varphi)|} - \frac{V_{nx}(\theta, \varphi)w_n(\theta, \varphi, \eta)}{|v(\theta, \varphi)|} \right]. \tag{4.68}$$

Like the quantity v_0 in formula (4.65), v_a also has the dimensions of velocity. The values of the angles $\theta = \pi/2$ and $\varphi = 0$ correspond to the vertex of the electron lens.

The second term, which contains the contribution to the conductivity from the locally flat region at the vertex of the lens, can be represented in the form

$$\sigma_{axx}^{(2)}(\omega, q) = \sigma_{xx}^{(0)}(q)\frac{8}{\pi^2}u_0 \sum_n V_{n0}^x \int_{-\infty}^0 e^\eta d\eta$$
$$\times \int_0^{\pi/2} \frac{w_{n0}(\eta)\sin\theta d\theta}{u_0^2 w_{n0}^2(\eta) - \cos^2\theta} \int_0^{\pi/2} \frac{\cos\varphi d\varphi}{|K(\theta, \varphi)|}. \tag{4.69}$$

Here, $u_0 = \omega/q|v(\pi/2, 0)|$, $V_{n0}^x = V_{nx}(\pi/2, 0)$, and $w_{n0}(\eta) = w_n(\pi/2, 0, \eta)$.

The main contribution to the integral with respect to θ and φ comes from the neighborhood of the lens vertex where the Fermi surface curvature vanishes in view of the assumption made above. Hence, while carrying out integration in (4.69), we can use an approximate expression describing the curvature of the Fermi surface corresponding to the dispersion relation (4.59) in the vicinity of the flattening region

$$K(\theta, \varphi) = \frac{1}{p_a^2}(\sin^2 \varphi + \cos^2 \theta)^s. \qquad (4.70)$$

Here, $p_a^2 = p_1^2(s+1)[4(s+1)^2 m_1^2 |v(\pi/2, 0)|^2/s^2 p_1^2]^{s+1}$ is a constant having the dimensions of the square of momentum, $s = 1 - 1/(2l - 1)$, and $0 < s < 1$ for $l > 1$. The closer the value of s to unity, the more flattened the Fermi surface in the vicinity of the point where $\theta = \pi/2$, $\varphi = 0$. For $s \to 1$, the Fermi surface is transformed into a plane in the vicinity of this point.

To within terms of the order of $\sigma^{(1)}$, the term $\sigma_{axx}^{(2)}$ can be described by the expression

$$\sigma_{xx}^{(2)}(\omega, q) = \sigma_{xx}^0(q)\rho_s \sum_n \left\{[1 - i \cot(\pi s)]\overline{(u_0 w_{n0})}^{1-2s} + \frac{2i}{\pi}\overline{(u_0 w_{n0})}\right\}, \qquad (4.71)$$

where the following notation has been used

$$\rho_s = \frac{2}{\pi^{1/2}} \frac{p_a^2}{p_0^2} \frac{\Gamma\left(s - \frac{1}{2}\right)}{\Gamma(s)}, \qquad (4.72)$$

$$\overline{(u_0 w_{n0})}^\beta = V_{n0}^x \int_{-\infty}^{0} e^\eta [u_0 w_{n0}(\eta)]^\beta d\eta, \qquad (4.73)$$

and $\Gamma(x)$ is the gamma function.

The obtained results show that the flattening of the Fermi surface at the electron lens vertices leads to the emergence of an additional term (the first term in (4.71)) in the asymptotic expansion of $\sigma_{xx}(\omega, q)$. In order of magnitude this term exceeds all the remaining corrections to the principal term in the expansion. The expansion of σ_{yy} does not contain any such term since in this case the electron velocity component v_y vanishes at the points of flattening of the Fermi surface.

Under the conditions of the anomalous skin effect we can confine ourselves to the calculation of the surface impedance under the assumption that electrons are reflected specularly at the metal surface. In this case, the components $Z_{\alpha\beta}(\alpha, \beta = x, y)$ of the surface impedance for a semi-infinite metal occupying the half-space $z < 0$ are defined by the expression

$$Z_{\alpha\beta} = \frac{8i\omega}{c^2} \int_0^\infty \left(\frac{4\pi i\omega}{c^2}\sigma - q^2 E\right)_{\alpha\beta}^{-1} dq. \qquad (4.74)$$

Here $E_{\alpha\beta} = \delta_{\alpha\beta}$. Substituting the obtained expression for the conductivity components into (4.74) we can evaluate the first terms in the expansion of Z_{xx} in the inverse powers of the anomaly parameter ξ:

$$Z_{xx} = \frac{8\omega\delta}{c^2}\left\{ \frac{\pi}{3\sqrt{3}}(1-i\sqrt{3}) - \frac{2\pi}{9\sqrt{3}}\frac{v_m}{\xi}(1+i\sqrt{3}) \right.$$

$$\left. \times \sum_n \left(\left\langle \left| \frac{w_n}{v_0} \right| \right\rangle + \left\langle \left| \frac{w_n}{v_a} \right| \right\rangle \right) - \rho_s b_s \sum_n \left(\frac{\overline{w_{n0}}}{\xi} \right)^{1-2s} \right\}. \qquad (4.75)$$

Here the complex constant b_s is defined as

$$b_s = \frac{2(1-s)}{\sqrt{3}} \left(\frac{v_m}{|v(\pi/2;0)|} \right)^{1-2s} \frac{\cot\left[2\pi/3(1-s)\right]}{\sin(\pi s)} \left[1 - i\tan\left(\frac{2\pi}{3}(1-s) \right) \right]. \qquad (4.76)$$

The results show that the existence of flattening points on an effective strip of the Fermi surface leads to the emergence of an additional term in the expression for impedance, in the same way as in the case when one of the effective strips coincides with the line of parabolic points or the line where one of the principal radii of curvature of the Fermi surface vanishes. The emergence of this contribution to the surface impedance (the last term in (4.75)) is associated with a broadening of the strip of effective electrons due to the presence of flattening points on it.

The extent to which the locally flat regions of the Fermi surface affects the surface impedance is determined by the value of the parameter ρ_s characterizing the relative number of the effective electrons associated with the flattened regions, and the parameter s characterizing the degree of flattening of the Fermi surface in the vicinity of the zero-curvature point. If s differs considerably from zero and ρ_s is not too small, this contribution may turn out to be significant and lead to the emergence of noticeable resonance-type singularities in the frequency (field) dependences of the surface impedance. For the case $s > \frac{1}{2}$, the resonance peak is observed in the impedance itself, while for a less pronounced flattening of the Fermi surface ($0 < s < \frac{1}{2}$) the resonance singularity is manifested only in the field and frequency dependences of the derivative of impedance.

The experimental results obtained in [47] and [48] for cadmium and zinc indicate that in all probability, the latter of the above possibilities is realized in these metals. In other words, the flattening of the Fermi surface at vertices of the electron lens is moderate ($0 < s < \frac{1}{2}$). However, the first possibility ($0 < s < \frac{1}{2}$) cannot be ruled out either, since a strong flattening of the Fermi surface is in keeping with a relatively small value of the parameter ρ_s. In this case, the resonance singularity of the surface impedance may turn out to be too weak to be manifested clearly in the experiment. However, the stronger singularity in the derivative of the impedance for the same ρ_s is exhibited quite clearly.

The derivative of the real part of the contribution to the impedance from a locally flat region is defined as

$$\frac{dR}{dB} = \frac{\rho_s a_s}{\xi^{2(1-s)}} Y_s(w), \qquad (4.77)$$

where

$$a_s = \frac{16}{\sqrt{3}} \frac{\delta|e|}{(\pi/2)c^3} V_{10}^x \left(\frac{v_m}{|v(\pi/2,0)|}\right)^{1-2s} \frac{1-s}{\sin(\pi s)\sin\left[2\pi(1-s)/3\right]}, \qquad (4.78)$$

and the function $Y_s(w)$, which describes the shape of the resonance curve, has the form

$$Y_s(w) = |w|^{-2s} \cos\left\{2s \arctan\left(\frac{w''}{w'}\right) + \frac{2\pi}{3}(1-s) + \pi s\theta(\Omega - \omega)\right\}. \qquad (4.79)$$

Here

$$\theta(x) = \begin{cases} 0, & x \le 0, \\ 1, & x > 0, \end{cases}$$

$$w = w' + iw'' = 1 - \frac{\Omega(\pi/2;0)}{\omega} + i\left(\frac{1}{\omega\tau} - \frac{\Delta(\pi/2;0)}{\omega}\overline{\eta}\right).$$

While deriving formula (4.77), we have taken into account only the term with $n = 1$ in the sum over n in the expression for the resonance term in formula (4.75), since this result describes the field dependence of dR/dB in the region of the magnetic field corresponding to the closeness of Ω and ω. The contribution from the harmonics of cyclotron resonance can be taken into account by considering other terms in the sum over n in (4.75). While considering the dependence of dR/dB in the vicinity of one of the resonance values of the magnetic field ($n\Omega = \omega$), it is sufficient to be confined to the term corresponding to this harmonic in the sum over n. Finally, the quantity $\overline{\eta}$ in the expression for w is defined by the equality

$$w(\overline{\eta}) = \int_{-\infty}^{0} e^{\eta} w(\eta) \, d\eta. \qquad (4.80)$$

The resonance nature of the dependence of dR/dB is manifested under the condition of smallness of the imaginary part of w. Apart from the obvious condition $\omega\tau \gg 1$, the inequality $|\Delta(\pi/2, 0)\overline{\eta}| \ll \omega$ must also be satisfied in order to keep the value of the above quantity low. If the Fermi surface flattening points are located at the vertices of an electron lens, the effective electron velocity vector varies slightly during its motion in the part of the cyclotron orbit passing through the flat region of the lens. Hence the quantity $\Delta(\pi/2, 0)$ can be assumed to be small under these conditions. However, if the direction of the magnetic field is changed so that the point with coordinates $\theta = \pi/2$, $\varphi = 0$ on the Fermi surface no longer coincides with the flattening point, the value of $\Delta(\pi/2, 0)$ may increase considerably, and the resonance can no longer be observed. This is one of the reasons behind the weakening of the resonance as the magnetic field deviates from the direction corresponding to the maximum amplitude of the resonance observed in the experiments of [47], [48].

Figure 4.6 shows the field dependence of the function $Y_s(w)$ for certain values of the parameter s. A comparison with Figure 4.7, which contains the recording for dR/dB obtained for cadmium in [47] shows that the shape of the resonance

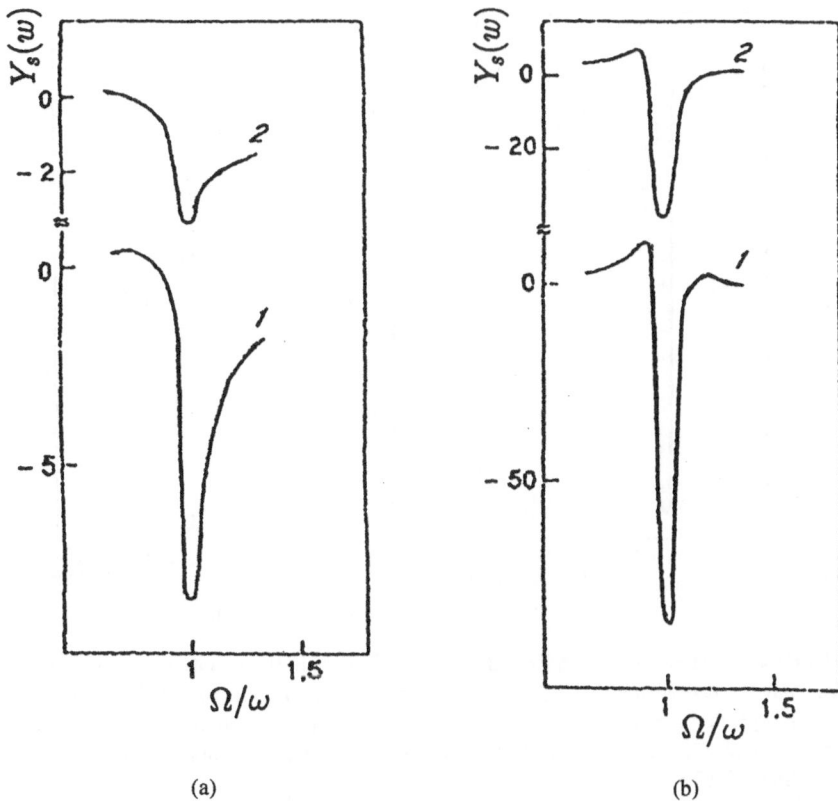

FIGURE 4.6. Plots of the function $Y_s(w)$ versus the magnetic field for (a) moderate $\left(0 < s < \frac{1}{2}\right)$ and (b) strong $\left(\frac{1}{2} < s < 1\right)$ flattening of the Fermi surface near vertices of the lenses. Curves are plotted for $\omega\tau = 20$, $\Delta(\pi/2, 0) = 0$, and (a) $s = \frac{3}{8}$ (curve 1), $s = \frac{1}{4}$ (curve 2); and (b) $s = \frac{3}{4}$ (curve 1), $s = \frac{5}{8}$ (curve 2).

lines described by (4.77) is in fairly good agreement with the experimental results. The variation of the amplitude and shape of the resonance lines (Figure 1.10) with decreasing s is found to be in accord with the variation of the experimental resonance lines upon an increase in the angle formed by the magnetic field with the axis [11$\bar{2}$0]. This can be seen, for example, from a comparison of the curves corresponding to $s = \frac{3}{4}$ and $s = \frac{1}{4}$ in Figure 4.6 with curves 1 and 2 in Figure 4.7(a).

The above similarity is due to the fact that as the magnetic field deviates from the axis [11$\bar{2}$0], the effective cross-section of the electron lens no longer passes through the flattening points at the vertices of the lens but still remains quite close to them for small angles of deviation. Hence it can be assumed that if the angle Φ between **B** and [11$\bar{2}$0] does not exceed a certain critical value Φ_0 characterizing the size of the flattening region on the Fermi surface, the parameter s decreases with increasing Φ, but remains nonzero. The value $s = 0$ corresponds to angles $\Phi \geq \Phi_0$,

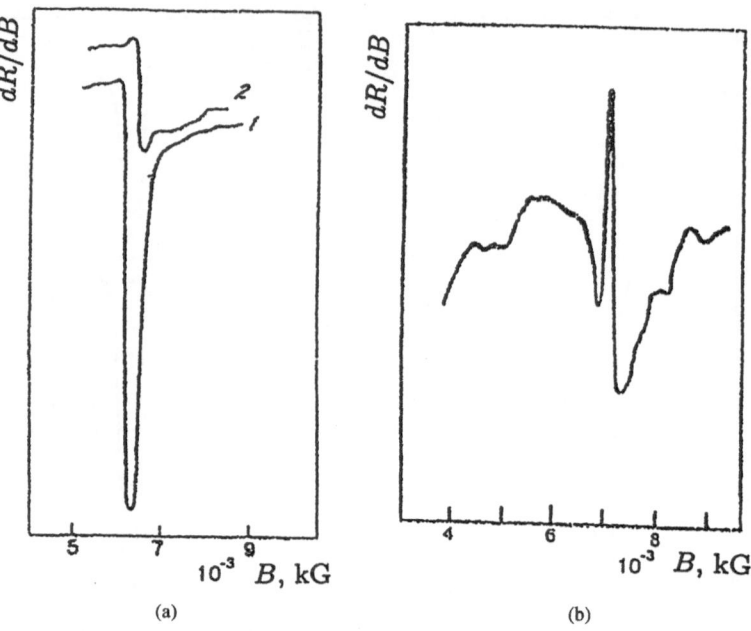

FIGURE 4.7. Cyclotron-resonance induced singularities in the magnetic field dependence of dR/dB for (a) cadmium and (b) zinc found in the experiments [47], [48] for $\mathbf{n}||[11\bar{2}0]$. Curve 1 is plotted for $\mathbf{B}||\mathbf{n}$ and curve 2 corresponds to the case when the angle Φ between $\mathbf{B}$ and $\mathbf{n}$ equals 8°. Amplification at the recording of curve 2 in comparison with curve 1 is approximately 10 : 1.

when the cyclotron orbit of the effective electrons no longer passes through the locally flat region on the Fermi surface. According to the data presented in [48], the values of Φ_0 for the electron lens in cadmium and zinc are close and amount to about 10°.

CHAPTER 5

Local Geometry of the Fermi Surface and Magnetoacoustic Oscillations in Metals

5.1   Magnetoacoustic Oscillations in Metals with a Nearly Cylindrical Fermi Surface

The absorption coefficient and velocity of sound propagating in a metal at right angles to the applied magnetic field **B** in the region of moderately strong magnetic fields for which the inequalities $\Omega\tau \gg 1$ and $qR \gg 1$ are satisfied simultaneously ($2R$ is the characteristic diameter of the cyclotron orbit and q is the wave vector of the acoustic wave) oscillate as a result of variation of the magnetic field. These oscillations, which are known as geometrical oscillations, are generated as a result of the periodic reproduction of the most favorable conditions for the absorption of the acoustic wave energy by electrons moving along the wave front. Their period is determined by the extremal diameter $2R_{ex}$ of the Fermi surface of the metal.

It follows from the theory of the geometrical oscillations expounded in [15], [123], [124] that the amplitude of oscillations in a simple metal with a closed Fermi surface has an order of magnitude smaller, by a factor of $1/\sqrt{qR_{ex}}$, than smooth components of the absorption coefficient and the velocity of sound. However, in the presence of certain peculiarities in the geometry of the Fermi surface, the number of electrons effectively participating in absorption can increase significantly, leading to an enhancement of the oscillations. Kirichenko and Peschanskii [125] proved, for example, that a sharp increase in the amplitude of geometrical oscillations must take place in a conductor with the Fermi surface in the form of a slightly corrugated cylinder under the condition that the magnetic field is directed along the cylinder axis. An increase in the number of electrons participating in the absorption of the acoustic wave energy, under the condition of geometrical resonance, can also be due to local anomalies in the Fermi surface curvature.

If one or both principal curvatures of the Fermi surface vanish at the points corresponding to stationary points of the cyclotron orbit of the extremal diameter, this also increases the number of electrons effectively absorbing the acoustic wave energy. Thus, the presence of flattening points or parabolic points on the Fermi surface can lead to an enhancement of geometrical oscillations. Let us consider a longitudinal acoustic wave propagating in a metal along the y-axis of the coordinate system whose z-axis is directed along the magnetic field $\mathbf{B}$ and coincides with a high-order symmetry axis of the crystal. Assume that the elastic displacement of the lattice $\mathbf{u}(\mathbf{r},t)$ is proportional to $\exp(iqy - i\omega t)$.

The force exerted by the electrons on the lattice contains a contribution originating from their interaction with the electromagnetic field accompanying the sound wave and the deformation contribution. Correspondingly, the magnitude of this force $F_{q\omega}$ can be written as follows:

$$F_{q\omega} = iq\left(\gamma_\alpha - \frac{iNe}{q}\delta_{\alpha y}\right)E_{q\omega}^{\prime\alpha} + i\omega q^2 \lambda u_{q\omega}, \tag{5.1}$$

where $\mathbf{E}_{q\omega}' = \mathbf{E}_{q\omega} + (i\omega/c)[\mathbf{u}_{q\omega} \times \mathbf{B}] + (m/e)\omega^2\mathbf{u}_{q\omega}$ and $\mathbf{E}_{q\omega}$ is the amplitude of the electric field accompanying the wave.

The amplitude $\mathbf{E}_{q\omega}$ satisfies the Maxwell equations

$$-[\mathbf{q} \times [\mathbf{q} \times \mathbf{E}_{q\omega}]] = \frac{4\pi i\omega\mathbf{J}_{q\omega}}{c^2}. \tag{5.2}$$

This expression contains the amplitude of the total density of current $\mathbf{J}_{q\omega}$ induced by the passage of an acoustic wave. The components of $\mathbf{J}_{q\omega}$ are given by

$$J_{q\omega}^\alpha = \sigma_{\alpha\beta}E_{q\omega}^{\prime\beta} + \omega q\left(\gamma_\alpha - \frac{iNe}{q}\delta_{\alpha y}\right)u_{q\omega}. \tag{5.3}$$

The electron kinetic coefficients λ and σ have the form

$$\lambda = \frac{i}{2\pi^2\hbar^3}\int dp_z m_\perp \sum_n \frac{U_{-n}(p_z, -q)U_n(p_z, q)}{\omega + i/\tau - n\Omega}, \tag{5.4}$$

$$\sigma_{\alpha\beta} = \frac{ie^2}{2\pi^2\hbar^3}\int dp_z m_\perp \sum_n \frac{v_{-n}^\alpha(p_z, -q)v_n^\beta(p_z, q)}{\omega + i/\tau - n\Omega}, \tag{5.5}$$

where $m_\perp$ is the cyclotron mass and $U_n(p_z, q)$ is the Fourier transform in the expansion in an azimuthal angle specifying the position of an electron on the cyclotron orbit

$$U_n(p_z, q) = \frac{1}{2\pi}\int_0^{2\pi} U_n(p_z, \psi, q)\exp(in\psi)\,d\psi, \tag{5.6}$$

where

$$U_n(p_z, \psi, q) = U(p_z, \psi)\exp\left[in\psi - \frac{iq}{\Omega}\int_0^\psi v_y(p_z, \psi')\,d\psi'\right],$$

$$U_n(p_z, \psi) = \Lambda_{yy}(p_z, \psi) - \frac{\langle\Lambda_{yy}\rangle - N}{g}. \tag{5.7}$$

Here, $\Lambda_{yy}(p_z, \psi)$ and $v_y(p_z, \psi)$ are the corresponding components of the deformation potential tensor and the electron velocity, N is the electron concentration, the symbol $\langle \cdots \rangle$ denotes the averaging over the Fermi surface, and g is the density of states on the Fermi surface.

The Fourier transforms of the electron velocity components in the expansion in the angle ψ are determined by relations similar to (5.6):

$$v_n^\alpha(p_z, q) = \frac{1}{2\pi} \int_0^{2\pi} v_\alpha(p_z, \psi) \exp\left[in\psi - \frac{iq}{\Omega} \int_0^\psi v_y(p_z, \psi') d\psi'\right] d\psi. \quad (5.8)$$

For a multiply connected Fermi surface, the integration with respect to p_z in (5.4) must be supplemented with summation over all the cavities of the Fermi surface. In this case, the values of $U_n(p_z, q)$ are calculated separately for each cavity.

In (5.1), (5.3) the term $(iNe/q)\delta_{\alpha y}$ has to be replaced by

$$\frac{ie}{q} \sum_k N_k \frac{e_k}{|e|}, \quad (5.9)$$

where summation has to be performed over the cavities of the Fermi surface, N_k is the concentration of charge carriers for the kth cavity, e_k and is their charge. When a considered metal has an equal number of electrons and holes, the term (5.9) is equal to zero, and the corresponding addends in the expressions for $\mathbf{F}_{q\omega}$ and $\mathbf{J}_{q\omega}$ vanish. We can obtain the expression for the kinetic coefficient γ_α replacing $U_{-n}(p_z, -q)$ by $ev_{-n}^\alpha(p_z, -q)$ in (5.4). To obtain the expression for $\overline{\gamma}_\alpha$ we have to replace $U_n(p_z, q)$ by $ev_n^\alpha(p_z, q)$.

To determine the wave vector of the acoustic wave propagating in metal we have to solve the equation for the amplitude of the elastic displacement of the lattice together with the Maxwell equations. As a result we arrive at the formula

$$q^2 = \frac{\omega^2}{s^2} - \frac{i\omega q^2}{\rho_m s^2}\left(\lambda^* + \frac{\gamma^*(\overline{\gamma}^* - Bcq/4\pi\omega)}{\sigma^* - c^2 q^2/4\pi i\omega}\right). \quad (5.10)$$

Here,

$$\lambda^* = \lambda - \left[\gamma_y - \frac{ie}{q}\sum_k N_k \frac{e_k}{|e|}\right]^2 / \sigma_{yy},$$

$$\gamma^* = \gamma_x - \left[\gamma_y - \frac{ie}{q}\sum_k N_k \frac{e_k}{|e|}\right]\frac{\sigma_{yx}}{\sigma_{yy}}, \quad (5.11)$$

$$\overline{\gamma}^* = \overline{\gamma}_x - \left[\overline{\gamma}_y - \frac{ie}{q}\sum_k N_k \frac{e_k}{|e|}\right]\frac{\sigma_{yx}}{\sigma_{yy}},$$

$$\sigma^* = \sigma_{xx} + \sigma_{yx}^2/\sigma_{yy}.$$

For small amplitudes of acoustic waves, the wave vector is described by the expression

$$q = \omega/s + \Delta q. \quad (5.12)$$

The increment Δq linear in $u_{q\omega}$, which emerges as a result of the interaction with electrons, in the case under investigation has the form

$$\Delta q = \frac{iq^2}{2\rho_m s}\left(\lambda^* + \frac{\gamma^*\left(\bar{\gamma}^* - Bcq/4\pi\omega\right)}{\sigma^* - c^2 q^2/4\pi i\omega}\right). \tag{5.13}$$

The wave vector q on the right-hand side of (5.13) is assumed to be equal to ω/s.

In the region under investigation, where $qR \gg 1$, the main contribution to the integral with respect to ψ in expressions (5.6), (5.8) for $U_n(p_z, q)$ and $v_n^\alpha(p_z, q)$ comes from the neighborhoods of the stationary points on cyclotron orbits. Accordingly, estimating the integrals by the stationary phase method, we can obtain the following asymptotic expressions for $U_{\pm n}(p_z, \pm q)$:

$$U_{\pm n}(p_z, \pm q) = \frac{1}{\pi} U_0(p_z)\exp\left[\pm iqR(p_z) \pm i\pi\frac{n}{2}\right] \tag{5.14}$$
$$\times \left\{\cos\left(qR(p_z) - \pi\frac{n}{2}\right)V(p_z) - \sin\left(qR(p_z) - \pi\frac{n}{2}\right)W(p_z)\right\},$$

where $U_0(p_z) = U(p_z, \psi_1) = U(p_z, \psi_2)$, $2R$ is the diameter of a cyclotron orbit of electrons in the direction of propagation of the acoustic wave, ψ_1 and ψ_2 are the values of the angle ψ corresponding to stationary points on the cyclotron orbit, and $\psi_1 - \psi_2 = \pi$. The form of the functions $V(p_z)$ and $W(p_z)$ is determined by singularities of the energy-momentum relation for electrons in the vicinity of stationary points.

Suppose that the Fermi surface of a metal includes a double-convex lens axisymmetric about the z-axis, for which the electron energy-momentum relation has the form

$$E(\mathbf{p}) = \frac{p_1^2}{2m_1}\left(\frac{p_x^2 + p_y^2}{p_1^2}\right) + \frac{p_2^2}{2m_2}\left(\frac{p_z^2}{p_2^2}\right)^k. \tag{5.15}$$

Here p_1 and p_2 are quantities having the dimensions of momentum and characterizing the diameter and thickness of the lens. For $k = 1$, the lens becomes ellipsoidal, and m_1 and m_2, in this case, are the principal values of the effective mass tensor associated with the region of the Fermi surface under investigation. If $k > 1$, the surface of the lens contains a line of parabolic points coinciding with the central cross-section of the lens by a plane perpendicular to its axis. In the vicinity of this cross-section, the shape of the lens surface is close to cylindrical, the degree of closeness increases when the value of the parameter k increases.

The functions $V(p_z)$ and $W(p_z)$ for the electrons of the lens have the form

$$V(p_z) = W(p_z) = \frac{1}{\sqrt{\pi q R_{\mathrm{ex}} f(p_z/p_2)}}, \tag{5.16}$$

where R_{ex} is the maximum value of the radius of the cyclotron orbit. The substitution of the expression (5.16) for $k = 1$ into (5.14) leads to the well-known result for a spherical Fermi surface with the Fermi momentum equal to p_2.

In this model, in the limit $qR \gg 1$, the Fourier transforms of the electron velocity components perpendicular to the magnetic field $\mathbf{B}$ have the form (for electrons associated with the lens):

$$v_{\pm n}^{x}(p_z, \pm q) = \pm i \frac{p_1}{m_1} \sqrt{\frac{2f(p_z/p_2)}{\pi q R_{ex}}} \sin\left[q R_{ex} f\left(\frac{p_z}{p_2}\right) - \frac{\pi n}{2} - \frac{\pi}{4}\right], \quad (5.17)$$

$$v_{\pm n}^{y}(p_z, \pm q) = \frac{n}{q R_{ex}} \frac{p_1}{m_1} \sqrt{\frac{2}{\pi q R_{ex} f(p_z/p_2)}} \cos\left[q R_{ex} f\left(\frac{p_z}{p_2}\right) - \frac{\pi n}{2} - \frac{\pi}{4}\right],$$

where $f(x) = \sqrt{1 - x^{2k}}$.

Using the asymptotics (5.14), (5.16), and (5.17), we can obtain asymptotic expressions for the kinetic coefficients appropriate in the limit of large qR. For example, the contributions of lens electrons to the electrical conductivity tensor components are given by

$$\sigma_{xx} = \frac{\sigma_0}{ql} a \coth\left[\pi \frac{1 - i\omega\tau}{\Omega\tau}\right] - \frac{\sigma_0}{ql} b \sqrt{\frac{p_2}{p_1}} \frac{\sin(2q R_{ex} - \pi/4k)}{(q R_{ex})^{1/2k} \sinh\left[\pi(1 - i\omega\tau)/\Omega\tau\right]}, \quad (5.18)$$

$$\sigma_{xy} = -\sigma_{yx} = b \frac{\sigma_0}{(ql)^2} \frac{1 - i\omega\tau}{\sinh\left[\pi(1 - i\omega\tau)/\Omega\tau\right]} \frac{\cos(2q R_{ex} - \pi/4k)}{(q R_{ex})^{1/2k}}, \quad (5.19)$$

$$\sigma_{yy} = d \frac{\sigma_0}{(ql)^2} (1 - i\omega\tau). \quad (5.20)$$

Here $\sigma_0 = Ne^2\tau/\sqrt{m_1 m_2}$, $l = \tau\sqrt{p_1 p_2/m_1 m_2}$, and the dimensionless constants a, b, and d are given, respectively, by

$$a = \frac{(1 + 1/2k)\sqrt{\pi}\Gamma(1/2k)}{(k + 1)\Gamma(\frac{1}{2} + 1/2k)},$$

$$b = \frac{2k + 1}{2k^2}\Gamma(1/2k) \quad \text{and} \quad d = 1 + 1/2k.$$

It follows from the obtained expressions that for $qR \gg 1$, the quantity σ^*, defined by (5.13) in our model, can be assumed to be equal to σ_{xx}. Similarly, we can prove that under these conditions the difference between the quantities λ^* and λ, γ^* and γ_x, and $\overline{\gamma}^*$ and $\overline{\gamma}_x$ is insignificant. Hence we can assume that for large qR the contributions of lens electrons to the kinetic coefficients λ^*, γ^*, and $\overline{\gamma}^*$ are described by the asymptotic expressions

$$\lambda^* = \lambda_1 + \lambda_2, \quad (5.21)$$

where

$$\lambda_1 = Q\frac{Np_1}{q} d \coth\left[\frac{\pi}{\Omega\tau}(1 - i\omega\tau)\right],$$

$$\lambda_2 = \frac{Np_1}{q} b\overline{U}_0^2(0)\frac{\sin(2q R_{ex} - \pi/4k)}{(q R_{ex})^{1/2k}} \left\{\sinh\left[\frac{\pi}{\Omega\tau}(1 - i\omega\tau)\right]\right\}^{-1}, \quad (5.22)$$

$$\gamma^* = \frac{iNe}{q} b \frac{\overline{U}_0(0)}{(q R_{ex})^{1/2k}} \exp\left[-iq R_{ex} + \frac{i\pi}{4k}\right]\Phi(q R_{ex})\left\{\sinh\left[\frac{\pi}{\Omega\tau}(1 - i\omega\tau)\right]\right\}^{-1}, \quad (5.23)$$

$$\overline{\gamma}^* = -\frac{iNe}{q} b \frac{\overline{U}_0(0)}{(q R_{ex})^{1/2k}} \exp\left[iq R_{ex} - \frac{i\pi}{4k}\right]\Phi^*(q R_{ex})\left\{\sinh\left[\frac{\pi}{\Omega\tau}(1 - i\omega\tau)\right]\right\}^{-1}. \quad (5.24)$$

Here

$$\overline{U}_0(p_z) = \frac{2U_0(p_z)}{p_1 p_2}\sqrt{m_1 m_2}, \qquad Q = \int_0^1 \frac{\overline{U}_0^2(p_2 t)}{\sqrt{1-t^{2k}}}dt.$$

The function $\Phi(q R_{ex})$ for $k \geq 2$ is described by the following asymptotic expression

$$1 - \exp\left(-\frac{i\pi}{4k}\right)\sin\left(2q R_{ex} - \frac{\pi}{4k}\right),$$

$\Phi^*(q R_{ex})$ is the quantity which is complex-congugate to $\Phi(q R_{ex})$.

If the electron lens (5.15) is ellipsoidal in shape ($k = 1$), the second term in formula (5.13) for the dynamic correction Δq for large qR is smaller than the first term by a factor of $1/\sqrt{qR}$ according to (5.22)–(5.24). In this case, the main contribution to Δq is associated with the deformation interaction of electrons with the acoustic wave

$$\Delta q = \Delta q_1 + \Delta q_2 = -\frac{iq^2}{2\rho_m s}(\lambda_1 + \lambda_2). \tag{5.25}$$

This formula is also applicable in the case when a line of parabolic points ($k > 1$) passes through the edge of the lens, but the curvature anomaly is manifested moderately. If the value of k is not too large, the parameter $(q R_{ex})^{-1/2k}$ can be regarded as small for $(q R_{ex}) \gg 1$, and the conclusion concerning the dominance of the deformation contribution in expression (5.13) remains in force.

We assume that the shape of the lens in the vicinity of its central cross-section is very close to cylindrical ($k \gg 1$), and the parameter $(q R_{ex})^{-1/2k}$ cannot be regarded as small any longer. In this case, the relation between the first and second terms in (5.13) depends on the relative number of conduction electrons connected with the lens. When the contributions from all parts of the Fermi surface are taken into account, the smooth components of the kinetic coefficients λ and σ_{xx} are proportional to the total concentration N_0 of the conduction electrons (for $\omega\tau < 1$), and their oscillating components are proportional to the concentration N of the electrons associated with the lens having a quasi-cylindrical belt. The lens contribution determines the main terms of the asymptotic expressions for the coefficients γ^* and $\overline{\gamma}^*$ in the region $qR \gg 1$, therefore they are also proportional to N.

When the major part of the conduction electrons of the metal in question is associated with the lens ($N/N_0 \sim 1$), the enhancement of the geometrical oscillations of the kinetic coefficients makes the second term in (5.13) as significant as the first term. Thus, the role of the electric fields accompanying the acoustic wave increases sharply under these conditions. For $N/N_0 \ll 1$, the second term in (5.13) is much smaller than the first term, i.e., the contribution from the interaction with the electric field of the acoustic wave is much smaller than the deformation contribution and can be neglected.

The effect of the electric field of the acoustic wave on the geometrical oscillations of the velocity of sound and ultrasonic absorption is analyzed in [125]. The model of the Fermi surface used in [125] satisfies both conditions ($k \gg 1$ and $N/N_0 \sim 1$) whose simultaneous fulfillment is required for the contribution from the interaction with the electric field of the acoustic wave to be comparable with the contribution from the deformation interaction. This model can be applied to organic layered conductors with a quasi-two-dimensional energy spectrum of charge carriers. In such compounds, the enhancement of the geometrical oscillations of electroacoustic coefficients leads to the resonant effect predicted in [125].

The Fermi surfaces for most of the real metals does not satisfy the conditions under which the interaction with the electric field of the acoustic wave can strongly affect the formation of magnetoacoustic oscillations ($k \gg 1$, $N/N_0 \sim 1$). However, the enhancement of the geometrical oscillations of the electroacoustic coefficient λ, due to local peculiarities of the Fermi surface geometry, leads no noticeable changes in the observables, even in the cases when the main contribution to the dynamic correction Δq to the wave vector of the acoustic wave is due to deformation interaction and is described by expression (5.25).

For moderate frequencies, the term Δq_1 in (5.25) does not exhibit an oscillatory dependence on the magnetic field and gives a smooth component of the contribution of the lens electrons to the damping coefficient and renormalization of the velocity of sound. Geometrical oscillations of the sound absorption coefficient Γ and the velocity shift $\Delta s/s$ are described by the term Δ_2:

$$\Gamma = \mathrm{Im}\, \Delta q_2 = \Gamma_0 b \overline{U}_0^2(0) \frac{\cos\left(2q R_{\mathrm{ex}} - \pi/2 - \pi/4k\right)}{(q R_{\mathrm{ex}})^{1/2k}}$$

$$\times \mathrm{Re}\left(\sinh\left[\pi \frac{1 - i\omega\tau}{\Omega\tau}\right]\right)^{-1}, \qquad (5.26)$$

$$\frac{\Delta s}{s} = \frac{s}{2\omega} \mathrm{Re}\, \Delta q_2 = \frac{N p_1}{4\rho_m s} b \overline{U}_0^2(0) \frac{\cos\left(2q R_{\mathrm{ex}} - \pi/2 - \pi/4k\right)}{(q R_{\mathrm{ex}})^{1/2k}}$$

$$\times \mathrm{Im}\left(\sinh\left[\pi \frac{1 - i\omega\tau}{\Omega\tau}\right]\right)^{-1}. \qquad (5.27)$$

Here $\Gamma_0 = N\omega p_1/2\rho_m s^2$.

Oscillations described by (5.26), (5.27) are formed by electrons associated with the line of parabolic points on the Fermi surface. The order of their amplitudes are larger by a factor of $(\sqrt{q R_{\mathrm{ex}}})^{1-1/k}$ than the amplitudes of the geometrical oscillations formed by electrons from other parts of the Fermi surface. Enhancement of oscillations is a consequence of the increase in the number of effective electrons on the quasi-cylindrical central belt of the Fermi surface under investigation. For $k \gg 1$, the enhancement of oscillations becomes significant.

The enhancement of geometrical oscillations considered above can be manifested only for a certain choice of the direction of the magnetic field relative to the symmetry axes of the crystal lattice. Like many other effects associated with local anomalies in the shape of the Fermi surface, this effect must strongly depend on the direction of the applied magnetic field (see, e.g., [79], [126], [127] and the previous chapter).

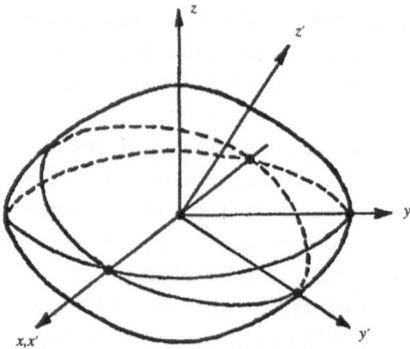

FIGURE 5.1. An axially symmetric Fermi surface corresponding to the energy-momentum relation (5.15). A line of zero curvature runs along the edge of the lens. An external magnetic field **B** is directed along the "z'''"-axis at an angle θ to the axis of symmetry of the lens ("z"-axis).

In this connection, let us consider the change in the amplitude of geometrical oscillations in a magnetic field tilted from the lens axis. We assume that the symmetry axis of the lens is turned relative to **B** through an angle θ. The most favorable situation for an analysis of the angular dependence of the amplitude of the geometrical oscillations, determined by the presence of a quasi-cylindrical belt on the surface of the electron lens, occurs when the angle θ lies in the yz-plane of the coordinate system under consideration. This is due to the following two circumstances.

First, if the rotation occurs in the yz-plane the extremal diameter of the electron cyclotron orbit, which determines the period of oscillations and considerably affects their amplitude, does not depend on the angle of rotation θ (Figure 5.1). Thus, the angular dependence of the oscillation amplitude is completely determined by the effect of the line of parabolic points on the Fermi surface.

Second, the stationary points of the cyclotron orbit remain parabolic points for any value of θ, although the line of parabolic points at the edge of the lens for $\theta \neq 0$ does not coincide any longer with the cross-section corresponding to the cyclotron orbit with the extremal diameter. The curvature of the Fermi surface in the vicinity of these points increases with θ at a higher rate than for $\theta = 0$, and hence the effect of local peculiarities of the geometry of the Fermi surface on magnetoacoustic oscillations must be suppressed with increasing θ, becoming vanishingly small as $\theta \to 90°$.

If the lens is rotated in some other plane, the points corresponding to stationary points of the cyclotron orbit "slip" from the line of parabolic points and fall into the region of the Fermi surface with a nonzero Gaussian curvature. The number of electrons which can participate in the absorption of the energy of sound waves in this case decreases more rapidly, and the effect of the line of parabolic points on the oscillation amplitude is manifested in a narrower range of the angle θ. Thus, a rotation in the yz-plane creates a more favorable situation for the experimental observation of the angular dependence under investigation.

Assuming that the rotational angle θ lives in the yz-plane, we can obtain the following expression for the oscillating component of the lens contribution to Δq_2:

$$\Delta q_2 = i\Gamma_0 d\overline{U}_0^2(0)\left[\sinh\left(\pi\frac{1-i\omega\tau}{\Omega\tau}\right)\right]^{-1}\left[\sin(2q\,R_{ex})S_k(\theta) - \cos(2q\,R_{ex})Y_k(\theta)\right],$$

$$(5.28)$$

where

$$S_k(\theta) = \int_0^\infty \cos\left[q\,R_{ex}\left(\frac{m_2}{m_1}\sin^2\theta y^2 + \cos^{2k}\theta y^{2k}\right)\right]dy,$$

$$Y_k(\theta) = \int_0^\infty \sin\left[q\,R_{ex}\left(\frac{m_2}{m_1}\sin^2\theta y^2 + \cos^{2k}\theta y^{2k}\right)\right]dy. \qquad (5.29)$$

The functions $S_k(\theta)$ and $Y_k(\theta)$ can be represented in the form of a power expansion in the parameter $\xi(\xi = (m_2/m_1)\tan^2\theta q\,R_{ex}^{1-1/k})$. For small values of θ, when $\xi \ll 1$, we obtain the following expression for $S_k(\theta)$:

$$S_k(\theta) = \left(2k\cos\theta(q\,R_{ex})^{1/2k}\right)^{-1}\sum_{r=0}^{\infty}\frac{(-1)^r}{r!}\xi^r\Gamma\left(\frac{2r+1}{2k}\right)\cos\left[\pi\frac{1-2r(k-1)}{4k}\right].$$

$$(5.30)$$

Here $\Gamma(x)$ is the gamma function.

As θ increases to the values ensuring the fulfillment of the inequality $\xi > 1$, the function $S_k(\theta)$ is described as follows:

$$S_k(\theta) = \frac{1}{2\sin\theta}\sqrt{\frac{m_1}{m_2}\frac{1}{q\,R_{ex}}}\sum_{r=0}^{\infty}\frac{(-1)^r}{r!}\xi^{-kr}\Gamma\left(kr+\tfrac{1}{2}\right)\cos\left[\frac{\pi}{4}(2r(k-1)+1)\right].$$

$$(5.31)$$

The expansions for $Y_k(\theta)$ can be obtained under similar conditions from (5.30) and (5.31) by replacing cosines by sines of the same arguments.

The oscillation amplitudes (5.27) are proportional to $F_k(\theta) = \left[S_k^2(\theta) + Y_k^2(\theta)\right]^{1/2}$. As $\theta \to 0$, the function $F_k(\theta)$ tends to the limit

$$F_k(0) = \Gamma\left(\frac{1}{2k}\right)\big/[2k(q\,R_{ex})^{1/2k}],$$

while with increasing θ it decreases and approaches the value $\sqrt{\pi m_1/(8m_2 q\,R_{ex})}$ which is a typical estimate for the amplitude of the magnetoacoustic geometrical oscillations in a metal with an ellipsoidal Fermi surface, as $\theta \to 90°$. The effect of the local anomaly of curvature on the amplitude of the geometrical oscillations remains significant in the range of angle θ between the axis of the lens and the magnetic field from zero to values of the order of $60°$. The results of the numerical calculations of the amplitude factor $F_k(\theta)$, for several values of the parameter k, are shown in Figure 5.2.

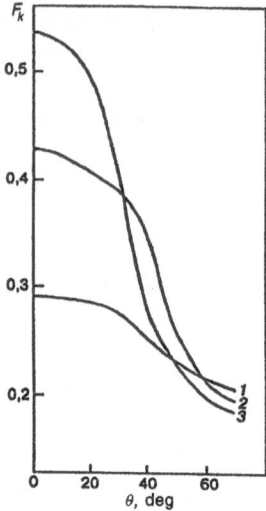

FIGURE 5.2. Dependence of the amplitude of the geometric oscillations on the angle θ between the magnetic field and the axis of symmetry of the electron lens. The curves are plotted for $q R_{ex} = 100$, $m_1/m_2 = 4$, $k = 2$ (curve 1), 3 (curve 2), 4 (curve 3).

Fermi surfaces of many real metals contain regions whose shape is close to cylindrical. The effect of enhancement of the geometrical oscillations of velocity and the absorption of ultrasonic waves can therefore be observed in many metals. The probability of its observation increases due to the existing dependence of the oscillation amplitude on the magnetic field orientation. The existence of such a dependence can serve as experimental evidence of the effect in question and can also give additional information on the geometrical parameters of the Fermi surface for some metals.

5.2   Local Flattening of the Fermi Surface and Magnetoacoustic Oscillations in Metals

A change in the number of electrons participating in the absorption of acoustic energy under the conditions of a Pippard geometrical resonance may be associated with the existence of local flattening of the Fermi surface at the points corresponding to the stationary points of a cyclotron orbit. Here we will analyze the effect of local flattening of the Fermi surface on the geometrical oscillations of absorption and the velocity of sound.

We assume that among cavities of a closed Fermi surface there is a biconvex lens, whose symmetry axis is the x-axis of a chosen coordinate system. We write the dispersion relation for the electrons associated with the lens in the form

$$E(\mathbf{p}) = \frac{p_1^2}{2m_1} \left(\frac{p_y^2 + p_z^2}{p_1^2} \right)^l + \frac{p_2^2}{2m_2} \left(\frac{p_x}{p_2} \right)^2. \qquad (5.32)$$

If the parameter l characterizing the shape of the lens assumes values greater than unity, then the Gaussian curvature of the surface vanishes at the points $(\pm p_2; 0; 0)$, which coincide with the vertices of the lens. Because of the axial symmetry of the lens, the vertices are points where the surface of the lens is flattened. The lens will be flatter near its vertices, the greater the value of l. Electrons associated with the vicinities of the vertices of the lens will participate strongly in the absorption of the energy of the acoustic wave, when both the magnetic field and the acoustic wave vector are perpendicular to the axis of the lens. Correspondingly, we assume that $\mathbf{B}$ is parallel to the z-axis, and a sound wave with wave vector q and frequency ω propagates along the y-axis of the coordinate system fixed in the lens. An expression for the wave vector of a sound wave can be written in the form (5.12).

For longitudinal sound the dynamical correction Δq is determined mainly by the deformation interaction with the electrons

$$\frac{\Delta q}{q} = -\frac{\omega}{2\rho_m s^2} \frac{1}{2\pi^2 \hbar^3} \int dp_z m_\perp \sum_n \frac{U_{-n}(p_z, -q) U_n(p_z, q)}{\omega + i/\tau - n\Omega}. \qquad (5.33)$$

Here the quantities $U_n(p_z, q)$ are defined by the relation (5.7).

For $qR \gg 1$ the neighborhoods of the stationary points on the cyclotron orbits make the main contribution to the integrals over ψ in the expressions for $U_n(p_z, q)$. Correspondingly, the asymptotic expressions for these quantities have the form (5.14) and the functions $V(p_z)$ and $W(p_z)$ can be written as follows:

$$V(p_z) = \int_0^\infty \cos\left[q R(p_z) Q_l(y, p_z) \right] dy,$$

$$W(p_z) = \int_0^\infty \sin\left[q R(p_z) Q_l(y, p_z) \right] dy, \qquad (5.34)$$

where

$$Q_l(y, p_z) = \sum_{k=1}^{l} a_k(p_z) y^{2k} \left(\frac{m_\perp^2}{m_1 m_2} \right)^k. \qquad (5.35)$$

All dimensionless coefficients $a_k(p_z)$ except for $a_l(p_z)$ turn zero at $p_z = 0$; the latter is of the order of unity at this point. Specifically, for $l = 2$, we have

$$Q_l(y, p_z) = \frac{m_\perp^2}{m_1 m_2} a_1(p_z) y^2 + \left(\frac{m_\perp^2}{m_1 m_2} \right)^2 a_2(p_z) y^4, \qquad (5.36)$$

where

$$a_1(p_z) = \frac{p_z^2}{p_1^2}, \qquad a_2(p_z) = \frac{1}{2} \left(1 - \frac{4}{3} \frac{p_z^4}{p_1^4} \right). \qquad (5.37)$$

For small values of p_z, corresponding to a neighborhood of the center cross-section of the lens, for which

$$\frac{p_z^2}{p_1^2} < [q R(p_z)]^{-1}, \tag{5.38}$$

the leading term of the asymptotic expansion of the function $V(p_z)$, in inverse powers of $q R$ with $l = 2$, has the form

$$V(p_z) = \frac{\Gamma(\frac{1}{4})}{4} \frac{\sqrt{m_1 m_2}}{m_\perp^{ex}} \left(\frac{2}{q R_{ex}}\right)^{1/4} \cos\left(\frac{\pi}{8}\right). \tag{5.39}$$

Here, $\Gamma(x)$ is the gamma function, $m_\perp^{ex} = m_\perp(0)$, and $R_{ex} = R(0)$.

For sufficiently large values of p_z, where the inequality (5.38) is not satisfied, the following approximation can be used for $V(p_z)$:

$$V(p_z) = \frac{1}{2} \sqrt{\frac{\pi m_1 m_2}{m_\perp^2} \frac{p_1^2}{p_z^2}} \frac{\cos(\pi/4)}{\sqrt{q R(p_z)}}. \tag{5.40}$$

The asymptotic expressions for $W(p_z)$ in the corresponding ranges of p_z are obtained from (5.39) and (5.40) by replacing the cosine by a sine of the same argument.

In calculating the dynamic correction arising in the wave vector of the sound wave as a result of the interaction with the electrons of the lens, the range of integration over p_z in expression (5.33) must be divided into regions, with small and large values of p_z. When the integration over each region is performed, the corresponding asymptotic form must be used for the functions $V(p_z)$ and $W(p_z)$. The result is

$$\Delta q = \Delta q_1 + \Delta q_2. \tag{5.41}$$

The principal term in the expansion of the addend Δq_1 in inverse powers of the parameter $q R$ is

$$\Delta q_1 = i\gamma_0 \left\{ \frac{\Gamma^2(\frac{1}{4})}{2\sqrt{2\pi}} \overline{U}_0^2(0) \coth\left[\pi \frac{1 - i\omega\tau}{\Omega_{ex}\tau}\right] \right. $$
$$\left. + \sqrt{q R_{ex}} \int_1^{\sqrt{q R_{ex}}} \overline{U}_0^2\left(\frac{t p_1}{\sqrt{q R_{ex}}}\right) \frac{\coth\left[\pi \frac{1 - i\omega\tau}{\Omega_{ex}\tau}\right]}{t^2 \sqrt{(q R_{ex})^2 - t^4}} dt \right\}, \tag{5.42}$$

where

$$\overline{U}_0(p_z) = \frac{m_2 U_0(p_z)}{p_2 \sqrt{p_1 p_2}}, \qquad \gamma_0 = \frac{m_1 N \omega p_2}{2\pi \rho_m m_\perp^{ex} s^2}.$$

The quantity γ_0 is of the order of the attenuation rate of the ultrasound in the absence of the external magnetic field. For $q R \gg 1$ the integral over "t" in expression (5.42) is of the order of $1/\sqrt{q R_{ex}}$. Thus both addends in the expression (5.42) have the same order of magnitude.

For not too high frequencies ($\omega\tau < 1$), the first term Δq_1 does not exhibit an oscillating dependence on the magnetic field and gives the smooth part of the contribution of the lens electrons to the absorption and the velocity shift of the ultrasonic wave.

The magnetoacoustic oscillations are described by the addend Δq_2 whose leading term is

$$\Delta q_2 = i\gamma_0 \overline{U_0(0)}^2 b \cos\left(2q R_{\text{ex}} + \frac{\pi}{4}\right) \left(\sinh\left[\pi\frac{1 - i\omega\tau}{\Omega_{\text{ex}}\tau}\right]\right)^{-1}, \tag{5.43}$$

where $b = \Gamma^2(\frac{1}{4})/(4\sqrt{2}\pi^2)$.

The real and imaginary parts of the dynamic correction Δq determine the renormalization of the velocity and the energy absorption coefficient for the ultrasonic wave. It follows from (5.43) that the principal term of the oscillating contribution to the absorption coefficient is

$$\gamma = \operatorname{Im}\Delta q_2 = \gamma_0 \overline{U_0(0)}^2 b \cos\left(2q R_{\text{ex}} + \frac{\pi}{4}\right) \operatorname{Re}\left(\sinh\left[\pi\frac{1 - i\omega\tau}{\Omega_{\text{ex}}\tau}\right]\right)^{-1}. \tag{5.44}$$

The amplitude of oscillations described by expression (5.44) is of the same order of magnitude in $(qR)^{-1}$ as the nonoscillating contribution to the absorption coefficient $\operatorname{Im}\Delta q_1$. This is a direct consequence of the larger number of effective electrons as a result of the flattening of the electron lens at the neighborhoods of its vertices. In a simple metal whose Fermi surface is closed and convex everywhere, the oscillating correction to the sound absorption coefficient is small compared to the smooth part.

The expressions (5.42)–(5.44) were derived under the assumption that the parameter l characterizing the degree of flatness of the lens near its vertices is equal to 2. However, even a moderate flattening of the Fermi surface can result in an enhancement of the geometric oscillations. For $l > 2$, the amplification of the oscillations will be even more pronounced. For an arbitrary value of l, the function $V(p_z)$ in a neighborhood of $p_z = 0$ is described by the asymptotic expression

$$V(p_z) = \frac{1}{2l}\frac{\Gamma(1/2l)}{(qR_{\text{ex}})^{1/2l}}\sqrt{\frac{m_1 m_2}{m_{\perp}^{\text{ex}}(a_l(0))^{1/l}}}\cos(\pi/4l). \tag{5.45}$$

A similar expression can also be written for $W(p_z)$.

As a result, an expression which extends the result (5.44) to arbitrary values of the parameter l is obtained for the oscillating part of Δq:

$$\gamma = \gamma_0 \overline{U_0(0)}^2 b_l (2q R_{\text{ex}})^{(l-2)/2l} \cos\left(2q R_{\text{ex}} + \frac{\pi}{2l}\right) \operatorname{Re}\left(\sinh\left[\pi\frac{1 - i\omega\tau}{\Omega_{\text{ex}}\tau}\right]\right)^{-1}. \tag{5.46}$$

Here, $b_l = \Gamma^2(1/2l)/\left(\pi^2 l^2 [a_l(0)]^{1/l}\right)$.

At the same time the first term in (5.42) has to be replaced by the expression

$$i\gamma_0\frac{\Gamma^2(1/2l)}{\pi^2 l^2 [a_l(0)]^{1/l}}\overline{U}_0^2(0)(2q R_{\text{ex}})^{(l-2)/2l}\coth\left[\pi\frac{1 - i\omega\tau}{\Omega_{\text{ex}}\tau}\right]. \tag{5.47}$$

The second term in the expression for the nonoscillating contribution to the dynamical correction Δq keeps the former order of magnitude.

Therefore, even with greater flattening of the surface of the lens, the amplitude of the geometric oscillations of ultrasonic absorption (5.46) is of the same order of magnitude as the smooth part of the absorption coefficient and is much greater than (by a factor qR_{ex} for $l \gg 1$) the amplitude of the corresponding oscillations in a simple metal, whose Fermi surface has no local flattenings. The oscillating contribution to the velocity of the ultrasonic wave is expressed in terms of the real part of the oscillating contribution to the correction Δq:

$$\frac{\Delta s}{s} = \frac{s}{2\omega} \operatorname{Re} \Delta q_2 = \frac{m_1 N}{4\pi \rho_m s} \frac{p_2}{m_\perp^{ex}} \bar{U}_0^2(0) b_l (2q R_{ex})^{(l-2)/2l} \cos\left(2q R_{ex} + \frac{\pi}{2l}\right)$$

$$\times \operatorname{Im}\left(\sinh\left[\pi \frac{1 - i\omega\tau}{\Omega_{ex}\tau}\right]\right)^{-1}. \tag{5.48}$$

The amplification produced in the magnetoacoustic oscillations as a result of the increase in the number of electrons participating in the oscillations is manifested in the oscillations of the sound speed just as in the oscillations of the ultrasonic absorption which we analyzed above.

Since the amplification of the geometric resonances in the velocity and absorption of ultrasound is due to the local geometric characteristics of the Fermi surface, it can be observed only for a definite choice of the direction of the magnetic field with respect to the symmetry axes of the crystal lattice. When the magnetic field is tilted away from the direction for which the point of flattening of the Fermi surface falls on its section corresponding to the cyclotron orbit of the electrons participating effectively in the formation of the oscillations, the influence of this point vanishes and the amplitude of the oscillations decreases. Therefore, the amplification of the geometric oscillations should exhibit a pronounced dependence on the direction of the external magnetic field. Specifically, for the model of the Fermi surface (5.32) considered here, the amplitude of the geometric oscillations of the ultrasound velocity shift and the absorption coefficient will depend on the angle between the external magnetic field and a plane perpendicular to the axis of the lens. The range of variation of the amplitude of these oscillations with an increase of this angle is determined by the rate of flattening of the lens near its vertices.

5.3 Acoustic Cyclotron Resonance and Giant High-Frequency Magnetoacoustic Oscillations in Metals with the Locally Flattened Fermi Surface

The enhancement of magnetoacoustic oscillations due to the local flattening of the Fermi surface can also exhibit itself in a high-frequency range ($\omega\tau > 1$). For these frequencies the magnetoacoustic oscillations may be superimposed on the acoustic cyclotron resonance.

Assume that among the cavities of a multiconnected Fermi surface there is a biconvex lens corresponding to the energy momentum relation (5.32). When we use the geometry described in the previous section (both the magnetic field and the wave vector of the ultrasound wave are directed perpendicularly to the axis of the lens), stationary points on the cyclotron orbit fall into the flattened segments of the lens near its vertices.

Using the asymptotic expressions (5.14), (5.45) for $U_{\pm n}(p_z, \pm q)$ and similar asymptotics for $v^{\alpha}_{\pm n}(p_z, \pm q)$ we arrive at expressions for electron kinetic coefficients. Taking into account that the largest contribution to the integrals over p_z in the expressions (5.4), (5.5) originates from the range of small p_z, we can replace all smooth functions of p_z in the integrands by their values at $p_z = 0$. For $qR \gg 1$, the main contribution to the asymptotic expression for λ is associated with the electrons of the lens

$$\lambda = \frac{ig}{\omega} \frac{\mu}{(qR_{ex})^{1/l}} U_0^2(0) W(\omega). \tag{5.49}$$

Here $R_{ex} = R(0)$, a dimensionless contant μ equals

$$\mu = \frac{a_l^2}{4\sqrt{\pi}} \frac{\langle 1 \rangle}{g} \Gamma\left(\frac{l+1}{2l}\right) \bigg/ \sqrt{\int_0^1 \overline{m}_{\perp}(x)\,dx}. \tag{5.50}$$

We introduce a notation $\langle 1 \rangle$ for the electron density of states on the lens, $x = p_z/p_m$ and $\overline{m}_{\perp}(x) = m_{\perp}(x)/m_{\perp}(0)$. The frequency-dependent factor $W(\omega)$ in Eq.(5.49) has the form

$$W(\omega) = \int_{-1}^{1} Y(\omega, x)\,dx, \tag{5.51}$$

where

$$Y(\omega, x) = -i\pi \frac{\omega}{\Omega} \left\{ \coth\left[\pi \frac{1 - i\omega\tau}{\Omega\tau}\right] \right.$$
$$\left. + \cos\left(2qR + \frac{\pi}{2l}\right) \left(\sinh\left[\pi \frac{1 - i\omega\tau}{\Omega\tau}\right]\right)^{-1} \right\}. \tag{5.52}$$

We obtain asymptotic expressions for the remaining electroacoustic kinetic coefficients in a similar way. Specifically, we have

$$\gamma_y = \frac{le}{q} \sum_k N_k \frac{e_k}{|e|} + \frac{ieg}{q} \frac{\mu}{(qR_{ex})^{1/l}} U_0(0) W(\omega), \tag{5.53}$$

$$\sigma_{yy} = -\frac{ie^2}{q^2} \omega g \left(1 - \frac{\mu}{(qR_{ex})^{1/l}} W(\omega)\right). \tag{5.54}$$

Oscillating terms in (5.53), (5.54) are mainly determined by the contributions from the flattened electron lens. Contributions from the remaining (nonflattened) cavities of the Fermi surface are proportional to the small factor $1/qR$ and we can omit them.

In the high-frequencies range ($\omega\tau \gg 1$) the function $Y(\omega, x)$ has singularities at frequencies ω which are equal to the multiple cyclotron frequency Ω. These singularities arise due to the acoustic cyclotron resonance which was analyzed in [128], [129]. The second term in (5.52) also contains the factor $\cos(2qR + \pi/2l)$ describing the geometrical oscillations.

The main contribution to the integral (5.51) is from the region of small x where the cyclotron frequency is close to its extremum value Ω_{ex}. In this region which corresponds to the vicinity of the center cross-section of the lens, we can use the following approximation:

$$\Omega(x) = \Omega_{ex}(1 + \eta^2 x^2), \tag{5.55}$$

where

$$\eta^2 = \frac{1}{\sqrt{\pi}} \frac{\Gamma((l+1)/2l)}{\Gamma(1+1/2l)} \int_0^1 \frac{dz}{z^2} \left(\frac{1}{\sqrt{1-z^2}} - \frac{1}{\sqrt{1-z^{2l}}} \right). \tag{5.56}$$

When $l = 1$ and the lens is ellipsoidal in shape this parameter η^2 turns to zero. In this case, the cyclotron frequency is independent of p_z. For a flattened lens ($l \geq 2$) this parameter takes nonzero values which may be of the order of unity.

An asymptotic expression for the function $W(\omega)$ near the cyclotron resonance depends on the ratio of the parameters $2q R_{ex}$ and $(\omega\tau)^{1/2}$. Under considered conditions, both parameters are large compared to unity. Suppose that $2q R_{ex} \gg (\omega\tau)^{1/2}$. Under conditions of the acoustic cyclotron resonance in convenient metals the parameter $q R_{ex} \sim v_F/s \sim 10^3$ (v_F is the Fermi velocity for the electrons associated with the lens). For $\Omega\tau \sim 10^2$, this inequality can be satisfied when the lens is moderately flattened ($l < 2$). The asymptotic expression for the function $W(\omega)$ near the cyclotron resonance can be written as follows:

$$W(\omega) = \frac{\pi}{\eta} \frac{1}{\mho} \left[1 + \frac{(-1)^n b}{(q R_{ex})^{1/2l}} \frac{\cos(2q R_{ex} + \pi/4l)}{\mho} \right], \tag{5.57}$$

where

$$b = \frac{2\eta}{\pi} \Gamma(1 + 1/2l), \qquad \mho = \sqrt{1 - \frac{\omega}{n\Omega_{ex}} - \frac{i}{\omega\tau}}.$$

The principal term in the obtained expression for $W(\omega)$ is its first term. The second term in (5.57), which describes the geometrical oscillations, is significantly smaller in magnitude.

When $2q R_{ex} \gg (\omega\tau)^{1/2}$ the dynamical correction Δq near the acoustic cyclotron resonance remains small compared to the main approximation of the ultrasound wave vector ω/s. For longitudinal waves this correction is mainly determined by the deformation interaction of the sound wave with the electrons. The resonance contribution to the correction Δ from the electrons associated with the neighborhood of the central cross-section of the lens (5.32) equals

$$\Delta q_r = \gamma_0 \frac{1}{(q R_{ex})^{1/l}} \frac{q R_{ex}}{n} \frac{1}{\mho} \left[1 + \frac{b \cos(2q R_{ex} + \pi n + \pi/4l)}{\mho (q R_{ex})^{1/2l}} \right]. \tag{5.58}$$

Here

$$\gamma_0 = \frac{\pi N q \omega m_\perp(0)\mu}{2\eta\rho_m s^2 p_2} U_0^2(0)$$

is the quantity of the dimensions and the order of the attenuation rate for high-frequency ultrasound waves in the absence of the external magnetic field.

The real and imaginary parts of the correction Δq_r determine the resonance contributions from the electrons associated with the lens to the velocity shift and the attenuation rate of the ultrasound wave. For $l = 1$, the result (5.58) for the attenuation rate coincides with the corresponding result of [129], which is obtained under the assumption that the Fermi surface of a metal everywhere has a finite and nonzero curvature. When $l = 1$, the magnitude of the resonance feature in the attenuation rate is of the order of $\gamma_0\sqrt{\omega\tau}/n$. In this case, the magnitude of the geometrical oscillations is smaller, by a factor of $\sqrt{\omega\tau/q R_{ex}}$, than the magnitude of the resonance feature connected with the cyclotron resonance.

When $l > 1$, the effective strip on the Fermi surface passes through the flattened segments near the vertices of the lens. It gives the amplification of the acoustic cyclotron resonance. The dependence of the attenuation rate of the ultrasound on the magnetic field near the cyclotron resonance is shown in the Figure 5.3. The Fermi surface is assumed to be moderately flattened. The resonance contribution to the ultrasonic absorption coefficient increases $(q R_{ex})^{(l-1)/l}$ times. This amplification arises due to the increase in the number of electrons participating in the resonance absorption of the energy of the ultrasound wave.

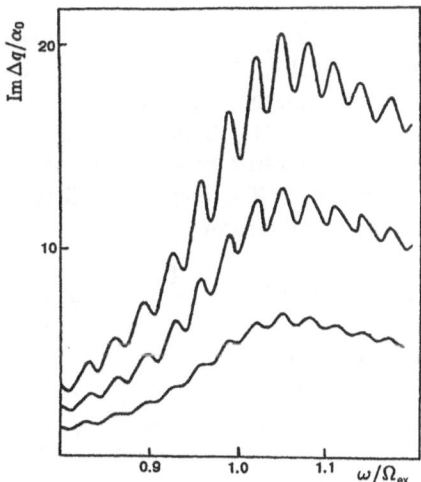

FIGURE 5.3. Attenuation of longitudinal ultrasound waves versus ω/Ω_{ex} in the vicinity of the cyclotron resonance for a moderate flattening of the Fermi surface near vertices of the electron lens. Curves are plotted for $\omega\tau = 10$, $q R_{ex} = 100$, $l = 1.25$ (curve 1), $l = 1.5$ (curve 2), $l = 1.75$ (curve 3).

This increase in the number of efficient electrons also leads to the amplification of the geometrical oscillations. The corresponding term in (5.58) is $(q R_{ex})^{(l-1)/2}$ times larger in magnitude than a similar term in the expression for Δq_r in a simple metal. When the flattening of the Fermi surface becomes stronger, the magnitude of the geometrical oscillations grows faster than the magnitude of the peak, corresponding to the acoustic cyclotron resonance. The larger l the relatively larger is the contribution from the term associated with the geometrical oscillations (the second term in expression (5.58)) to the correction Δq_r.

We can use expression (5.58) to describe the resonance part of the dynamic correction Δq only for moderate flattening of the electron lens and moderately large $\omega\tau$. When the flattening of the lens near its vertices is strong, the quantity $(\omega\tau)^{l/2}$ exceeds the parameter $2q R_{ex}$. Under the conditions of acoustic cyclotron resonance in typical metals the inequality $2q R_{ex} \ll (\omega\tau)^{l/2}$ can be satisfied for $l > 3$.

In this case, we have to use a new asymptotic expression for the function $W(\omega)$. This new asymptotic expression can be written as follows:

$$W(\omega) = \frac{\pi}{\eta} \frac{1}{\mho} \left[1 + (-1)^n \cos\left(2q R_{ex} + \frac{\pi}{2l} \right) \right]. \tag{5.59}$$

Both terms in this expression (5.59) are of the same order of magnitude. It critically changes the magnetic field dependence of the function $W(\omega)$ near the resonance ($\omega \approx n\Omega_{ex}$). When the asymptotic expression (5.59) is applicable, the factor $W(\omega)/(q R_{ex})^{1/l}$ in the expressions for the kinetic coefficients is not small compared to unity. In this connection, the contribution to the dynamic correction Δq arising due to the interaction with the electromagnetic field accompanying the sound wave becomes significant.

The effects originating from the coupling of electromagnetic and ultrasound waves are well known. Specifically, it is shown that the ultrasound wave propagating perpendicularly to the external magnetic field can couple to shortwave cyclotron waves (see [24], [130], [131]). In our geometry, longitudinal ultrasound waves couple to longitudinal cyclotron waves whose dispersion relation is determined by the equation $\sigma_{yy} = 0$. The dispersion curve of this mode near the frequency $n\Omega_{ex}$ can be written in the form

$$\omega_1 = n\Omega_{ex} \left[1 - \frac{1}{(q R_{ex})^{2/l}} f^2(q) \right]. \tag{5.60}$$

Here $f(q)$ is an oscillating function

$$f(q) = \frac{2\pi\mu}{\eta} \cos^2 \left[q R_{ex} + \frac{\pi n}{2} + \frac{\pi}{4l} \right]. \tag{5.61}$$

This cyclotron mode can propagate in a metal under the condition $2q R_{ex} \ll (\omega\tau)^{l/2}$. The shape of the dispersion curve of the considered cyclotron wave depends on the local geometry of the Fermi surface. Longitudinal cyclotron waves, similar to the mode described by (5.60), can propagate in a metal with a spherical Fermi surface under the condition $q R_{ex} < \omega\tau$. Their dispersion relation has the form (see [11]):

$$\omega_1 = n\Omega(1 + 1/2qR).$$ (5.62)

The difference in expressions (5.60) and (5.62) describing the dispersion curves of the longitudinal cyclotron waves are caused completely caused by the local flattening of the considered Fermi surface.

For a very strong flattening of the vicinities of the vertices of the electron lens $(2qR_{ex} \ll (\omega\tau)^{1/2})$ we can write the following expression for the resonance contribution to the dynamic correction Δq_r:

$$\Delta q_r = \gamma_0 \frac{qR_{ex}}{n} \frac{f^2(q)}{(qR_{ex})^{2/l}} \frac{\omega}{\omega_1 - \omega - i/\tau}.$$ (5.63)

Here ω_1 is the frequency of the longitudinal cyclotron wave described by formula (5.60). The frequency ω_1 corresponds to the resonance rather than the cyclotron frequency Ω_{ex}. The shift of the peak of the acoustic cyclotron resonance caused by the coupling ultrasound to the short-wave cyclotron wave was studied for the spherical and ellipsoidal Fermi surfaces. When the effective segments of the Fermi surface are locally flattened this shift is more pronounced and more available for experimental observations.

Besides the cyclotron mode described by (5.60), Fermi-liquid cyclotron waves can propagate in metals. Coupling to these Fermi-liquid modes can change the resonance contribution to the dynamical correction Δq near the acoustic cyclotron resonance. However, it is shown in [68] that these changes are not very significant because the coupling of the ultrasound to these Fermi-liquid modes is more weaker

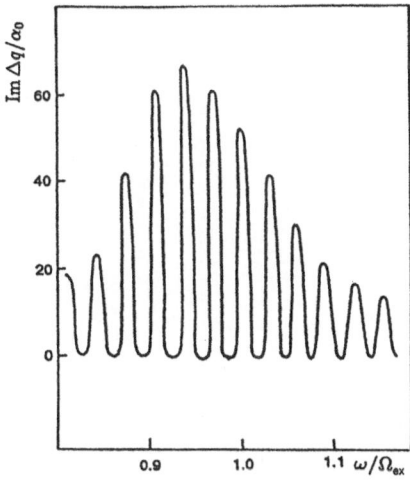

FIGURE 5.4. Giant geometric oscillations of the attenuation rate of longitudinal ultrasound waves in the vicinity of the cyclotron resonance for strong flattening of the Fermi surface near vertices of the electron lens ($l = 4$). The curve is plotted for $\omega\tau = 10$, $qR_{ex} = 100$, $\mu = 0.1$. Shift of the resonance occurs due to the coupling of the ultrasound to the cyclotron wave.

than to the mode analyzed above. It gives reasons to neglect the Fermi-liquid effects in the present consideration.

The factor $f^2(q)$ in expression (5.60) describes the geometrical oscillations which are superimposed on the peak corresponding to the acoustic cyclotron resonance. The amplitude of the geometrical oscillations sharply increases near the resonance. In order of magnitude, it is determined by the height of the resonance peak. Thus the geometrical oscillations of the ultrasonic absorption coefficient in metals with strongly flattened Fermi surfaces may be sufficiently amplified near the acoustic cyclotron resonance. Figure 5.4 illustrates this conclusion.

CHAPTER 6

Fermi-Liquid Cyclotron Doppleron Waves in Metals

6.1   Local Geometry of the Fermi Surface and Spectra of Cyclotron Doppleron Waves in Metals

It is known that cyclotron waves can propagate in metals along the direction of an external magnetic field. Waves of this kind exist due to the Fermi-liquid interaction of electrons. Such waves were predicted in [4] (see also the review article [21]).

Spectra of cyclotron waves and opportunities for their observation in experiments were repeatedly analyzed (see [14], [135]–[141]). It was shown that in metals with a spherical Fermi surface the dispersion curve of a cyclotron wave terminated on the boundary of the region of transparency. By virtue of such limitation of their frequencies the Fermi-liquid cyclotron waves can propagate in a rather narrow frequency and wave vector range. Their frequencies are close to the limiting value which corresponds to a wave vector tending to zero. The subsequent development of the theory for several concrete models of Fermi surfaces—a sherical belt, a surface close in shape to a sphere or ellipsoid—has also confirmed the above-mentioned limitation on the frequency of the waves [100], [142].

In the present section we analyze the character of spectra of the Fermi-liquid cyclotron waves for a wide class of axisymmetric Fermi surfaces. We show that when the Fermi surface has a nearly paraboloidal shape in the vicinities of limiting points or lines of inflection, the dispersion curve of the wave does not intersect the boundary of the region of transparency and the limitation on the values of frequency misses. A brief report about such an effect is contained in [143]. This effect arises due to the singularity of the dielectric constant on the boundary of the cyclotron absorption (or, in other words, when the frequency satisfies the condition for the Doppler-shifted cyclotron resonance) in metals with the paraboloidal Fermi surfaces. For the spherical Fermi surface this singularity is weaker. A similar effect was investigated in the theory of dopplerons [103]. The frequency range for the dopplerons which is very narrow for a spherical Fermi surface is essentially extended for a paraboloidal surface. This extension also appears because of the significant amplification of the singularity of the effective conductivity due to the Doppler-shifted cyclotron resonance. The detailed study has shown that segments of the Fermi surface, determining the response of the electron system under conditions of the Doppler-shifted cyclotron resonance for some metals, are much closer in shape to paraboloidal than to spherical (see [142]).

The character of dependence of the frequency ω of the Fermi-liquid cyclotron doppleron of its wave vector $\mathbf{q}$ is retrieved below. Under some conditions the wave frequency can be much smaller than the cyclotron frequency Ω. In this low-frequency range the dependence ω of q is similar to that for dopplerons. Thus in metals whose Fermi surfaces include nearly paraboloidal segments, we can expect a new type of collective exitation—Fermi-liquid cyclotron dopplerons—alongside the well-known dopplerons.

Let us assume that the magnetic field $\mathbf{B}$ is directed along the axis of an axisymmetric Fermi surface (the z-axis in the chosen coordinate system) and consider the transverse waves propagating along the magnetic field ($\mathbf{q} = (0, 0, q)$). To simplify the following calculations we shall first assume that the cyclotron frequency Ω does not depend on p_z. Circularly polarized waves can propagate in metals in the considered geometry. Their polarizations correspond to the circular components of a field $E_{\pm} = E_x \pm i E_y$. One of the polarizations (the component E_- in the case of electrons) corresponds to helicons, dopplerons, and cyclotron waves. Not being not interested here by the rather feeble influence of the Fermi-liquid interaction on the frequencies of helicons and dopplerons, we write the dispersion equation for the Fermi-liquid cyclotron wave. Using the expression (2.75) for the transverse conductivity of a metal with an axisymmetric Fermi surface and supposing that the cyclotron mass is a constant we can arrive at the equation

$$\Delta(u) = \frac{1 + A_2}{A_2}, \tag{6.1}$$

where

$$\Delta(u) = \frac{u}{Q_2} \left(\Phi_2(u) - \frac{\Phi_1^2(u)}{\Phi_0(u)} \right). \tag{6.2}$$

It is necessary to remark that the Fermi-liquid constant A_2, included in (6.1), can be positive as well as negative, and its magnitude can accept values about unity. According to existing knowledge, which is based on the experimental data obtained for alkali metals, the constant A_2 is small in magnitude compared to unity [24], [144]. However, we apparently have no basis to suppose this constant to be small in magnitude for metals with complex multiconnected Fermi surfaces.

In the derivation of (6.1) we did not take into account the displacement currents (i.e., values of the order $c^2 q^2 / \omega^2$), and also corrections of the order of $c^2 q^2 / \omega_p^2$, where ω_p is a plasma frequency for the quasi particles. Under these approximations the constant A_1 vanishes in the resulting dispersion equation.

In the limit of small q we can expand the function $\Delta(u)$ in inverse powers of the parameter u. Keeping the terms of the expansion no less than u^2 in the order of magnitude, we can derive from (6.1) the following expression for the frequency of the cyclotron wave (we omit the imaginary contribution arising due to the scattering, supposing that $\Omega\tau \gg 1$):

$$\omega = \Omega(1 + A_2)\left[1 + \frac{a}{A_2}\left(\frac{qv_m}{\Omega}\right)^2\right], \qquad (6.3)$$

where

$$a = \left[\int_{-1}^{1} \bar{s}(x)\bar{v}^2(x)x^2dx - \left(\int_{-1}^{1} \bar{s}(x)\bar{v}(x)xdx\right)^2\left(\int_{-1}^{1}\bar{s}(x)dx\right)^{-1}\right]$$

$$\times \left(\int_{-1}^{1}\bar{s}(x)x^2dx\right)^{-1}. \qquad (6.4)$$

Formula (6.3) is valid when the correction to the limiting frequency $\Omega(1 + A_2)$, which is proportional to q^2, is small in comparison with $A_2\Omega$. For a spherical Fermi surface we have $a = \frac{8}{35}$, and formula (6.3) coincides with the appropriate result obtained in [135].

When the wave vector increases, the frequency of a cyclotron wave remains greater than Ω for $A_2 > 0$, and it is less than Ω for $A_2 < 0$. For a spherical Fermi surface and other surfaces belonging to the same type (see below) the dispersion curve touchs the boundary of the region of transparency ($\omega = \Omega - qv_m$ for $A_2 < 0$ or $\omega = \Omega + qv_m$ for $A_2 > 0$) when q equals q_m. For the spherical Fermi surface the value q_m is determined in [135], $q_m = 5A_2\Omega/v_F(3 - A_2)$. In this case, the waves strongly damp when q is larger than q_m. We will show that for the Fermi surface close in shape to a paraboloid the dispersion curve of the cyclotron wave does not reach the region of cyclotron absorption, and the limitation on the frequency of the wave vanishes.

Let us first consider the Fermi surface which represents two paraboloidal "cups." Such a model was repeatedly used to study dopplerons in metals [90]–[92]. Within this model we have

$$E(\mathbf{p}) = \frac{p_x^2 + p_y^2}{m_\perp} + v_0|p_z|, \qquad (6.5)$$

where v_0 is a positive constant characterizing a magnitude of the longitudinal component of the velocity of electrons. To study the effects of qualitative character in the theory of dopplerons [90]–[92], [145] we have to consider contributions to the response functions from all parts of the Fermi surface (corresponding to electrons and holes). In particular, we have to consider all these contributions to take into account the compensation of the electron and hole contribution to the Hall currents in metals with equal numbers of electrons and holes. In the analysis of our problem it is possible to consider separately the effects originating from that segment of the Fermi surface which produces cyclotron dopplerons.

Calculating the function $\Delta(u)$ for the selected model of the Fermi surface (6.5) we obtain the following equation (neglecting contributions from electron scattering) which was derived in [143]:

$$3(\chi + \alpha_2)(1 - (u\chi)^2) = \alpha_2, \tag{6.6}$$

where $\chi = \Omega/\omega - 1$, $u = \omega/qv_m$, and $\alpha_2 = A_2/(1 + A_2)$. The considered waves correspond to solutions of (6.6) in the interval $\omega < \Omega - qv_m$ or $\omega > \Omega + qv_m$. In the long wavelength interval ($q^2v_m^2 \ll \alpha_2^2\Omega^2$) we can neglect unity in comparison with $(u\chi)^2 \approx \alpha_2^2\Omega^2/(qv_m)^2$. Then (6.6) will have a solution of the form (6.3) where the constant "a" equals $\frac{1}{3}$, as follows from expression (6.4).

When q increases and A_2 is positive, the frequency of the wave increases beyond all bounds. Supposing $\omega \gg \Omega$, we arrive at the following expression:

$$\omega = \Omega + qv_m\left\{1 + \frac{1}{2}\left[\sqrt{\left(1 - \frac{\alpha_2}{3}\right)^2 + \tfrac{4}{3}\alpha_2} - 1 + \frac{\alpha_2}{3}\right]\right\}^{1/2}. \tag{6.7}$$

For small positive values of the constant A_2 the dispersion curve runs nearby the boundary $\omega = \Omega + qv_m$:

$$\omega = \Omega + qv_m(1 + \alpha_2/6). \tag{6.8}$$

When the Fermi-liquid constant A_2 is negative the frequency of the wave tends to zero when $qv_m \to \Omega$. The dispersion dependence in the low-frequency range ($\omega \ll \Omega$) is described by the formula

$$\omega = (\Omega - qv_m)\frac{1 - |\alpha_2|}{6(1 - |\alpha_2|)}, \tag{6.9}$$

which results from (6.6) under the conditions $\omega \ll \Omega$ and $q \to \Omega/v_m$. For small $|A_2|$, formula (6.9) remains valid for rather small q as low as qv_m which is of the order of $\Omega|A_2|$. In this case, we can write a simple expression circumscribing all dispersion curve. For small $|A_2|$, the main role on the left-hand side of (6.9) is played by the factors $\chi + \alpha_2$ and $1 - (u\chi)^2$. Taking into account that both factors are small in magnitude (it follows from the assumption that $|A_2| \ll 1$) we can approximate the dependence ω of q to provide the validity of the asymptotics (6.3) and (6.9). Namely, assume that $(u\chi)^2 \approx 1 + \alpha_2^2\Omega^2/(qv_m)^2$ and $\omega \approx \Omega - \sqrt{(qv_m)^2 + \alpha_2^2\Omega^2}$. Then we obtain the following result:

$$\omega = \Omega - \frac{qv_m + |\alpha_2|\Omega}{2} - \sqrt{\frac{(qv_m - |\alpha_2|\Omega)^2}{4} + \frac{|\alpha_2|\Omega(qv_m)^2}{3(qv_m + \sqrt{(qv_m)^2 + |\alpha_2|^2\Omega^2})}}. \tag{6.10}$$

For $|A_2| \ll 1$, this formula gives a correct description for both limits $qv_m \ll |\alpha_2|\Omega$ and $qv_m \gg |\alpha_2|\Omega$ and the smoothly varying approximations for the dispersion curve the in intermediate interval. For arbitrary values of A_2 we can derive the relation corresponding to the dispersion curve as a result of equation (6.9).

These results show that for the paraboloidal Fermi surface there are no limitations on frequency of the Fermi-liquid cyclotron wave which exist for the spherical

Fermi surface. The only limitation on the wave frequency is stipulated by the increase in damping of the wave due to collisions near the boundary $\Omega - q v_m = \omega$. Taking into account the contribution into (6.6), from the electron scattering we can prove that the wave is weakly damping up to a magnitude of the wave vector of the order of $\Omega(1 - 1/|A_2|\Omega\tau)/v_m$. This value (especially for small A_2) is significantly larger than the value q_m for the spherical Fermi surface. Therefore, despite this limitation, the frequency of the Fermi-liquid cyclotron wave for negative A_2 can be much smaller than Ω (remaining larger than $1/\tau$), and its dependence of q is similar to that for dopplerons (see (6.9)). Thus, it is possible to predict that two dopplerons with identical circular polarization can propagate in some metals: the usual low-frequency doppleron and high-frequency Fermi-liquid wave with a doppleron-like spectrum.

Let us extend our consideration to include arbitrary axisymmetric Fermi surfaces. Its follows from (6.1) that the dispersion curve of the cyclotron wave will not intersect the boundary of the region of transparency when the function $\Delta(u)$ has a singularity at this boundary. The behavior of the function $\Delta(u)$ in the vicinity of the boundary is determined by the contributions to the integrals $\Phi_m(u)$ from the neighborhoods of those p_z, which correspond to the maximum value v_m of the longitudinal velocity $v_z(p_z)$. Therefore, to proceed, we have to define conditions for these contributions to tend to infinity when $\omega - \Omega \pm q v_m \to 0$. A similar analysis was carried out in the theory of dopplerons in [145]. It was shown that when the appropriate component of the tensor of conductivity (integral of a type of $\Phi_0(u)$) became infinite at the boundary of the cyclotron absorption (for the Doppler-shifted cyclotron resonance) it provides the propagation of the doppleron without damping in a wide frequency range.

For the spherical Fermi surface the longitudinal velocity $v_z(p_z)$ reaches its maximum in the limiting point $p_z = p_m$, and in the vicinity of this point it is a linear function of $p_z - p_m$. Just because of such a linear character of the dependence of v_z of $(p_z - p_m)$ the integrals $\Phi_m(u)$ remain finite when $\omega - \Omega \pm q v_m \to 0$. That is also valid for any nonspherical Fermi surface when the first derivative $\partial v_z/\partial p_z|_{p_z=p_m}$ is nonzero, and the greatest value of the longitudinal velocity $v_z(p_z)$ is reached at $p_z = p_m$. We shall assume now that the derivatives of the function $v_z(p_z)$ of the order l and above ($l > 1$) are nonzero at the point $p_z = p_m$ and its derivatives of the order less than l are equal to zero at this point. Then in the vicinities of the points $p_z = \pm p_m$, we have

$$v_z(p_z) \approx \pm v_m \left[1 - \left(\frac{p_m \mp p_z}{\Delta p} \right)^l \right], \tag{6.11}$$

$$S(p_z) \approx 2\pi m_\perp v_m (p_m \mp p_z). \tag{6.12}$$

Here, Δp is the parameter of dimensions of momentum describing a scale of change of the function $v_z(p_z)$. We can extract from the expressions for $\Phi_m(u)$ the integrals over small intervals $(p_m - \delta p, p_m), (-p_m, -p_m + \delta p)$, and use the asymptotic expressions (6.11), (6.12) in calculating these integrals. In the limit when $\omega - \Omega \pm q v_m$ goes to zero we shall discover singular parts of the functions

$\Phi_m(u)$. It appears that the logarithmic anomaly already occurs when $l = 2$, but the singular terms in the expression (6.2) for $\Delta(u)$ are canceled out. We obtain a similar result for $l = 3$, so the function $\Delta(u)$ has a singular part Δ_s only for $l \geq 4$. Neglecting the scattering ($1/\tau \to 0$) we obtain

$$\Delta_s = \pm\gamma \left(\frac{\Delta p}{p_m}\right)^4 \frac{\omega}{q v_m} \ln \frac{\Omega}{|\pm\omega \mp \Omega - q v_m|}, \qquad (l = 4), \qquad (6.13)$$

$$\Delta_s = \pm\eta_l \left(\frac{\Delta p}{p_m}\right)^4 \frac{\omega}{q v_m} \left(\frac{\Omega}{\pm\omega \mp \Omega - q v_m}\right)^{1-4/l}, \qquad (l > 4), \quad (6.14)$$

where γ and η_l are positive constants of the order of unity:

$$\gamma = \frac{\pi m_\perp v_m p_m^4}{2Q}, \qquad \eta_l = \frac{2\pi^2 m_\perp v_m p_m^4 \sin^2(\pi/l)}{Ql \sin(4\pi/l) \sin^2(6\pi/l)}. \qquad (6.15)$$

Thus when the Taylor series for the longitudinal velocity $v_z(p_z)$ in powers of $(p_z \pm p_m)$ does not contain terms of the first-, second- and third-order, the function $\Delta(u)$ tends to infinity when $\omega - \Omega \mp q v_m \to 0$, and the Fermi-liquid cyclotron doppleron waves can propagate in metals. The higher degree l corresponds to the first nonvanishing term of the Taylor series (6.11), the stronger is the singularity of the function Δ. When $l \to \infty$, which is appropriate to the paraboloidal shape of the Fermi surface, the function Δ has a simple pole, as well as in (6.6).

The maximal value of the longitudinal velocity $v_z(p_z)$ can be reached at an internal point p_1. Near these points $\pm p_1$ where $v_z(\pm p_1) = \pm v_m$ we can write the asymptotic expressions

$$v_z(p_z) \approx \pm v_m \left[1 - \left(\frac{p_1 \mp p_z}{\Delta_p}\right)^{2s}\right], \qquad (s \geq 1), \qquad (6.16)$$

$$S(p_z) \approx S(p_1) + 2\pi m_\perp v_m (p_1 \mp p_z). \qquad (6.17)$$

In this case, the function Δ tends to infinity when $\omega - \Omega \mp q v_m \to 0$ for $s > 1$. The singular part of this function Δ_s is equal to

$$\Delta_s = \pm v_s \left(\frac{\Delta p}{p_1}\right)^3 \frac{\omega}{q v_m} \left(\frac{q v_m}{\pm\omega \mp \Omega - q v_m}\right)^{1-3/2s}. \qquad (6.18)$$

Here, $v_s = \pi S(p_1) p_m^3 / Qs \sin(3\pi/2s)$ is a dimensionless coefficient of the order of unity. For large s the Fermi surface near $p_z = \pm p_1$ is close in shape to a paraboloidal belt and the function Δ described by the asymptotic (6.18) has a simple pole.

If the Fermi-liquid constant A_2 is small in comparison with unity, the expressions (6.13)–(6.14) and (6.18) for the singular part of the function $\Delta(u)$ immediately give us an asymptotic for the frequency of the cyclotron doppleron wave near the Doppler-shifted cyclotron resonance, since we can neglect the nonsingular part in the dispersion equation. In particular, using the formula (6.13) (a line of inflection on the Fermi surface) we obtain for small negative A_2:

$$\omega = \Omega - q v_m - \Omega \left[|\alpha_2| v_s \left(\frac{\Delta p}{p_1} \right)^2 \left(1 - \frac{q v_m}{\Omega} \right) \right]^{2s/(2s-3)} . \qquad (6.19)$$

It follows from this result that the frequency of the wave tends to zero when $\Omega - q v_m \to 0$ although its dependence on q is nonlinear in contrast to the paraboloidal model (see (6.9)). To take into account the scattering processes we have to add the term i/τ on the right-hand side of expression (6.19). The collisions restrict the frequency range of the wave from below.

Thus, assuming that Ω is independent of p_z, we showed that the Fermi-liquid cyclotron wave with a frequency much smaller than $\Omega(A_2 < 0)$ or much larger than $\Omega(A_2 > 0)$, could propagate in metals when their Fermi surfaces are close to paraboloids in the vicinities of the points of maxima of the longitudinal velocity of the electrons. To extend the frequency of the wave to zero (for $A_2 < 0$), or to infinity (for $A_2 > 0$), we have to suppose that all derivatives of the function v_z of the order no less than three turn to zero in the extreme points $p_z = \pm p_m$ and $p_z = \pm p_1$.

For a more realistic consideration we have to take into account that the cyclotron frequency depends on p_z. When Ω depends on p_z, the left-hand side of the dispersion equation (6.1) takes the form

$$\Delta = \frac{1}{Q_2} \left[\Gamma_-(v, v) - \frac{\Gamma_-(v, \pi) \Gamma_-(\pi, v)}{\Gamma_-(\pi, \pi)} \right]. \qquad (6.20)$$

Here the integrals $\Gamma_-(a, b)$ and Q_2 are the same as in the expression for the transverse conductivity of a metal with an axisymmetric Fermi surface (see Section 2.3).

Let us assume that the cyclotron frequency of the electrons in a magnetic field **B** accepts values in between Ω_{min} and Ω_{max}. The dispersion curves of weakly damping electromagnetic waves run along the region of transparency which now includes frequencies less than Ω_{min} and more than Ω_{max}. For definiteness, assume that $\omega < \Omega_{min}$, i.e., we will analyze solutions of the dispersion equation for negative values of the Fermi-liquid constant A_2.

Let us first consider the solution of (6.1) for small q, which corresponds to the Fermi-liquid cyclotron wave. Unlike the case analyzed before (Ω does not depend on p_z), now we have to consider contributions merely from electrons associated with a neighborhood of the cross-sections of the Fermi surface which correspond to the minimum cyclotron frequency. A limiting frequency of the Fermi-liquid cyclotron wave is determined with the character of dependence of Ω on p_z in these vicinities. Also it depends on the position of p_z corresponding to the extremum Ω in the interval of its possible values, namely whether it is an internal or a limiting point.

Assume that Ω_{min} is reached at the internal points ($p_z = \pm p_1$). Near these points we can write an asymptotic expression for $\Omega(p_z)$:

$$\Omega(p_z) = \Omega_{min} \left[1 + \xi^2 \left(\frac{p \pm p_z}{\Delta p} \right)^{2s} \right], \qquad (6.21)$$

where ξ^2 is a dimensionless positive constant.

For small q we can expand the function Δ, which is defined by the relation (6.20), in powers of q and solve (6.1). As a result we obtain the following expression for the frequency of the Fermi-liquid cyclotron wave

$$\omega = \Omega_{\min}\left[1 - (|\alpha_2|b_s)^{2s/(2s-1)}\right] - \frac{\Omega_{\min}\beta_s}{(|\alpha_2|b_s)^{2s/(2s-1)}}\left(\frac{qv_z(p_z)}{\Omega_{\min}}\right)^2. \qquad (6.22)$$

Here b_s and β_s are dimensionless positive constants of the order of unity

$$b_s = \frac{2\pi p_1^3 m_\perp(p_1)S(p_1)}{sQ\sin(\pi/2s)\xi^{1/s}},$$

$$\beta_s = \frac{1}{\pi}\sin\left(\frac{\pi}{2s}\right)\left\{B\left(\frac{1}{2s}, 3 - \frac{1}{2s}\right) - B^2\left(\frac{1}{2s}, 2 - \frac{1}{2s}\right)\right\},$$

$B(x, y)$ is the beta function.

We can show in a similar way that when $\Omega_{\min}$ is reached at the limiting points $p_z = \pm p_m$, the wave frequency depends on q as follows:

$$\omega = \Omega_{\min}\left[1 - (|\alpha_2|c_s)^{s/(s-1)}\right] - \frac{\Omega_{\min}\delta_s}{(|\alpha_2|c_s)^{s/(s-1)}}\left(\frac{qv_z(p_z)}{\Omega_{\min}}\right)^2. \qquad (6.23)$$

The dimensionless constants c_s and δ_s are accordingly equal

$$c_s = \frac{2\pi^2 v_z(p_m)m_\perp^2(p_m)p_m^4}{Q_s\sin^2(\pi/2s)\xi^{2/s}},$$

$$\delta_s = \frac{1}{\pi}\sin\left(\frac{\pi}{s}\right)\left\{B\left(\frac{1}{s}, 3 - \frac{1}{s}\right) - B^2\left(\frac{1}{s}, 2 - \frac{1}{s}\right)\right\}.$$

It follows from these results, (6.22), (6.23), that the dependence of the cyclotron frequency of p_z does not bring any qualitative changes for small q: the dependence ω from q still remains square-law. As well as in (6.3), the obtained formulas remain valid so long as the correction to the limiting frequency ω_0 depending on q remains small compared to $\Omega - \omega_0$.

It was mentioned above that the dispersion curve of the Fermi-liquid cyclotron wave has a doppleron-like extension in the low-frequency range when the function Δ tends to infinity on the boundary of the region of transparency. Assuming that the cyclotron mass does not depend on p_z we showed that it would take place when the Fermi surface is close to a paraboloid near those cross-sections where the longitudinal velocity of the electrons reaches its maxima. The absorption of energy of the wave on the boundary of the region of transparency is carried out by electrons corresponding to these cross-sections of the Fermi surface. Hence the possibility for this cyclotron doppleron wave to propagate in a metal is provided by the local geometry of the Fermi surface.

When Ω depends on p_z, electrons associated with various cross-sections of the Fermi surface participate in the absorption of the energy of the cyclotron wave at the boundary of the region of transparency. Therefore, to support tending of the function Δ to infinity on this boundary, we have to require that not merely the narrow strips near the lines of inflection or the vicinities of limiting points but rather the large segments of the Fermi surface are nearly paraboloidal. This condition is too stringent for the Fermi surfaces of real metals. Thus we can expect that the dispersion curve of the Fermi-liquid cyclotron wave will intersect the boundary of the region of transparency at rather small q.

However, the solution of the dispersion equation corresponding to the cyclotron doppleron can exist at low frequencies $\omega \ll \Omega$. The equations circumscribing the boundary of the region of transparency have the form

$$\begin{cases} R(\omega, q, p_z) = 0, \\ \\ \partial R(\omega, q, p_z)/\partial p_z = 0, \end{cases} \tag{6.24}$$

where $R(\omega, q, p_z) = \omega - \Omega(p_z) - q v_z(p_z)$. For small ω we have

$$\begin{cases} \Omega(p_z) \left(1 - \dfrac{cq}{2\pi |e| B} \dfrac{\partial S}{\partial p_z} \right) = 0, \\ \dfrac{\partial \Omega}{\partial p_z} \left(1 - \dfrac{cq}{2\pi |e| B} \dfrac{\partial S}{\partial p_z} \right) - \Omega(p_z) \dfrac{cq}{2\pi |e| B} \dfrac{\partial^2 S}{\partial p_z^2} = 0. \end{cases} \tag{6.25}$$

It follows from these equations that the absorption on the boundary for small ω is carried out by the electrons belonging to a neighborhood of particular cross-sections of the Fermi surface where extrema of the value $\partial S/\partial p_z$ are reached. It can be neighborhoods of limiting points or lines of inflection. Thus, in the low-frequency range $\omega \ll \Omega$, the dependence of the cyclotron frequency on p_z can be neglected and a singular part of the function Δ is still featured by the formulas (6.13), (6.14), (6.18). This means that the conditions for propagation of the cyclotron doppleron remain the same, as well as for the Fermi surface which is characterized by a stationary value of cyclotron mass.

The obtained results can be significant in the analysis of observation of such Fermi-liquid waves in experiments because the Fermi surfaces of some metals actually have segments close to paraboloids in shape. The predicted cyclotron doppleron waves have a wider frequency range than cyclotron waves observed before [135]–[142] and can be more available for observation in experiments.

6.2 Propagation of Cyclotron Doppleron Waves Through a Metal Plate

To clarify the possible manifestations of the Fermi-liquid cyclotron doppleron waves in the high-frequency properties of a metal plate, i.e., to determine the magnitude and shape of the appropriate size oscillations, we have to calculate the contribution of these waves to the surface impedance or transmission coefficient. In this section we consider the transmission coefficient for the cyclotron doppleron wave propagating through a metal plate. The Fermi-surface of the metal is assumed to be axisymmetric.

We consider a metal plate of width L in the presence of an applied magnetic field directed along a normal to the boundaries. The symmetry axis of the Fermi surface is parallel to the magnetic field (z-axis) and boundaries reflect the conduction electrons in a similar manner. Then the Maxwell equations inside the metal are reduced to two independent equations for the circular components of an electrical field $E_\pm(z)\exp(-i\omega t)(E_\pm = E_x \pm i E_y)$. High-frequency properties of the plate are featured by the surface impedances $Z^0_\pm$ and $Z^1_\pm$, combining the magnitudes of the electrical and magnetic fields ($E_\pm(z)$ and $b_\pm(z)$) on the boundaries $z = 0$ and $z = L$:

$$E_\pm(0) = \frac{c}{4\pi}\big[Z^{(0)}_\pm b_\pm(0) - Z^{(1)}_\pm b_\pm(L)\big], \tag{6.26}$$

$$E_\pm(L) = \frac{c}{4\pi}\big[Z^{(1)}_\pm b_\pm(0) - Z^{(0)}_\pm b_\pm(L)\big]. \tag{6.27}$$

When the circularly polarized electromagnetic wave of a magnitude E_i falls on the boundary $z = 0$, the magnitude of the transmitted wave $E_t = E_\pm(L)$ is determined by the surface impedance $Z^{(1)}_\pm$. Taking into account the relations

$$b_\pm(z) = \frac{c}{i\omega}E'_\pm(z), \tag{6.28}$$

$$2E_i = E_\pm(0) + b_\pm(0), \qquad E_\pm(L) = b_\pm(L), \tag{6.29}$$

we can write the following expression for the transmission coefficient

$$T_\pm = \frac{E_\pm}{E_t} = \frac{E_\pm(L) + b_\pm(L)}{E_\pm(0) + b_\pm(0)} \approx \frac{c}{2\pi}Z^{(1)}_\pm. \tag{6.30}$$

This result is valid when $(c/4\pi)Z^{(1,0)}_\pm \ll 1$. The magnitude of the reflected wave (or reflection coefficient) is determined by the impedance $Z^{(0)}_\pm$. In the following consideration we omit, for simplicity, the indices $\pm$.

To calculate the impedances $Z^{(0)}$ and $Z^{(1)}$ we have to solve the Maxwell equations for the field inside the plate. To proceed we can expand the magnitude of the field intensity $E(z)$ in Fourier series:

$$E(z) = \frac{2}{L}\sum_{N=0}\big(1 - \tfrac{1}{2}\delta_{N0}\big) E_N \cos(q_N z) = \frac{2}{L}\sum_{N=1} \tilde{E}_N \sin(q_N z), \tag{6.31}$$

where

$$E_N = \int_0^L E(z)\cos(q_N z)\,dz,$$

$$\tilde{E}_N = \int_0^L E(z)\sin(q_N z)\,dz,$$

$$q_N = \pi N/L.$$

We can use a similar expansion for the current density $j(z)$. Fourier components E_N satisfy the equation

$$-\frac{c^2 q_N^2}{4\pi i\omega}E_N + j_N = \frac{c}{4\pi}\big[b(0) - (-1)^N b(L)\big]. \tag{6.32}$$

The Fourier components of the current density j_N are connected to E_N by the relations

$$j_N = \sigma_N E_N + \sum_{N'=0} \left(1 - \tfrac{1}{2}\delta_{N'0}\right) \sigma_{NN'} E_{N'}. \tag{6.33}$$

Here σ_N is the bulk conductivity of a metal calculated for $q = q_N$, and $\sigma_{NN'}$ is the surface contribution to the conductivity. Due to the axial symmetry of the Fermi surface components, j_N with even N are expressed in terms of the components E_N also having even numbers. Fourier components with odd numbers N are also connected in a similar way. Therefore we can write (6.32) separately for even and odd N and define the impedances for symmetric and antisymmetric excitation as follows:

$$Z^{(s)} = \frac{2}{L} \sum_{n=0} \left(1 - \tfrac{1}{2}\delta_{n0}\right) Z_{2n}^{(s)},$$

$$Z^{(a)} = \frac{2}{L} \sum_{n=0} Z_{2n+1}^{(a)}. \tag{6.34}$$

Here the quantities $Z_{2n}^{(s)}$ and $Z_{2n+1}^{(a)}$ satisfy the equations [146], [147]:

$$-\frac{c^2 q_{2n}^2}{4\pi i\omega} Z_{2n}^{(s)} + j_{2n} = 1, \tag{6.35}$$

$$-\frac{c^2 q_{2n+1}^2}{4\pi i\omega} Z_{2n+1}^{(a)} + j_{2n+1} = 1. \tag{6.36}$$

Thus the impedances $Z^{(0)}$ and $Z^{(1)}$ can be expressed in terms of $Z^{(s)}$ and $Z^{(a)}$ as follows:

$$Z^{(0)} = Z^{(s)} + Z^{(a)},$$

$$Z^{(1)} = Z^{(s)} - Z^{(a)}. \tag{6.37}$$

We consider the transmission coefficient for the specular reflection of electrons from the boundaries, assuming that the polarization of the field corresponds to the cyclotron doppleron wave propagating in the plate. We start from the following expression for the transmission coefficient:

$$T = \frac{4i\omega^2}{Lc} \sum_{n=0} (-1)^{n+1} \left(1 - \tfrac{1}{2}\delta_{n0}\right) f(q_n), \tag{6.38}$$

where

$$f(q_n) = \left[q_n^2 - \frac{4\pi i\omega\sigma_n}{c^2} \right]^{-1}, \tag{6.39}$$

where $\sigma_n = \sigma_{xx}(\omega, q_n) - i\sigma_{yx}(\omega, q_n)$ is the Fourier component of the conductivity. Using the Poisson sum formula

$$\sum_{n=0}^{\infty} f(q_n) = \sum_{r=-\infty}^{\infty} \int_0^{\infty} f\left(\frac{\pi}{L}x\right) \exp(2\pi i r x)\, dx, \tag{6.40}$$

we convert the expression for the transmission coefficient and arrive at the result

$$T = \frac{2}{\pi} \frac{\omega}{c} \int_{-\infty}^{\infty} \text{sign}(q) \, \text{cosec} \, (Lq) f(q) dq. \tag{6.41}$$

An important contribution to the integral (6.41) comes from the vicinities of poles of the function $f(q)$, i.e., the roots of the dispersion equation

$$q^2 - 4\pi i \omega \sigma / c^2 = 0. \tag{6.42}$$

It is shown previously that when the Fermi surface of a metal includes nearly paraboloidal segments, equation (6.42) has a solution corresponding to the cyclotron doppleron. For low frequencies ($\omega \ll \Omega$), the dispersion dependence of this wave is determined mainly by the shape of the Fermi surface near those p_z which correspond to the maxima of the quantity $\partial S / \partial p_z$. These values of p_z correspond to limiting points or lines of inflection of the Fermi surface. The dispersion curve of the cyclotron doppleron runs above the line $\omega = \Omega$ for the Fermi-liquid coefficient $A_2 > 0$, and lower than this line when $A_2 < 0$. In further consideration, we will analyze the case when the parameter A_2 is negative.

Assume that the extreme values of the longitudinal velocity $v_z(p_z)$ are reached at the limiting points $p_z = \pm p_m$. The appropriate expression for the wave vector q (we denote it by q_1) for rather low frequencies $\tau^{-1} \ll \omega \ll \Omega$ and a small negative Fermi-liquid constant A_2 has the form

$$q_1 v_m = \Omega - \omega - \Omega \left[|\alpha_2| \eta_l \frac{\omega}{\Omega} \left(\frac{\Delta p}{p_m} \right)^4 \right]^{l/(l-4)}. \tag{6.43}$$

When the maximum v_z is reached at $p_z = \pm p_1$ on a line of inflection, the wave vector of the wave for $\omega \ll \Omega$ and small negative A_2 equals

$$q_1 v_m = \Omega - \omega - \Omega \left[|\alpha_2| v_s \frac{\omega}{\Omega} \left(\frac{\Delta p}{p_1} \right)^3 \right]^{2s/(2s-3)}. \tag{6.44}$$

When the frequency ω is close to the cyclotron frequency (supposing that $m_\perp = \text{const.}$), the root of the dispersion equation q_1 is approximately equal

$$q_1 = \frac{\Omega}{v_m} \sqrt{-\frac{\alpha_2}{a} \frac{\omega}{\Omega} \left(1 - \frac{\omega}{\omega_0} + \frac{i}{\Omega \tau} \right)}, \tag{6.45}$$

which corresponds to the Fermi-liquid cyclotron wave whose limiting frequency is $\omega_0 = \Omega(1 + A_2)$.

The possible frequencies of the cyclotron doppleron have to satisfy the inequality $(|A_2| \tau)^{-1} \ll \Omega < \omega_0$. For $\tau < 10^{-9}$ s the frequency ω cannot be lower than 10^5–10^6. Correspondingly, the anomaly parameter $\xi = 4\pi \omega N v_m^2 m_\perp c / |e| B^3$, in magnetic fields of the order of 30–50 kG, is of the order of 10^3–10^4. Thus, for all frequency ranges of the cyclotron doppleron, the skin effect is of anomalous character. Under these conditions the solution q_1 is a unique solution of the dispersion equation (6.42) which corresponds to a weakly damping electromagnetic wave which can propagate through the plate.

The contribution from the cyclotron doppleron to the transmission coefficient is equal to a residue from the appropriate pole of the integrand in expression (6.41). Taking into account only the singular parts of the integrals, $\Gamma_-(v, v)$, $\Gamma_-(\pi, \pi)$, $\Gamma_-(v, \pi)$ in calculation of the conductivity, we arrive at the following expression for the magnitude of the transmission coefficient of the cyclotron doppleron for $\omega \ll \Omega$:

$$
|T_1| = \frac{\beta_l}{\xi} \frac{v_m}{c} \frac{\omega}{\Omega(p_m)} \left(|\alpha_2| \frac{\omega}{\Omega(p_m)} \right)^{(2l-2)/(l-4)} \left(1 - \frac{\omega}{\Omega(p_m)} \right)^{-6l/(l-4)}
$$
$$
\times \left(\frac{\Delta p}{p_m} \right)^{6l/(l-4)} \left[\sin^2 \left(\frac{L}{v_m} [\Omega(p_m) - \omega] \right) + \sinh^2 \left(\frac{L}{v_m \tau} \right) \right]^{-1/2}. \quad (6.46)
$$

This expression is fair when the maxima of the derivative $\partial S/\partial p_z$ are reached at the limiting points ($p_z = \pm p_m$).

When $\partial S/\partial p_z$ gets its maximum value at $p_z = \pm p_1$, the magnitude of the transmission coefficient of the cyclotron doppleron equals

$$
|T_1| = \frac{\rho_s}{\xi} \frac{v_m}{c} \frac{\omega}{\Omega(p_1)} \left(|\alpha_2| \frac{\omega}{\Omega(p_1)} \right)^{(4s-3)/(2s-2)} \left(1 - \frac{\omega}{\Omega(p_1)} \right)^{-3/(2s-3)}
$$
$$
\times \left(\frac{\Delta p}{p_1} \right)^{6s/(2s-3)} \left[\sin^2 \left(\frac{L}{v_m} [\Omega(p_1) - \omega] \right) + \sinh^2 \left(\frac{L}{v_m \tau} \right) \right]^{-1/2}. \quad (6.47)
$$

Here β_l, ρ_s are dimensionless factors of the order of unity

$$
\beta_l = \frac{l}{l-4} \frac{Q_0}{Q_2} \sec \left(\frac{2\pi}{l} \right) (\eta_l)^{(l+2)/(l-4)},
$$
$$
\rho_s = \frac{2s}{2s-3} \frac{Q_0}{Q_2} \frac{p_1^2}{p_m^2} (v_s)^{2s/(2s-3)}. \quad (6.48)
$$

When the frequency ω is close to the cyclotron frequency Ω we obtain instead of (6.46), (6.47) the following results for the magnitude of the transmission coefficient:

$$
|T_1| = \frac{\mu}{a^{3/2}} \left[|\alpha_2| \left(\frac{\omega}{\Omega} \right) \right]^2 \frac{B^2}{m_\perp v_m N c} \frac{1}{\xi} \sqrt{\frac{1-z}{z}}
$$
$$
\times \left[\left(\frac{L\Omega|\alpha_2|}{v_m} \sqrt{\frac{1-z}{az}} \right) + \sinh^2 \left(\frac{L}{2v_m \tau} \sqrt{\frac{z}{a(1-z)}} \right) \right]^{-1/2}, \quad (6.49)
$$

where $z = (\Omega/\omega - 1)/|\alpha_2|$, $m_\perp$ is assumed to be a constant.

The magnitude of the transmission coefficient of the cyclotron doppleron increases when its frequency ω approachs the cyclotron frequency and the width of the plate decreases. The cyclotron doppleron waves are available for observation in thin films when $L \ll v_m \tau$. Considering this condition to be satisfied, we can obtain the following estimates for $|T_1|$ in a typical metal in a magnetic field of the order of 50 kG:

$$|T_1| < (10^{-7} - 10^{-8}) \left(\frac{\Delta p}{p_m}\right)^{6l/(l-4)} \frac{v_m \tau}{L}, \tag{6.50}$$

or

$$|T_1| < (10^{-7} - 10^{-8}) \left(\frac{\Delta p}{p_1}\right)^{6s/(2s-3)} \frac{v_m \tau}{L}. \tag{6.51}$$

When $\Delta p/p_1 \sim 0.05$, $v_m \tau/L \sim 10^3$, and $s \gg 1$, these estimates give that $|T_1| \sim (10^{-8}$–$10^{-9})$. The values of such an order can be measured in experiments on transmission of the electromagnetic waves through metal films.

Under considered conditions, the transmission coefficient also includes contributions from electrons corresponding to the vicinities of those cross-sections of the Fermi surface where the longitudinal component of their velocity tends to zero. These contributions always exist under conditions of the anomalous skin effect. They are produced by electrons which determine the asymptotic of the conductivity for large $q(qv_m/\omega \gg 1)$.

The asymptotic expression for the transverse conductivity for large q depends on the geometry of the Fermi surface near those cross-sections where $v_z = 0$. When the Fermi surface everywhere has a finite and nonzero curvature the main term in the expansion of the conductivity in inverse powers of q is featured by formula (3.8). Using this expression, we can obtain the following approximations for the roots of the dispersion equation (6.42), corresponding to the abnormal skin effect:

$$q_2 = i\frac{\omega}{v_m}\xi, \qquad q_3 = i\frac{\omega}{v_m}\exp(i\pi/6)\xi. \tag{6.52}$$

Assume that the effective cross-sections of the Fermi surface include one of a zero curvature. The number of the effective electrons associated with this cross-section can be large enough to make their contribution to the conductivity predominant. In this case, expression (3.8) does not give the correct asymptotic for the conductivity. Under these conditions the main term in the asymptotic expression for the conductivity takes the form (see (3.17)):

$$\sigma = \sigma_0 R \left(1 - i\tan\frac{\pi\beta}{2}\right)\left(\frac{\omega}{qv_m}\right)^\beta \left(1 - \frac{\Omega}{\omega}\right)^\beta, \qquad -1 < \beta < 0. \tag{6.53}$$

The value of the dimensionless parameter R is determined by the relative number of electrons associated with the nearly cylindrical effective cross-section. Here it is of the order of unity. When $\beta \to -1$ the shape of the Fermi surface near the effective cross-section is very close to cylindrical, and the roots of the dispersion equations q_2 and q_3 are featured by the expression

$$q_2 \sim (R\xi^3)^{1/2}\left(1 - \frac{\Omega}{\omega} + \frac{i}{\omega\tau}\right)^{-1/2}\left(1 + i\frac{\pi}{4}(\beta + 1)\right),$$

$$q_3 \sim (R\xi^3)^{1/2}\left(1 - \frac{\Omega}{\omega} + \frac{i}{\omega\tau}\right)^{-1/2}\left(1 + 3i\frac{\pi}{4}(\beta + 1)\right). \tag{6.54}$$

Thus, when the Fermi surface everywhere has a finite nonzero curvature (e.g., it is a sphere) the "skin" roots $q_{2,3}$ are proportional to ξ, and for the surface having nearly cylindrical segments $q_{2,3} \sim \xi^{3/2}$. The indicated cases do not exhaust all possibilities. For example, in a model of the Fermi surface, shaped as a paraboloidal lens (such a surface has infinite curvature on its cross-section at $p_z = 0$), $q_{2,3} \sim \xi^{3/4}$. Further we assume that the Fermi surface has finite curvature in every point. It provides the most favorable conditions to discover the contribution of the cyclotron doppleron wave into the transmission coefficient.

The contribution to the transmission coefficient from electrons associated with the effective cross-sections is equal to the sum of residues from poles of the function $f(q)$, which correspond to the roots of the dispersion equation. The most favorable conditions for the observation of the cyclotron doppleron wave in experiments are provided when the contribution from this wave to the transmission coefficient is no smaller than other contributions. It is provided when $L\omega\xi > v_m$. When the Fermi surface everywhere has a finite nonzero curvature, the expressions for $|T_{2,3}|$ can be written as follows:

$$|T_2| \approx \frac{4}{3} \frac{v_m}{c} \frac{1}{\xi} \exp\left(-\frac{L\omega\xi}{v_m}\right);$$ (6.55)

$$|T_3| \approx \frac{4}{3} \frac{v_m}{c} \frac{1}{\xi} \exp\left(-\frac{2L\omega\xi}{v_m}\right).$$ (6.56)

In magnetic fields, ~ 50 kG s and for $L\omega \sim v_m$, the contributions from the skin components have the order of 10^{-10}–10^{-11}, i.e., the predominance of the contribution from the cyclotron doppleron can be reached.

We can easily separate the contribution from the cyclotron doppleron to the transmission coefficient when the Fermi surface is nearly paraboloidal in shape. In this case, the roots of the dispersion equation are featured by formulas (6.54) which give the following expressions for $|T_{2,3}|$:

$$|T_2| \approx \frac{2}{3} \frac{v_m}{c} \sqrt{\frac{1 - \Omega/\omega}{R\xi^3}} \exp\left[-\frac{\pi(\beta+1)}{4} \frac{L\omega}{v_m}\xi\sqrt{\frac{R\xi}{1-\Omega/\omega}}\right],$$ (6.57)

$$|T_3| \approx \frac{2}{3} \frac{v_m}{c} \sqrt{\frac{1 - \Omega/\omega}{R\xi^3}} \exp\left[-\frac{3\pi(\beta+1)}{4} \frac{L\omega}{v_m}\xi\sqrt{\frac{R\xi}{1-\Omega/\omega}}\right].$$ (6.58)

For $\beta + 1 \sim 0.1$ and $L\omega \sim v_m$ we have that $|T_{2,3}| \sim 10^{-13}$, i.e. the contribution from the cyclotron doppleron to the transmission coefficient will exceed the contributions from the skin components.

Alongside the partial contributions from the poles of the function $f(q)$, the transmission coefficient includes contributions from branch points of this function appropriate to size oscillations described by V.F. Gantmakher and E.A. Kaner [8]. The transverse conductivity $\sigma(\omega, q)$ has two branch points in the upper part of a complex plane, $q = \pm(\Omega - \omega)/v_m + i/v_m\tau$. These points cause the Gantmakher–Kaner size oscillations of the transmission coefficient. For $L\Omega > v_m$, their contribution has a magnitude of the order of 10^{-9}–10^{-10} or less.

Thus, according to the present estimates, we can conclude that size oscillations of the transmission coefficient, arising due to the Fermi-liquid cyclotron doppleron under typical conditions, have the same magnitude as other contributions, to the transmission coefficient or more. It creates a rather favorable possibility for their observation in experiments.

6.3   Influence of the Surface Scattering of Electrons on the Excitation of Cyclotron Dopplerons

It is known (see, e.g., [148], [149]) that the surface scattering of electrons can play a considerable role in the excitation of waves in a metal plate. Therefore we have to analyze the influence of surface scattering on the excitation of the cyclotron doppleron. To study the effects of a qualitative character we can restrict our considerations to the simple model of a paraboloidal Fermi surface and assume is the diffusivity coefficient a constant.

Taking into account surface scattering of the electrons in calculations of the current density (6.33) we have to compute a surface part of the conductivity $\sigma_{NN'}$. For an axisymmetric Fermi surface the dependence of the nonequilibrium part of the electronic distribution function on an azimuthal angle Φ can be explicitly expressed as follows:

$$g(p_\perp, p_z, \Phi, z) = \tfrac{1}{2}\{g_+(p_\perp, p_z, z)\exp(i\Phi) + g_-(p_\perp, p_z, z)\exp(-i\Phi)\}. \quad (6.59)$$

Starting from this expression we can write kinetic equations for circular components

$$\left(\omega + \frac{i}{\tau}\right)g_\pm(p_\perp, p_z, z) - \left(\pm\Omega + iv_z\frac{\partial}{\partial z}\right)g_\pm^e(p_\perp, p_z, z) - \frac{ep_\perp}{m_\perp}\frac{\partial f_p}{\partial E_p}E_{\pm\omega}(z) = 0. \quad (6.60)$$

Using the approximation (2.71) for the Fermi-liquid kernel we obtain a relation connecting the functions $g_\pm(p_\perp, p_z, z)$ and $g_\pm^e(p_\perp, p_z, z)$:

$$g_\pm^e(p_\perp, p_z, z) = g_\pm(p_\perp, p_z, z) - \frac{\partial f_p}{\partial E_p}\varphi_{10}p_\perp G_1^\pm(z) - \frac{\partial f_p}{\partial E_p}\varphi_{11}p_\perp p_z G_2^\pm(z), \quad (6.61)$$

where $G_m^\pm(z) = \sum_p p_\perp p_z^m g_\pm(p_\perp, p_z, z)$.

We have to solve (6.60) together with the boundary conditions which describe a change of the electronic distribution function due to the electron scattering on the surfaces of the plate. Introducing the functions $g_{1,2}^\pm(p_\perp, p_z, z)$ and $g_{1,2}^{\pm e}(p_\perp, p_z, z)(g_1^\pm = g^\pm|_{v_z>0}, g_2^\pm = g^\pm|_{v_z<0}, g_1^{\pm e}$ and $g_2^{\pm e}$ are defined similarly) and assuming that the diffuse-reflection factor is a constant we can write the following boundary conditions for the functions $g^{\pm e}(p_\perp, p_z, z)$ (see [150]):

$$g_1^{\pm e}(p_\perp, p_z, 0) = (1 - P)g_2^{\pm e}(p_\perp, -p_z, 0),$$
$$g_2^{\pm e}(p_\perp, -p_z, L) = (1 - P)g_1^{\pm e}(p_\perp, p_z, L), \quad (6.62)$$

Here P is the diffuse-reflection factor. To proceed we expand the distribution functions in Fourier series in cosines and sines:

$$g_{\pm}(p_{\perp}, p_z, z) = \frac{2}{L} \sum_{N=0} \left(1 - \tfrac{1}{2}\delta_{N0}\right) g_N^{\pm}(p_{\perp}, p_z) \cos(q_N z),$$

$$g_{\pm}(p_{\perp}, p_z, z) = \frac{2}{L} \sum_{N=1} \tilde{g}_N^{\pm}(p_{\perp}, p_z) \sin(q_N z),$$

$$g_{\pm}^e(p_{\perp}, p_z, z) = \frac{2}{L} \sum_{N=0} \left(1 - \tfrac{1}{2}\delta_{N0}\right) g_N^{e\pm}(p_{\perp}, p_z) \cos(q_N z),$$

$$g_{\pm}^e(p_{\perp}, p_z, z) = \frac{2}{L} \sum_{N=1} \tilde{g}_N^{e\pm}(p_{\perp}, p_z) \sin(q_N z).$$

$$(6.63)$$

Going in (6.60) to Fourier transforms in a space variable z, we arrive at the set of equations for Fourier components of the functions $g_{\pm}(p_{\perp}, p_z, z)$ and $g_{\pm}^e(p_{\perp}, p_z, z)$ in their expansions in cosines and sines:

$$(\omega + i/\tau)g_N^{\pm} \mp \Omega g_N^{e\pm} - iq_N v_z \tilde{g}_N^{e\pm} - \frac{ep_{\perp}}{m_{\perp}} \frac{\partial f_p}{\partial E_p} E_N^{\pm} = iv_z \left[g_{\pm}^e(L)(-1)^N - g_{\pm}^e(0)\right],$$

$$(6.64)$$

$$(\omega + i/\tau)\tilde{g}_N^{\pm} \mp \Omega \tilde{g}_N^{e\pm} - iq_N v_z g_N^{e\pm} - \frac{ep_{\perp}}{m_{\perp}} \frac{\partial f_p}{\partial E_p} \tilde{E}_N^{\pm} = 0.$$

Here the Fourier components $g_N^{\pm}$ and $g_N^{e\pm}$ are connected by a relation

$$g_N^{e\pm} = g_N^{\pm} - \frac{\partial f_p}{\partial E_p} \varphi_{10} p_{\perp} G_{1N}^{\pm} - \frac{\partial f_p}{\partial E_p} \varphi_{11} p_{\perp} p_z G_{2N}^{\pm},$$

$$G_{mN}^{\pm} = \sum_p p_{\perp} p_z^m g_N^{\pm}(p_{\perp}, p_z).$$

$$(6.65)$$

Fourier components $\tilde{g}_N^{\pm}$ and $\tilde{g}_N^{e\pm}$ in the expansions in sines are connected in a similar way.

To simplify further calculations we introduce the functions $g_N^{(s)}(p_{\perp}, p_z)$, $g_N^{(a)}(p_{\perp}, p_z)$, defined by the relations (for brevity we omit symbols of polarization):

$$g_N^{\binom{s}{a}} = \frac{g_{N1}(p_{\perp}, p_z) \pm g_{N2}(p_{\perp}, p_z)}{2}.$$

$$(6.66)$$

The functions $\tilde{g}_N^{\binom{s}{a}}(p_{\perp}, p_z)$, $g_N^{e\binom{s}{a}}(p_{\perp}, p_z)$, $\tilde{g}_N^{e\binom{s}{a}}(p_{\perp}, p_z)$, can be defined similarly. The Fourier components in expansion (6.33) of the current density in cosines are expressed in terms of $g_N^{(s)}$:

$$j_N = \frac{2e}{m_{\perp}} \sum_p p_{\perp} g_N^{(s)}(p_{\perp}, p_z) = \frac{2e}{m_{\perp}} G_{1N}^{(s)}.$$

$$(6.67)$$

These quantities can be found from equations which follow from (6.64):

$$(\omega + i/\tau)g_N^{(s)} \mp \Omega g_N^{e(s)} - iq_N v_z g_N^{e(a)} - \frac{ep_{\perp}}{m_{\perp}} \frac{\partial f_p}{\partial E_p} E_N$$

$$-iv_z\big[g^{e(a)}(L)(-1)^N - g^{e(a)}(0)\big] = 0,$$

$$\big(\omega + i/\tau\big)g_N^{(a)} \mp \Omega g_N^{e(a)} + iq_N v_z g_N^{e(s)} = 0, \qquad (6.68)$$

where

$$g_N^{e(s)} = g_N^{(s)} - 2\varphi_{10}\frac{\partial f_p}{\partial E_p}G_{1N}^{(s)}p_\perp,$$

$$g_N^{e(a)} = g_N^{(a)} - 2\varphi_{11}\frac{\partial f_p}{\partial E_p}G_{2N}^{(a)}p_\perp p_z. \qquad (6.69)$$

The quantities $(-1)^N g^{e(a)}(L) - g^{e(a)}(0)$ included in (6.68) can be presented in the form

$$(-1)^N g^{e(a)}(L) - g^{e(a)}(0) = \frac{P}{2}D(p_z)\big[1+(-1)^N\big]+\frac{P}{2}F(p_z)\big[1-(-1)^N\big], \quad (6.70)$$

where the functions $D(p_z)$ and $F(p_z)$ are accordingly equal:

$$D(p_z) = \tfrac{1}{2}\big[g_2^e(0, p_\perp, -p_z) + g_1^e(L, p_\perp, p_z)\big],$$

$$F(p_z) = \tfrac{1}{2}\big[g_2^e(0, p_\perp, -p_z) - g_1^e(L, p_\perp, p_z)\big]. \qquad (6.71)$$

Using the boundary conditions and expanding $g^e(p_\perp, p_z, z)$ in a Fourier series in z, we can express $D(p_z)$ and $F(p_z)$ in terms of $g_N^{e(s)}$:

$$(2 - P)D(p_z) = \frac{2}{L}\sum_{N=0}\big(1 - \tfrac{1}{2}\delta_{N0}\big)\, g_N^{e(s)}\big[1 + (-1)^N\big],$$

$$(2 - P)F(p_z) = \frac{2}{L}\sum_{N=0}\Big(1 - \frac{1}{2}\delta_{N0}\Big)\, g_N^{e(s)}\big[1 - (-1)^N\big]. \qquad (6.72)$$

Equations (6.67), (6.68), (6.70), and (6.72) together are the set of equations which enables us to find j_N. We have to solve the equations separately for even and odd N, which determine the surface impedances for antisymmetric and symmetric excitation. As follows from expressions (6.70) and (6.72) the functions $F(p_z)$ and $D(p_z)$ determine, correspondingly, the surface impedances for antisymmetric $(Z^{(a)})$ and symmetric $(Z^{(s)})$ excitation.

For the antisymmetric excitation (N are odd) the system of equations has the form

$$(\omega \mp \Omega + i/\tau)g_N^{(s)} - iq_N v_z g_N^{(a)} - 4iv_z\rho\sum_N g_N^s$$

$$+\frac{\partial f_p}{\partial E_p}\left[\pm 2\Omega\varphi_{10}p_\perp G_{1N}^s + 2iq_N\varphi_{11}p_\perp p_z G_{2N}^a - \frac{ep_\perp}{m_\perp}E_N\right] = 0,$$

$$-(\omega \mp \Omega + i/\tau)g_N^{(a)} - iq_N v_z g_N^{(s)}$$

$$+\frac{\partial f_p}{\partial E_p}\big[\pm \Omega\varphi_{11}p_\perp p_z G_{2N}^a + iq_N v_z\varphi_{10}p_\perp G_{1N}^s\big] = 0, \qquad (6.73)$$

where $\rho = P/(2 - P)$.

The solution of this system is given by the following expressions for the volume and surface conductivity (the Fermi surface is supposed to be paraboloidal):

$$\sigma_N^{\pm} = \pm \frac{ine^2}{m_\perp \omega} \frac{\chi \theta_N^* \mp \frac{2}{3}\alpha_2 k_N^2}{\theta_N^* \bar{\theta}_N - \frac{2}{3}\alpha_1 \alpha_2 k_N^2}, \tag{6.74}$$

$$\sigma_{NN'}^{\pm} = \pm \frac{ne^2}{m_\perp \omega} \rho \frac{v_m}{\omega L} \frac{\chi^2}{U T_N T_N'} \{\theta_N^* \theta_{N'}^* (1 \mp \rho M_1)$$

$$+ 2 \frac{\chi^{*4}}{\chi^4} \theta_N \theta_{N'} (1 \mp \rho M_2) - \frac{2}{3}\rho \frac{\chi^{*2}}{\chi^2} (\theta_N^* \theta_{N'} + \theta_{N'}^* \theta_N) M_3\}. \tag{6.75}$$

Here

$$U = (1 \mp \rho M_1)(1 \mp \rho M_2) - \frac{2}{9}\rho^2 M_3^2,$$

$$\chi = 1 \mp \Omega/\omega + i/\omega\tau, \qquad k_N = q_N v_m/\omega,$$

$$\theta_N = \chi^2 - k_N^2, \qquad \theta_N^* = \chi^2 \mp \alpha_2 \chi - k_N^2 \equiv \chi^{*2} - k_N^2,$$

$$\bar{\theta}_N = \chi^2 \mp \alpha_1 \chi - k_N^2 \equiv \bar{\chi}^2 - k_N^2,$$

$$T_N = \theta_N^* \bar{\theta}_N - \frac{2}{3}\alpha_1\alpha_2 k_N^2, \qquad \alpha_{1,2} = \frac{A_{1,2}}{(1 + A_{1,2})},$$

$$\begin{bmatrix} M_1 \\ M_2 \\ M_3 \end{bmatrix} = -\frac{4v_m}{L\omega} \sum_N \frac{i}{T_N} \begin{bmatrix} (\chi \mp \alpha_2)(\theta_N^* \mp \chi\alpha_2 \mp \frac{1}{3}\chi\alpha_1) \\ \chi(\bar{\theta}_N \mp \chi\alpha_2 \pm \frac{1}{3}\chi\alpha_1) \\ \chi^{*2} \end{bmatrix}.$$

We can analyze the qualitative features of the surface impedance arising due to the cyclotron doppleron, taking into account only one of the Fermi-liquid parameters, namely A_2. When accepting this parameter nonzero value the dispersion equation (6.42) has a solution corresponding to the Fermi-liquid cyclotron doppleron wave. Taking into account the Fermi-liquid parameter A_1 we do not achive any qualitative changes in the results but make the finite formulas much too unwieldy. Therefore, in further calculations, we assume that $A_1 = 0$. This enables us to simplify the expressions for σ_N and $\sigma_{NN'}$:

$$\sigma_N = \frac{\pm ine^2}{3m_\perp \omega} \chi \left(\frac{1}{\chi^2 - k_N^2} + \frac{2\chi^{*2}/\chi^2}{\chi^{*2} - k_N^2} \right), \tag{6.76}$$

$$\sigma_{NN'} = \frac{4}{3} \frac{ne^2}{m_\perp \omega} \rho \frac{v_m}{\omega L} \chi^2 \left(\frac{1}{1 \mp \rho s^a} \frac{1}{\theta_N \theta_{N'}} + \frac{2\chi^{*4}/\chi^4}{1 \mp \rho s^{*a} \chi^*/\chi} \frac{1}{\theta_N^* \theta_{N'}^*} \right). \tag{6.77}$$

Here,

$$s^a = -\frac{4v_m}{\omega L} \sum_N \frac{i\chi}{\chi^2 - k_N^2} = i \tan \left(\frac{L\Omega}{v_m} \chi \right), \tag{6.78}$$

$$s^{*a} = -\frac{4v_m}{\omega L} \sum_N \frac{i\chi^*}{\chi^{*2} - k_N^2} = i \tan \left(\frac{L\Omega}{v_m} \chi^* \right). \tag{6.79}$$

Using the obtained results for the Fourier components of the current density we can write the equation for the surface impedance of the plate at antisymmetric excitation

$$\frac{(\chi_1^2 - q_N^2 d^2)(\chi_2^2 - q_N^2 d^2)(\chi_3^2 - q_N^2 d^2)}{(\chi^2 - q_N^2 d^2)(\chi^{*2} - q_N^2 d^2)} Z_N^{(a)}$$

$$+ \xi\rho \frac{\chi^2}{3} \left[\frac{1}{1 \mp \rho s^a} \frac{1}{\chi^2 - q_N^2 d^2} \frac{4id}{L} \sum_{N'} \frac{Z_{N'}^{(a)}}{\chi^2 - q_{N'}^2 d^2} \right.$$

$$\left. + \frac{2}{1 \mp \rho s^{*a}} \frac{\chi^{*4}/\chi^4}{\chi^*/\chi} \frac{4id}{\chi^{*2} - q_N^2 d^2} \frac{Z_{N'}^{(a)}}{L} \sum_{N'} \frac{Z_{N'}^{(a)}}{\chi^{*2} - q_{N'}^2 d^2} \right] = \frac{4\pi i\omega d^2}{c^2}. \quad (6.80)$$

The parameter ξ is the anomaly parameter of the skin effect, and χ_1, χ_2, χ_3 are the solutions of the dispersion equation (6.42) expressed in terms of $q_N v_m / \Omega \equiv q_N d$.

Solving (6.30) for the electromagnetic field we can find the surface impedance of the plate for the antisymmetric excitation:

$$\frac{Z^{(a)}}{Z_0} = X - \tfrac{1}{3}\xi\rho\chi^2$$

$$\times \frac{Y^2 + 2(\chi^{*2}/\chi^2)\Gamma^2 + \tfrac{2}{3}\xi\rho\chi[Y^2 W - 2Y\Gamma R + \Gamma^2 V](\chi^{*4}/\chi^4)}{1 + \tfrac{1}{3}\xi\rho\chi^2[V + 2W(\chi^{*4}/\chi^4)] + \tfrac{2}{9}\xi^2\rho^2\chi^4[WV - R^2](\chi^{*4}/\chi^4)}.$$

$$(6.81)$$

Here

$$X = \frac{\gamma_i}{\chi_i} s_i^{(a)}, \qquad Y = \frac{\gamma_i}{\beta_i} \frac{s_i^{(a)}}{\chi_i}, \qquad \Gamma = \frac{\gamma_i}{\beta_i^*} \frac{s_i^{(a)}}{\chi_i}, \qquad V = \frac{\gamma_i}{\beta_i^2} \frac{s_i^{(a)}}{\chi_i},$$

$$W = \frac{\gamma_i}{\beta_i^{*2}} \frac{s_i^{(a)}}{\chi_i}, \qquad R = \frac{\gamma_i}{\beta_i\beta_i^*} \frac{s_i^{(a)}}{\chi_i}, \qquad Z_0 = \frac{2\pi v_m}{c^2}, \qquad \gamma_i = \frac{\beta_i\beta_i^*}{\beta_{ji}\beta_{ki}},$$

$$j, k \neq i, \qquad j \neq k, \qquad \beta_i = \chi^2 - \chi_i^2,$$

$$\beta_i^* = \chi^{*2} - \chi_i^2, \qquad \beta_{ki} = \chi_k^2 - \chi_i^2, \qquad \chi_i^2 = k_i^2 \omega^2 \Omega^2. \quad (6.82)$$

Summation over $i(i = 1, 2, 3)$ has to be performed in the expressions for X, Y, Γ, V, W, R. The quantities s_i are defined by the relations (6.78), (6.79), where the quantity χ has to be replaced by k_i.

The set of equations determining the surface impedance for symmetric excitation is similar to (6.80) with the difference that the summation has to be carried out over even N. This leads to the formula for the surface impedance for symmetric excitation similar to (6.81) where the functions defined by (6.78), (6.79) are replaced by the following functions:

$$s^{(s)} = -\frac{4v_m}{\omega L} \sum_N (1 - \tfrac{1}{2}\delta_{N0}) \frac{i\chi}{\chi^2 - k_N^2} = i \cot\left(\frac{L\omega}{v_m}\chi\right), \quad (6.83)$$

$$s^{*(s)} = -\frac{4v_m}{\omega L} \sum_N (1 - \tfrac{1}{2}\delta_{N0}) \frac{i\chi^*}{\chi^{*2} - k_N^2} = i \cot\left(\frac{L\omega}{v_m}\chi^*\right). \quad (6.84)$$

Let us consider the excitation of the cyclotron doppleron which corresponds to the root k_3. We assume the inequalities $\mathrm{Im}(\chi_{1,2}/d) \gg 1$ to be satisfied so the contributions to the transmission coefficient from the roots $k_{1,2}$ are exponentially small. Keeping in (6.81) the terms of the order of small parameter $(\xi)^{-1}$ we have

$$\frac{Z^{(a)}}{Z_0} = \frac{1}{2k_1} + \frac{1}{2k_2} + \frac{1}{6k_1k_2} \frac{\mp \rho \chi [1 + 2\chi^{*4}/\chi^4]k_3^2/\chi^{*2} + \rho^2\chi^2 s_3^{(a)}/k_3}{1 \mp \rho s_3^{(a)} k_3/\chi}. \quad (6.85)$$

To analyze the frequency dependence of the surface impedance we transform this formula as follows:

$$\frac{Z^{(a)}}{Z_0} = -\frac{i}{2\xi^{1/4}\mho_1^{\mp}} \left\{ 1+i \right.$$

$$\left. + \frac{1}{3}\frac{\rho}{(\xi\mho_1^{\pm})^{1/4}} \frac{[\mp 1 + (\omega + i/\tau)/\Omega]^2 + 2(\mho_2^{\pm})^2 \mp \rho s_3^{(a)}\sqrt{\mho_2^{\pm}}(\mho_1^{\pm})^{3/2}}{\mho_1^{\pm} + \rho s_3^{(a)}\sqrt{\mho_2^{\pm}\mho_1^{\pm}}} \right\}. $$

$$(6.86)$$

Here $\mho_1^{\pm}$, $\mho_2^{\pm}$ are defined as follows:

$$\mho_1^{\pm} = \mp 1 + \frac{\omega}{\Omega_1} + \frac{i}{\Omega\tau}, \qquad \mho_2\pm = \mp 1 + \frac{\omega}{\Omega_2} + \frac{i}{\Omega\tau},$$

$\Omega_1 = \Omega(1 + A_2)(1 + A_2/3)$, and $\Omega_2 = \Omega(1 + A_2)$.

Dependence of the surface impedance on the thickness of the plate L is determined by the function $s_3^{(a)}$. When L increases $s_3^{(a)} \to 1$, and in the limit $L \to \infty$, this expression (6.86) passes into the expression for the surface impedance of a massive metal. The first term in (6.86), which is proportional to $(\xi)^{-1/4}$, determines the asymptotic of the surface impedance in this limit. This term has a singularity at the point $\omega = \Omega$. However, this result is an inference of our choice of the model of the Fermi surface. Fermi surfaces of typical metals include many sheets and the paraboloidal model used before cannot correspond to the surface as a whole but only to a part of the surface. Therefore, formulas (6.85), (6.86) should be changed to include the contributions from the remaining segments of the Fermi surface. We showed that the main term of the surface impedance of the metal does not exhibit a resonance at the cyclotron frequency when the Fermi surface has effective cross-sections of a finite nonzero curvature. These conditions are typical for most of the metals.

Let us discuss in more detail that size oscillations of the transmission coefficient stipulated by the excitation of that cyclotron doppleron wave. Starting from the relations (6.30), (6.37), and (6.85), we obtain

$$T = \frac{ic}{9\pi} Z_0 \frac{\alpha_2^2 k_3^3}{\chi^{*2}\chi^2 k_1 k_2} \left(\rho^2 + 2\frac{\chi^2 k_2}{k_1^3} \right)$$

$$\times \left\{ \left(1 + \rho^2\frac{k_3^2}{\chi^2} \right) \sin\left(L\omega\frac{k_3}{v_m} \right) \mp 2i\rho\frac{k_3}{\chi}\cos\left(L\omega\frac{k_3}{v_m} \right) \right\}^{-1}. \quad (6.87)$$

We will consider a minus polarization and we will assume that the collision frequency $1/\tau$ is small in comparison with the various combinations of the frequencies ω and Ω. Formula (6.87) includes the following combinations of these frequencies:

$$\Delta_0 = \Omega - \omega, \tag{6.88}$$

$$\Delta_1 = \Omega(1 + A_2) - \omega, \tag{6.89}$$

$$\Delta_2 = \Omega(1 + A_2) - \omega(1 + A_2/3). \tag{6.90}$$

Within the cylotron doppleron frequency range the appropriate values of Δ_0, Δ_1, and Δ_2 are of the same order.

Under considered conditions the anomaly parameter ξ is significantly more than unity and can reach values of the order of 10^2. When the inequality

$$\rho^2 \sqrt{\xi} \Omega^{3/2} \frac{\sqrt{\Delta_2}}{\Delta_0^2} \gg 1 \tag{6.91}$$

is satisfied or $\rho^2 \sqrt{\xi} \gg 1$, which is the same, the magnitude of the transmission coefficient is determined by the surface scattering of electrons and equals

$$|T| = \tfrac{2}{9}\rho^2 \frac{A_2^2}{\sqrt{1 + A_2}} \frac{v_m}{c\sqrt{\xi}} \frac{\sqrt{\Delta_1}}{\Delta_2^2} \frac{\omega^2}{\sqrt{\Omega}} \left\{ \left(1 - \rho^2 \frac{\Delta_1}{\Delta_2}\right)^2 \sin^2\left(\frac{L\Delta_0}{d\Omega}\sqrt{\frac{\Delta_1}{\Delta_2}}\right) \right.$$
$$+ \left[\left(1 + \rho^2 \frac{\Delta_1}{\Delta_2}\right) \sinh\left(\frac{L}{l}\sqrt{\frac{\Delta_1}{\Delta_2}}\left(1 - \tfrac{1}{6}A_2(1 + A_2)\frac{\omega\Delta_0}{\Delta_1\Delta_2}\right)\right) \right.$$
$$\left. \left. + 2\rho \frac{\Delta_1}{\Delta_2} \cosh\left(\frac{L}{l}\sqrt{\frac{\Delta_1}{\Delta_2}}\left(1 - \tfrac{1}{6}A_2(1 + A_2)\frac{\omega\Delta_0}{\Delta_1\Delta_2}\right)\right) \right]^2 \right\}^{-1/2}. \tag{6.92}$$

It is known from the theory of dopplerons that for the diffusive reflection of electrons the contribution from the wave to the surface impedance can be significantly larger than for their specular reflection. We obtain similar results for the contribution to the transmission coefficient from our Fermi-liquid cyclotron doppleron wave.

When the electron mean free path l does not exceed the width of a film, the sizes of oscillations do not appear noticeably in the transmission coefficient which is rather small in itself. Starting from formula (6.92) we obtain the following asymptotic expression for the transmission coefficient which is valid in the case of a very thin film ($L \ll l$):

$$|T| = \tfrac{2}{9}\rho^2 \frac{A_2^2}{\sqrt{1 + A_2}} \frac{v_m}{c\sqrt{\xi}} \frac{\sqrt{\Delta_1}}{\Delta_2^2} \frac{\omega^2}{\sqrt{\Omega}} \left\{ \left(1 - \rho^2 \frac{\Delta_1}{\Delta_2}\right)^2 \sin^2\left(\frac{L\Delta_0}{d\Omega}\sqrt{\frac{\Delta_1}{\Delta_2}}\right) \right.$$
$$\left. + \left[\left(1 + \rho^2 \frac{\Delta_1}{\Delta_2}\right) \frac{L}{l}\sqrt{\frac{\Delta_1}{\Delta_2}}\left(1 - \tfrac{1}{6}A_2(1 + A_2)\frac{\omega\Delta_0}{\Delta_1\Delta_2}\right) + 2\rho \frac{\Delta_1}{\Delta_2} \right]^2 \right\}^{-1/2}.$$
$$\tag{6.93}$$

When the surface scattering is strong ($\rho \sim 1$) the term $2\rho\Delta_1/\Delta_2$ in square brackets exceeds the term proportional to L/l and the magnitude and shape of the size oscillations are determined by surface scattering. In the opposite limit when the diffuse-reflection factor is small ($\rho \ll 1$), it produces some opportunities for the resonance excitation so that the transmission coefficient has peaks. The positions of these peaks are determined by the equality

$$\frac{\Delta_0}{\Omega}\sqrt{\frac{\Delta_1}{\Delta_2}} = \pi n \frac{d}{L}, \tag{6.94}$$

and their shape is featured by the formula

$$|T| = \tfrac{2}{9}\rho^2 \frac{A_2^2}{\sqrt{1+A_2}} \frac{v_m}{c\sqrt{\xi}} \frac{\sqrt{\Delta_1}}{\Delta_2^2} \frac{\omega^2}{\sqrt{\Omega}} \left\{ \left(\frac{L\Delta_0}{d\Omega}\sqrt{\frac{\Delta_1}{\Delta_2}} - \pi n \right)^2 \right.$$
$$\left. + \left(\frac{\Omega}{\Delta_0}\pi n \frac{L}{d} \right)^2 \left[\frac{L}{l}\left(1 - \tfrac{1}{6}A_2(1+A_2)\frac{\omega\Delta_0}{\Delta_1\Delta_2} \right) + 2\pi\rho n \frac{d}{L}\frac{\Omega}{\Delta_0} \right]^2 \right\}^{-1/2} . \tag{6.95}$$

The width of peaks is determined by both the volume and surface scattering. There are two additive contributions of the order of L/l and ρ, correspondingly. Both contributions are small in the regime of resonance excitation. When the inequality $\rho^2\sqrt{\xi} \gg 1$ is satisfied, the quantity $|T_n|$ is proportional to ρ^2. It follows from (6.91) that this inequality can be satisfied for $\rho \ll 1$ because of the large value of the anomaly parameter ξ. For small $\rho (\rho^2\sqrt{\xi} \ll 1)$ the results correspond to a case of the specular reflection of electrons analyzed in [151].

A direct observation of cyclotron waves, propagating along a magnetic field in experiments concerning transmission of the electromagnetic waves through a thin film of alkali metal, is possible when the parameter $\Omega\tau$ takes values of about several hundreds (see [14]). However, this conclusion of [14] concerns the special case of alkali metals for which the resonant frequency of cyclotron waves is very close to the cyclotron frequency. According to the results expounded here, the difference between the resonant frequency of the cyclotron wave and the cyclotron frequency can be comparable to the latter. Therefore these experiments on the transmission of cyclotron dopplerons through metal films can be carried out when $\Omega\tau$ is of the order of 10–20. The resonance transmission can be displayed in very thin films when L is small compared to the electron mean free path.

CHAPTER 7

The Influence of the Local Geometry of Fermi Surfaces on Quantum Oscillations of the Velocity of Sound in Metals

7.1 The Effect of Curvature-Related Features of the Fermi Surface on Quantum Oscillations of the Velocity of Sound in Metals in the Low-Frequency Range

Intensive studies of the effect of quantizing the orbital motion of electrons in strong magnetic fields on the observable characteristics of ultrasound waves propagating in metals were started from the theoretical prediction (see [20]) of giant quantum oscillations of the ultrasonic absorption coefficient and the following observation of such oscillations in experiments. Some important aspects of the theory of quantum oscillations of the velocity and absorption coefficient of ultrasound were analyzed in [126], [152]–[157].

However, the effects of the local features of the shape of Fermi surfaces in the vicinity of its extremal sections, on the amplitude and shape of quantum oscillations of the elasticity moduli and sound velocity, have not been studied in much detail. Studies of the sound propagation in metals with complex Fermi surfaces revealed, however, that the presence of flat or cylindrical portions, or even separate points at which the Gaussian curvature of the Fermi surface turns zero, may lead to qualitative changes in the frequency and temperature dependence of the attenuation rate and velocity shift of sound waves. Therefore we can expect that under certain conditions, local curvature anomalies at the extremal sections of the Fermi surface will noticeably affect the characteristics of the quantum oscillations of the sound velocity and attenuation.

The electron contribution to the static elastic moduli of a lattice is determined by the locally equilibrium component of the force exerted by the electrons on the lattice. Fourier transforms, for this part of the force in its expansion in space variables, are given by formula (2.31) which can be transformed as follows:

$$F_{q\omega}^{L\alpha} = q_\beta q_\delta \lambda_{\alpha\beta\gamma\delta} u_{q\omega}^\gamma. \tag{7.1}$$

Using this expression (7.1) we can derive the expression for the electron contribution to static elastic moduli

$$\lambda_{\alpha\beta\gamma\delta} = \frac{[(\Lambda_{\alpha\beta}^*, n) - N\delta_{\alpha\beta}][(\Lambda_{\gamma\delta}, n^*) - N\delta_{\gamma\delta}]}{(n^*, n)} - (\Lambda_{\alpha\beta}^*, \Lambda_{\gamma\delta}) \tag{7.2}$$

$$- 4\pi \left(D_{g\beta\alpha} - \frac{N\delta_{\alpha\beta}}{(n^*, n)} \frac{\partial M^g}{\partial \zeta} \right) (1 - 4\pi\chi)_{g\kappa}^{-1} \left(D_{\kappa\delta\gamma} - \frac{N\delta_{\gamma\delta}}{(n^*, n)} \frac{\partial M^\kappa}{\partial \zeta} \right).$$

In (7.2), N is the electron concentration, $\mathbf{M}$ is the magnetic moment, χ is the magnetic susceptibility tensor, ζ is the chemical potential of electrons, and D is a tensor determined by the relation

$$\frac{e}{mc} \left[(P_\gamma^*, \Lambda_{\alpha\beta}) - \frac{(P_\gamma^*, n)}{(n^*, n)} (n^*, \Lambda_{\alpha\beta}) \right] A_q^\gamma = -b_q^\gamma D_{\gamma\beta\alpha}, \tag{7.3}$$

where $\mathbf{b}_q$ is the Fourier transform of the magnetic component of the self-consistent electromagnetic field and A_q is its vector potential.

In order to avoid cumbersome expressions, (7.2) and (7.3) are written using the (A, B) symbols defined as

$$(A, B) = -\sum_{\nu\nu'} \frac{f_\nu - f_{\nu'}}{E_\nu - E_{\nu'}} A_{-q}^{\nu\nu'} B_q^{\nu'\nu}. \tag{7.4}$$

The notation $n_{\nu'\nu}(\mathbf{q})$ and $\mathbf{P}_{\nu'\nu}(\mathbf{q})$ is introduced to represent the Fourier components of the matrix elements of operators of the electron density and electron kinetic momentum; $\Lambda_{\nu'\nu}^{\alpha\beta}(\mathbf{q})$ is the Fourier transform of the tensor operator of the deformation potential density; and the quantities $n_{\nu'\nu}^*(\mathbf{q})$, $\mathbf{P}_{\nu'\nu}^*(\mathbf{q})$, and $\Lambda_{\nu'\nu}^*(\mathbf{q})$ are the Fourier components of the matrix elements of the corresponding effective operators renormalized due to the Fermi-liquid interaction. The connection between unrenormalized and renormalized operators is given by relations such as (2.12), (2.26). The first two terms in (7.2) are retained in the semiclassical limit. In the semiclassical limit, these terms give us the well-known result obtained by Silin in [96]. The term $N^2/(n^*, n)$ describes the contribution of electronic compressibility to the elastic moduli, and all other terms are due to deformation interaction between the electrons and sound.

The last term in (7.2) is of a purely quantum nature and vanishes on going to the semiclassical limit. This term represents the effects of magnetostriction, that is, the appearance of an additional magnetic field $\mathbf{b}$ on the deformation of a metal in the presence of a quantizing magnetic field. Under typical conditions, when the quantity $\theta = 2\pi^2 T/\hbar\Omega$ (T is the temperature expressed in energy units and Ω is the cyclotron frequency) is not too small in comparison with unity, the contribution of this term is smaller than that of the other terms in (7.2) and can be neglected. In the low-temperature range ($\theta \ll 1$), the magnetostriction contribution to the elastic moduli becomes significant [157], [158] and must be allowed for in order to provide an adequate description of their quantum oscillations.

In calculating the quantum corrections to observables under the usual experimental conditions, when the number of Landau levels under the Fermi surface is sufficiently large $(\hbar\Omega \ll \zeta)$, the matrix elements included in (7.2) and (7.3) can be replaced by their semiclassical analogs. In particular, $\Lambda_{\nu'\nu}^{\alpha\beta}(q)$ is replaced by

$$\Lambda_n^{\alpha\beta}(p_z, \mathbf{q}) = \frac{1}{2\pi} \int_0^{2\pi} d\psi \Lambda^{\alpha\beta}(p_z, \psi) \exp\left[in\psi - \frac{i}{\Omega} \int_0^\psi q_z \Delta v_z(p_z, \psi') d\psi'\right],$$
(7.5)

where $\psi = \Omega t$, t is the time of motion of an electron over the cyclotron orbit, and $\Delta v_z = v_z(p_z, \psi) - \langle v_z \rangle$.

Furthermore, under the considered conditions when the wavelength of the ultrasound critically exceeds the radius of the cyclotron orbit, we can neglect the dependence of the matrix elements of $\mathbf{q}$ supposing that $\mathbf{q} = 0$. As a result we can write the following expression for the principal oscillating terms in the expansions of the quantities (7.4) with respect to a small parameter $\gamma^{-1} = (\hbar\Omega/\zeta)^{1/2}$:

$$(A^\sim, B) = \sum_i A_i(\zeta) B_i(\zeta) \Delta_i,$$
(7.6)

where the oscillating function Δ_i has the form

$$\Delta_i = \frac{m_{\perp i}}{\pi^2 \hbar^3 g} \sqrt{\frac{2\pi\hbar|e|B}{c|\partial^2 S/\partial p_z^2|_i}} \sum_{r=1}^\infty \frac{\psi_r(\theta)}{r^{1/2}} \cos\left(\frac{rcS_i(\zeta)}{\hbar|e|B} \pm \frac{\pi}{4} - \pi r\delta\right) \cos\left(\pi r \frac{m_{\perp i}}{m}\right).$$
(7.7)

Here g is the density of the electron states on the Fermi surface in the absence of the magnetic field, $\psi_r(\theta) = r\theta/\sinh r\theta$, $S_i(\zeta)$ is the area of the extremum cross-section of the Fermi surface, and $m_{\perp i}$ is the cyclotron mass corresponding to the ith extremum cross-section. Summation over i in (7.6) is performed over the extrema of the function $S(\zeta, p_z)$. The signs "plus" corresponds to its minima and "minus" to its maxima.

Expression (7.7) for Δ_i is valid provided only that the Gaussian curvature of the Fermi surface is everywhere finite and nonzero. If the Fermi surface contains separate points or lines on which the curvature turns to zero, then (7.7) fails to be valid.

Consider, for example, a metal with an axisymmetric Fermi surface. The curvature turning to zero at some extremal section implies that the expansion of the function $S(p_z)$ in the vicinity of the corresponding point $p_z = p_i$ has the form

$$S(p_z, \zeta) = S_i(\zeta) + \frac{1}{(2l)!} \left(\frac{d^{2l}S}{dp_z^{2l}}\right)_{\substack{E=\zeta \\ p_z=p_i}} (p_z - p_i)^{2l} + \cdots,$$
(7.8)

with $l > 1$. In this case (7.7), for the oscillating function Δ_i, has to be replaced by

$$\Delta_i = \frac{m_{\perp i}}{\pi^2 \hbar^3 g} \left(\frac{\Gamma(2l+1)\hbar|e|B}{c|\partial^{2l}S/\partial p_z^{2l}|_{\substack{E=\zeta \\ p_z=p_i}}}\right)^{1/2l} \frac{\Gamma(1/2l)}{l}$$

$$\times \sum_{r=1}^\infty \frac{\psi_r(\theta)}{r^{1/2l}} \cos\left(\frac{crS_i(\zeta)}{\hbar|e|B} \pm \frac{\pi}{4l} - \pi r\delta\right) \cos\left(\pi r \frac{m_{\perp i}}{m}\right),$$
(7.9)

where $\Gamma(x)$ is the gamma-function.

Oscillations described by (7.9) and (7.7) differ both in phase and amplitude. The amplitude of convenient oscillations given by (7.7) is of the order of $\gamma^{-1}\theta^{-1/2}$, while (7.9) yields a magnitude of the order of $\gamma^{-1/l}\theta^{1/2l-1}$. Therefore, the amplitude of oscillations related to the extremal section of zero curvature is approximately $(\gamma\sqrt{\theta})^{(l-1)/l}$ times that of the usual quantum oscillations. This must be especially well pronounced at low temperatures, i.e., when $\theta \ll 1$. In the low-temperature range, a contribution due to an extremal section of zero curvature can be cosiderably (more than tenfold) greater than the contributions due to the other extremal sections.

Thus, the local features of the shape of the Fermi surface near the extremal sections may strongly affect both the amplitude and shape of the quantum oscillations. This is a quite natural result. Indeed, all oscillating corrections to observables originate from the nearest vicinity of the extremal sections, hence, the local geometric features of the Fermi surface in these sections must influence the characteristics of oscillations.

Using (7.6) we can obtain from (7.2) the following expressions for the oscillating corrections to the components of the static tensor of the elastic moduli

$$\tilde{\lambda}_{\alpha\beta\gamma\delta} = -\sum_i L_i^{\alpha\beta} L_i^{\gamma\delta} \Delta_i, \tag{7.10}$$

where the summation is carried out over the extremal sections

$$L_i^{\alpha\beta} = \Lambda_i^{*\alpha\beta} - \frac{\langle \Lambda_{\alpha\beta}^* \rangle}{g} + \frac{N\delta_{\alpha\beta}}{g}, \tag{7.11}$$

the symbol $\langle \cdots \rangle$ denotes averaging with the derivative of the Fermi distribution function over the Fermi surface in the absence of an external magnetic field.

The electron–electron interaction manifests itself in this result (7.10) only as a renormalization of the electron quantities included in (7.11). Neglecting this renormalization and also the deformation terms we arrive at the simple result

$$\tilde{\lambda}_{\alpha\beta\gamma\delta} = -\frac{N^2}{g}\delta_{\alpha\beta}\delta_{\gamma\delta}\sum_i \Delta_i. \tag{7.12}$$

We can readily express the electron contribution to the velocity of sound in low-frequency range in terms of the elastic moduli $\lambda_{\alpha\beta\gamma\delta}$. Specifically, we can obtain the following expression for the oscillating correction to the velocity of the longitudinal ultrasound wave propagating along the external magnetic field

$$\tilde{s}^2 = -\frac{N^2}{\rho_m}\frac{1}{g}\sum_i \Delta_i. \tag{7.13}$$

This formula was derived in [154].

The principal terms in the expressions for the electron contribution to the velocity of ultrasound waves for not too low temperatures ($\theta \simeq 1$) are described by formulas (7.10), (7.11). It follows from these results that oscillations connected with any extremum cross-section of anomalous (zero or anomalously large) curvature have to exhibit special characteristic features determined by a special form of the oscillating function (7.9).

However, these features will vanish on changing the orientation of the applied magnetic field with respect to the crystallographic axes of a single-crystal sample, because this change will violate the matching of the extremal section with the line of parabolic points. Thus, the amplitude and phase of the sound velocity oscillations related an "anomalous" extremal section of the Fermi surface must strongly depend on the magnetic field direction. The detection of this dependence in experiments might be evidence for the existence of an "anomalous" line on the corresponding part of the Fermi surface of the metal studied.

The oscillating correction $\tilde{s}$ to the velocity s of sound, propagating in a metal at low frequencies, can be written as follows:

$$\frac{\tilde{s}}{s} = \sum_i A_i \Delta_i. \tag{7.14}$$

The values of dimensionless amplitude factors A_i in a particular metal may differ, depending on the polarization of the ultrasonic wave and the direction of its propagation relative to the magnetic field. Hence A_i will generally change upon a variation of the direction of the external magnetic field $\mathbf{B}$. However, these variations can be disregarded while solving the problem under consideration, since the dependence of the amplitude of oscillations at the extremal cross-section on the direction of the magnetic field $\mathbf{B}$ is primarily determined by the behavior of the oscillating factors Δ_i describing the oscillations of the density of states of electrons in a quantizing magnetic field.

In order to find the character of the dependence under investigation, it is sufficient to consider a simple model of a metal with a closed simply connected axially symmetric Fermi surface having a unique extremal cross-section for any direction of the magnetic field. In this case, the oscillating function Δ is described by the expression

$$\Delta = \sum_{r=1}^{\infty} (-1)^r \psi^r(\theta) \cos\left(\pi r \frac{m_{\perp i}}{m}\right) \int_0^1 \cos\left(\frac{rcS(x)}{\hbar|e|B}\right) dx, \tag{7.15}$$

where $x = p_z/p_m$ (p_m is the maximum value of p_z), and $S(x)$ is the area of cross-section of the Fermi surface. Further, we assume that the dependence $S(x)$ of x can be presented in the form

$$S(x) = S_{ex}(1 - a^2 x^2 - b^2 x^{2k}), \qquad k > \tfrac{1}{2}. \tag{7.16}$$

Here b^2 is a dimensionless positive constant, and the quantities a^2 and S_{ex} are functions of the angle Φ between the magnetic field direction and the Fermi surface symmetry axis. The quantity $a^2(0) = 0$, and the value of a^2 increases quite rapidly with Φ so that for a certain value of Φ_0 it becomes and stays much larger than the constant b^2 in order of magnitude.

Thus, for $\Phi = 0$ (the magnetic field is directed along the symmetry axis), the central cross-section of the Fermi surface in the chosen model is quasi cylindrical for $k > 1$. If, however, $\frac{1}{2} < k < 1$, one of the principal radii of curvature vanishes at all points of the central cross-section of the Fermi surface. When $\mathbf{B}$ deviates from the symmetry axis by an angle of the order of, or larger than, Φ_0, the term proportional to x^2 becomes the principal term depending on x on the right-hand side of (7.16). In other words, the extremal cross-section of the Fermi surface passes through that segment of the surface where its curvature is finite and nonzero.

Substituting (7.16) into (7.15), we obtain

$$\Delta = \sum_{r=1}^{\infty} (-1)^r \psi_r(\theta) \cos\left(\pi r \frac{m_\perp}{m}\right) \left[\cos(\pi r \gamma^2) U_r(a) + \sin(\pi r \gamma^2) W_r(a)\right]. \quad (7.17)$$

Here,

$$U_r(a) = \int_0^1 \cos\left[\pi r \gamma^2 (a^2 x^2 + b^2 x^{2k})\right] dx,$$

$$W_r(a) = \int_0^1 \sin\left[\pi r \gamma^2 (a^2 x^2 + b^2 x^{2k})\right] dx. \quad (7.18)$$

Under normal experimental conditions, when the number of Landau levels under the Fermi surface is large, the parameter γ^2 assumes a value much larger than unity.

The principal terms in the expansions of the integrals (7.18), in powers of the small parameter γ^{-1}, can easily be obtained in two limiting cases $a^2 \ll b^2$ and $a^2 \gg b^2$. The first inequality is satisfied for small values of the angle between the magnetic field and the symmetry axis, when the extremal cross-section of the Fermi surface nearly coincides with the line at whose points one of the principal radii of curvature vanishes or tends to infinity. The asymptotic expression for the oscillating function Δ will have the form

$$\Delta = \frac{\eta_k}{\gamma^{1/k}} \sum_{r=1}^{\infty} \frac{(-1)^r}{r^{1/2k}} \psi_r(\theta) \cos\left(\pi r \gamma^2 - \frac{\pi}{4k}\right) \cos\left(\pi r \frac{m_\perp}{m}\right), \quad (7.19)$$

where

$$\eta_k = \frac{\Gamma(1/2k)}{2k(b\sqrt{\pi})^{1/k}}.$$

In the opposite limit $a^2 \gg b^2$, which corresponds to the situation when the angle Φ is sufficiently large, the principal term in the expansion of Δ in powers of γ^{-1} becomes

$$\Delta = \frac{1}{\gamma} \sum_{r=1}^{\infty} \frac{(-1)^r}{\sqrt{r}} \psi_r(\theta) \cos\left(\pi r \gamma^2 - \frac{\pi}{4}\right) \cos\left(\pi r \frac{m_\perp}{m}\right). \quad (7.20)$$

Thus, in this case, the function Δ describes ordinary oscillations of the electron density of states in a quantizing magnetic field. Expressions (7.19) and (7.20) can be straightforwardly derived from (7.9) and (7.10) when the appropriate model of the Fermi surface is chosen.

The analysis of the dependence of the amplitude of oscillations on Φ in the intermediate region of values of Φ, where a^2 and b^2 have the same order of magnitude, is a more complex problem. In this region, the integrals (7.18) can be presented in the form of expansions in powers of the parameter $w (w = \pi r \gamma^2 a^2 / (\pi r \gamma^2 b^2)^{1/k})$. In particular, the expression for U_r in the case $w < 1$ has the form

$$U_r(a) = \frac{(\pi r \gamma^2 b^2)^{-1/2k}}{2k} \sum_{n=0}^{\infty} \frac{(-1)^n}{n!} w^n \Gamma \left(\frac{2n+1}{2k} \right) \cos \left[\pi \frac{n(1-k)}{2k} + \frac{\pi}{4k} \right],$$

(7.21)

while for $w < 1$ the expression assumes the form

$$U_r(a) = \frac{1}{2\pi a \gamma \sqrt{r}} \sum_{n=0}^{\infty} \frac{(-1)^n}{n!} w^{-n} \Gamma \left(kn + \tfrac{1}{2} \right) \cos \left[\pi \frac{n(k-1)}{2} + \frac{\pi}{4} \right]. \quad (7.22)$$

Expansions for $W_r(a)$ are obtained from (7.21) and (7.22) by replacing cosines with sines of the same arguments.

The results of the numerical computation of the oscillation amplitude (7.17) $Y (Y = \sqrt{U^2 + W^2})$ as a function of w, obtained by means of the expansions (7.21), (7.22) of U_r and W_r in powers of w, are presented in Figures 7.1 and 7.2 for several values of the parameter k. Calculations were made under the assumption that $\theta \approx 1$, i.e., only the first term was taken into consideration in the sum over n in (7.17). The parameter $\pi b^2 \gamma^2$ was assumed to be equal to 10^3. Figure 7.1 shows the dependence of Y on w calculated for $k = 2$ and $k = 4$. The central Fermi surface cross-section for $\Phi = 0$ is found to be quasi cylindrical in this case. The oscillation amplitude assumes the largest value when $\Phi = 0$ and decreases monotonically when w increases. Both curves asymptotically approach the straight line $Y = 1/\gamma$ for $w \gg 1$. As expected, the scale of amplitude variation and the rate of its decrease are much larger for the curve constructed for larger values of the parameter k. This

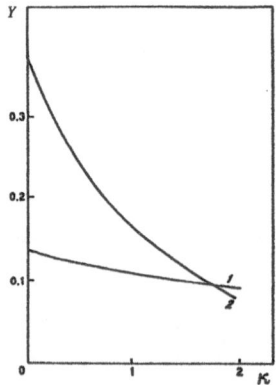

FIGURE 7.1. The amplitude of the quantum oscillations of the velocity of sound as a function of the orientation of the external magnetic field with respect to the axis of the axially symmetric Fermi surface for $k > 1$. The curves are plotted for $\pi b^2 \gamma^2 = 10^3$, $k = 2$ (curve 1), $k = 4$ (curve 2).

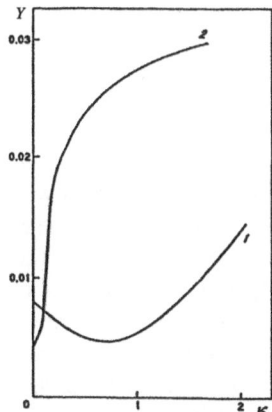

FIGURE 7.2. The amplitude of the quantum oscillations of the velocity of sound as a function of the orientation of the magnetic field with respect to the axis of symmetry of the Fermi surface for $\frac{1}{2} < k < 1$. The curves are plotted for $\pi b^2 \gamma^2 = 10^3$, $k = \frac{3}{4}$ (curve 1), $k = \frac{5}{8}$ (curve 2).

corresponds to the assumption that the Fermi surface is closer to the cylindrical shape in the vicinity of its central cross-section for $\Phi = 0$.

The behavior of the amplitude of quantum oscillations of the velocity of sound upon a variation of the angle Φ is shown in Figure 7.2 under the assumption that the curvature at the points of the central cross-section of the Fermi surface becomes infinite for $\Phi = 0$. The curves shown in the figure are constructed for $k = \frac{3}{4}$ and $k = \frac{5}{8}$. The second value of k corresponds to a stronger anomaly of the Fermi surface curvature at the extremal cross-section. For $k = \frac{5}{8}$, the oscillation amplitude increases monotonically with w and tends to attain the value $1/\gamma$. For a less strongly manifested anomaly of the curvature at the central cross-section ($k = \frac{3}{4}$), the dependence of Y on w is nonmonotonic. However, in this case also, the decrease in the value of Y for small values of w is replaced by an increase.

Finally, let us consider the quantity Φ_0 describing the real width of the range of variation of the angle Φ corresponding to a well-manifested dependence of the magnitude of the oscillations on the angle Φ. While the parameter w characterizes the shape of the Fermi surface in the vicinity of its extremal cross-section, Φ_0 describes the width of the Fermi surface strip in which the curvature anomalies are manifested. The order-of-magnitude estimate for Φ_0 can easily be obtained for the corresponding equation (7.16) at $k = 2$. The dependence $w(\Phi)$ in this case has the form

$$w(\Phi) \approx 2\sqrt{\hbar}\frac{m_\perp}{m_\parallel} \sin^2 \Phi \cos \Phi, \qquad (7.23)$$

where $m_\parallel$ is the longitudinal mass at the reference points ($p_z = \pm p_m$). Since the dependence $Y(w)$ for $k = 2$ becomes weak for $w \geq 2$, we can obtain the following estimate:

$$\Phi_0 \approx \sqrt{m_{||}/m_{\perp}\gamma}.$$

Thus, if the ratio $m_{||}/m_{\perp}$ accepts values of the order of unity and $\gamma^2 \approx 10^4$, Φ_0 can be of the order of $5°$–$10°$. We can expect that the quantity Φ_0 does not exceed a few degrees in the general case. This is in agreement with the experimental results of [126], [159] obtained for copper, as well as with the results of cyclotron resonance studies in a magnetic field normal to the metal surface of [47], [48].

Qualitative features of the quantum oscillations of static elastic moduli in the low-temperature range ($\theta \ll 1$) are manifested, even in the simplest model of an isotropic metal. It should be noted that the influence of the magnetostriction on the quantum oscillations of elastic moduli depends considerably on the orientation of the deformation gradient with respect to the external magnetic field (see [158]). The most noticeable effect of magnetostriction will take place when the deformation gradient is perpendicular to the applied field **B**. On the contrary, if the deformation is nonuniformly distributed along **B**, the effect of magnetostriction becomes insignificant. Therefore, the features of the quantum oscillations of the elastic moduli are conveniently studied in a frame of reference whose y-axis is directed along the deformation gradient, where it suffices to consider only the components λ_{yyyy} and λ_{xyxy}. It follows from (7.2) that the oscillating correction to λ_{yyyy} is expressed as

$$\tilde{\lambda}_{yyyy} = -\frac{N}{g}L_{yy}^2 \frac{\Delta}{1 + (1 + V - 4\pi\kappa\gamma^4)\Delta}, \tag{7.24}$$

where

$$L_{yy} = 3\sum_{j=2}(2j+1)\frac{\Lambda_{j0}^{yy}}{1 + A_j}P_j(0) + 1, \tag{7.25}$$

where Λ_{j0}^{yy} are dimensionless coefficients in the expansion of the corresponding component of the deformation potential tensor in spherical harmonics, $P_j(x)$ are the Legendre polynomials, and V is a constant describing the Fermi-liquid interaction, expressed in terms of the Fermi-liquid coefficients A_j, B_j:

$$V = \sum_{j=0}(2j+1)\left(\frac{A_j}{1 + A_j} + \frac{B_j}{1 + B_j}\right)[P_j(0)]^2 - \frac{A_0}{1 + A_0}, \tag{7.26}$$

Δ is a function of (7.20) describing quantum oscillations of the electron density of states in a metal with a spherical Fermi surface, and $\kappa = e^2 p_F/\hbar(2\pi c)^2 m(1 + A_1)$. An expression for the oscillating part of λ_{xyxy} is obtained from (7.24) by replacing L_{yy} by

$$L_{xy} = 3\sum_{j=2}(2j+1)\frac{\Lambda_{j0}^{xy}}{1 + A_j}P_j(0). \tag{7.27}$$

In deriving (7.24), only the principal terms of expansions of the quantities (7.4) in powers of the small parameter γ^{-1} were taken into consideration, while the terms which are less significant for qualitative description of the quantum oscillations were omitted.

The effect of magnetostriction is determined by the term $-4\pi\kappa\gamma^4\Delta$ in the denominator of (7.24). This term may become significant only at low temperatures, when the amplitude of oscillations of the function Δ is sufficiently large. It is also important that the magnetic field must not be too strong, because the value of $4\pi\kappa\gamma^4$ decreases with growing B and may become much lower than unity at large B.

The denominator of the right-hand part of (7.24) may tend to zero at the peaks of the quantum oscillations, which will lead to the vanishing of the corresponding elastic constants near these peaks, showing evidence for the instability of the lattice. If the quantity $1 + (1 + V - 4\pi\kappa\gamma^4)\Delta$ tends to zero, then $1 - 4\pi\chi_{||} \to 0$ (where $\chi_{||} = \kappa\gamma^4\Delta/[1 + (1 + V)\Delta]$ is the longitudinal magnetic susceptibility). The expression for the magnetic susceptibility used above was derived in [160]. Thus, a possible instability of the moduli λ_{yyyy} and λ_{xyxy} is related to the instability of magnetization and is due to the magnetostriction effects. An essential role belongs to the Fermi-liquid interaction, because at $4\pi\kappa\gamma^4 < 1$ the instability can only arise only when $V < 0$.

It seems that the instability of the lattice due to the quantum oscillations of the elastic moduli can occur under conditions when the magnetostriction effect is insignificant. Specifically, the oscillating term in the expression for λ_{zzzz} is

$$\tilde{\lambda}_{zzzz} = -\frac{N^2 L_{zz}^2}{g} \frac{\Delta}{1 + (1 + V)\Delta}, \tag{7.28}$$

where

$$L_{zz} = 3\sum_{j=2}(2j + 1)\frac{\Lambda_{j0}^{zz}}{1 + A_j} P_j(0) + 1.$$

The denominator in (7.28) may tend to zero when $V < -1$, and the magnitude of the oscillations described by the function Δ is large enough ($T \to 0$). This instability was predicted in [161]. However, on increasing Δ the quantity $1 + (1 + V - 4\pi\kappa\gamma^4)\Delta$ reaches zero sooner than the quantity $1 + (1 + V)\Delta$. Therefore, the phase transition arising due to the magnetostriction effect will prevent the appearance of the instability of the lattice connected with the singularity of Λ_{zzzz}.

The instability of the elasticity moduli can be observed provided that the amplitude of oscillations of the function Δ, which is of the order of $\gamma^{-1}\theta^{-1/2}$, is comparable with unity. This is a stringent condition. For example, at $\hbar\Omega/\zeta \sim 10^{-3}$ it can be satisfied if the value $T/\hbar\Omega$ is of the order of 10^{-2}, which implies that at a magnetic field of about 100 kG the temperature must not exceed several tenths of 1 K. The conditions for observation of the effects of this kind will be much more favorable if the Fermi surface has lines of zero curvature.

Consider a metal with an axisymmetric Fermi surface and assume that one of its extremal sections in the magnetic field directed along the symmetry axis represents a zero-curvature line. The matrix elements $\varphi(\mathbf{q}, Nlp_z; N_1, l_1 p_{z1})$ and $\psi(\mathbf{q}, Nlp_z; N_1, l_1 p_{z1})$ included in the Fermi-liquid kernel can be replaced by their semiclassical analogs calculated in the limit of small $q(q \rightarrow 0)$. For the components of the tensor of the deformation potential, we may use an approximation

$$\Lambda^{\alpha\beta}_{Nlp_z} \approx \Lambda_{\alpha\beta} n_{Nlp_z} + G\Pi^{\alpha\beta}_{Nlp_z}, \tag{7.29}$$

where $\Lambda^{\alpha\beta}_{Nlp_z}$ is a tensor whose elements are constants with the dimensionality of energy, G is a dimensionless constant, and $\Pi^{\alpha\beta}_{Nlp_z}$ is the tensor of the electron momentum flux density.

This approximation leads to the following expressions for the oscillating corrections to the elastic moduli λ_{yyyy} and λ_{xyxy}:

$$\tilde{\lambda}_{yyyy} = -\frac{N^2}{g} \frac{(1+G^*)\Delta}{1+(1+V-4\pi\kappa\gamma^4)\Delta}, \tag{7.30}$$

$$\tilde{\lambda}_{xyxy} = -G^{*2}\frac{N^2}{g} \frac{\Delta}{1+(1+V-4\pi\kappa\gamma^4)\Delta}. \tag{7.31}$$

Here $G^* = G/(1+\tilde{A}_2)$, $V = \alpha - \bar{\alpha}_0 + \beta$, and the dimensionless coefficients $\alpha, \alpha_0, \bar{\alpha}_0, \beta$ are described by the relations

$$\alpha = \frac{A_0^*}{1+\tilde{A}_0}, \qquad \beta = \frac{B_0^*}{1+\tilde{B}_0},$$

$$\alpha_0 = \frac{A_0}{1+\tilde{A}_0}, \qquad \bar{\alpha}_0 = \frac{\bar{A}_0}{1+\tilde{A}_0}, \tag{7.32}$$

where

$$A_0 = \left(\frac{1}{p_0}\int P^0_{00}(p_z)\,dp_z\right)^2, \qquad \bar{A}_0 = \frac{P^0_{00}(0)}{p_0}\int P^0_{00}(p_z)\,dp_z,$$

$$\tilde{A}_0 = \frac{1}{p_0}\int \left[P^{|0|}_{00}(p_z)\right]^2 dp_z, \qquad \tilde{B}_0 = \frac{1}{p_0}\int \left[Q^{|0|}_{00}(p_z)\right]^2 dp_z, \tag{7.33}$$

$$A_0^* = \left[P^{|0|}_{00}(p_z)\right]^2, \qquad B_0^* = \left[Q^{|0|}_{00}(p_z)\right]^2,$$

and the constant p_0 has the dimensions of the momentum and equals $2\pi^2\hbar^3 g/m_\perp$.

The function Δ describing the quantum oscillations is now given by (7.9) and the magnitude of the oscillations is of the order of $\gamma^{-1/l}\theta^{(1/2l)-1}$. The amplitude of oscillations may become comparable with unity at somewhat higher temperatures. For example, at $l = 2$, $\hbar\Omega/\zeta \sim 10^{-3}$, and $B \sim 100$ kG, the condition $\gamma^{-1/l}\theta^{(1/2l)-1} \sim 1$ will be satisfied at temperatures of the order of 1 K. Therefore, the presence of a quasi-cylindrical extremal section on the Fermi surface would provide more favorable conditions for the manifestation of the above-described features of the low-temperature quantum oscillations of the static elasticity moduli of metals.

7.2   Quantum Oscillations of the Velocity of Sound Propagating in a Nonlocal Regime

The oscillations of the ultrasound velocity in a quantizing magnetic field for a so-called nonlocal regime when the ultrasound wavelength does not exceed the electron mean free path ($ql > 1$) were investigated far less thoroughly than the oscillations in a low-frequency range. Yet, it is precisely under these conditions that the quantum oscillations of the sound velocity are more complicated in character.

The basic qualitative features of the quantum oscillations of sound velocity in this nonlocal propagation regime and in the rf range are caused by collisionless absorption and will be analyzed below. At first we consider the geometry when an ultrasound wave propagates perpendicularly to the magnetic field. In this geometry the collisionless absorption cannot appear and the electron renormalization of the elastic moduli is described by (7.1) and (7.2) in an extended frequency range $\omega < \Omega$. Now, however, we have to take into account the dependence of the matrix elements included in (7.4), and on q to calculate the averages. For not too strong magnetic fields ($qR \geq 1$) this dependence cannot be neglected.

The oscillating contribution to the elastic modulus λ_{yyyy} which determines the electron renormalization of the velocity of a longitudinal ultrasound wave propagating perpendicularly to the magnetic field in a metal with a spherical Fermi surface, can be represented as follows:

$$\tilde{\lambda}_{yyyy} = -\frac{N^2}{8} \frac{L_{yy}^2(q)\Delta}{1 + [V_q + J_0^2(qR) - 4\pi\kappa\gamma^4]\Delta}, \tag{7.34}$$

where

$$V_q = \sum_{j=0}\sum_{|m|\leq j} c_{jm}\left(\frac{A_j}{1+A_j} + \frac{B_j}{1+B_j}\right)[P_j^{|m|}(0)J_{-m}(qR)]^2 - \frac{A_0}{1+A_0}J_0^2(qR), \tag{7.35}$$

$$L_{yy}(q) = 3\sum_{j=0}\sum_{|m|\leq j} c_{jm}\frac{\Lambda_{jm}^{yy}}{1+A_j}P_j^{|m|}(0)J_{-m}(qR) + J_0(qR). \tag{7.36}$$

Here

$$c_{jm} = (2j+1)\frac{(j-|m|)!}{(j+|m|)!},$$

$P_j^{|m|}(x)$ are the associated Legendre functions, and $J_{-m}(x)$ are the Bessel functions. In the limit of small $q(qR \ll 1)$ this expression (7.34) coincides with (7.23).

The oscillating correction to the λ_{xyxy} component of the elastic moduli is described by the expression similar to (7.34) where $L_{yy}(q)$ is replaced by $L_{xy}(q)$:

$$L_{xy}(q) = 3\sum_{j=0}\sum_{|m|\leq j} c_{jm}\frac{\Lambda_{jm}^{xy}}{1+A_j}P_j^{|m|}(0)J_{-m}(qR). \tag{7.37}$$

If the temperature is not too low, the amplitude of the oscillations of the function Δ is small compared to unity. In this case, the oscillating term in the denominator of (7.34) may be neglected, and instead of (7.34), we obtain

$$\tilde{\lambda}_{yyyy} = -\frac{N^2}{g} L_{yy}^2(q)\Delta, \tag{7.38}$$

for the longitudinal wave and

$$\tilde{\lambda}_{xyxy} = -\frac{N^2}{g} L_{xy}^2(q)\Delta, \tag{7.39}$$

for the transverse wave.

These equations describe the quantum oscillations in the velocities of the longitudinal and the transverse (linear polarized along the x-axis) sound waves that propagate normally to the external magnetic field under typical experimental conditions. As the functions $L_{yy}(q)$ and $L_{xy}(q)$ depend on the parameter qR, the oscillation amplitudes have a sufficiently complicated dependence on the external magnetic field B. This dependence becomes more simple in the limiting case of $qR \gg 1$. Using the asymptotical representation of the Bessel functions, one may obtain

$$L_{yy}(q) \approx 3\sqrt{\frac{2}{\pi qR}} \sum_{j=2} \sum_{|m|\leq j} c_{jm} \frac{\Lambda_{jm}^{yy}}{1+A_j} P_j^{|m|}(0) \cos\left(qR - \frac{\pi}{4} - \frac{\pi}{2}m\right)$$

$$+ \sqrt{\frac{2}{\pi qR}} \cos\left(qR - \frac{\pi}{4}\right), \tag{7.40}$$

$$L_{xy}(q) \approx 3\sqrt{\frac{2}{\pi qR}} \sum_{j=2} \sum_{|m|\leq j} c_{jm} \frac{\Lambda_{jm}^{xy}}{1+A_j} P_j^{|m|}(0) \cos\left(qR - \frac{\pi}{4} - \frac{\pi}{2}m\right). \tag{7.41}$$

Hence, at $qR \gg 1$, the oscillation amplitude is periodic in the reciprocal magnetic field strength. This is a manifestation of the geometrical oscillations in the interaction of the ultrasound waves with the electron system. In this regime, the quantum oscillations are caused by the behavior of the electron density of states in the magnetic field, as they are in the local propagation regime. The ratio of the quantum oscillation period to the period of the geometrical oscillations is of the order of $\hbar q/p_F \ll 1$.

If the sound wave propagates in a direction perpendicular to the external magnetic field, the low-temperature quantum oscillations of the sound velocity are substantially influenced by the oscillating term in the denominator of (7.34), which results from the magnetostriction. Comparison of (7.34) with (7.23) shows that all the characteristic features of the oscillations arising due to the magnetostriction effect also appear in the nonlocal propagation regime.

To study the quantum oscillations of the ultrasound velocity for $(\mathbf{q} \cdot \mathbf{B}) \neq 0$ we have to take into account the collisionless damping which critically changes their character. In a quantizing magnetic field the collisionless damping is suppressed in some frequency intervals. It produces giant quantum oscillations in the absorption of the ultrasonic waves and is more complicated than in the low-frequency range, the structure of the quantum oscillations of the ultrasound velocity is as shown in [27].

To analyze the electron renormalization of the velocity of ultrasound in a nonlocal regime, we will start from the kinetic equation for the electron density matrix (2.20). At first we confine consideration to the case of a longitudinal ultrasound wave propagating along the magnetic field **B** in a metal with a spherical Fermi surface. Nevertheless, this simplified model enables us to expose characteristic features of these quantum oscillations arising due to the collisionless damping. The oscillating correction to the velocity of sound can be represented as follows:

$$\tilde{s}^2 = \frac{N^2}{g\rho_m} L_{zz}^2 \, \mathrm{Re} \, \Delta^*. \tag{7.42}$$

Here Δ^* is the oscillating part of the function F_q:

$$F_q = \frac{1}{(2\pi\hbar)^2} \frac{|e|B}{cg} \sum_{n,\sigma} \int dp_z \frac{f_{np_z}^{\sigma} - f_{np_z-\hbar q}^{\sigma}}{E_{np_z-\hbar q}^{\sigma} - E_{np_z}^{\sigma} + \hbar\omega + i\hbar/\tau}, \tag{7.43}$$

where n, p_z, σ are the quantum numbers of electrons in a magnetic field. There is a pole in the integrand in (7.43), and the presence of this pole produces some problems in the calculation of the integral over "p_z". Taking into account only the principal terms in the expansion of the function Δ^* in powers of γ^{-1} we arrive at the result

$$\Delta^* = X + X'. \tag{7.44}$$

Here the first term X is the contribution from the pole

$$X = \frac{1}{2qR} \sum_{r=1} \frac{(-1)^r}{r} \psi_r(\theta) \cos\left(\pi r \frac{m_\perp}{m}\right)$$

$$\times \left\{ \exp\left[-\pi r\gamma^2 \left(2\left|\frac{\omega}{qv_F} + \frac{\hbar q}{2p_F}\right| \frac{1}{ql} + i\left(1 - \left[\frac{\omega}{qv_F} + \frac{\hbar q}{2p_F}\right]^2\right)\right)\right] \right.$$

$$\left. - \exp\left[-\pi r\gamma^2 \left(2\left|\frac{\omega}{qv_F} - \frac{\hbar q}{2p_F}\right| + i\left(1 - \left[\frac{\omega}{qv_F} - \frac{\hbar q}{2p_F}\right]^2\right)\right)\right] \right.$$

$$\times \mathrm{sign}\left[\frac{\omega}{qv_F} - \frac{\hbar q}{2p_F}\right] \right\}. \tag{7.45}$$

This term appears in the expressions for the velocity and absorption coefficient of ultrasound waves in the nonlocal propagation regime and the high-frequency range. It describes giant quantum oscillations of the absorption and the corresponding contribution to the oscillations of the velocity of ultrasound waves. Under certain conditions this term can also describe coupling of the ultrasound waves to the longitudinal quantum waves (see [27], [162]).

The second term X' has the form:

$$X' = \frac{1}{2\gamma} \sum_{r=1} \frac{(-1)^r}{\sqrt{r}} \psi_r(\theta) \cos\left(\pi r \frac{m_\perp}{m}\right)$$

$$\times \left\{ \exp\left[i\left(r\gamma^2 - \frac{\pi}{4} \right) \right] G_r^- + \exp\left[-i\left(r\gamma^2 - \frac{\pi}{4} \right) \right] G_r^+ \right\}. \quad (7.46)$$

Here

$$G_r^\pm = \pm \frac{i}{2} \int_0^\infty \exp[\pm iy] \exp\left[-\left(\frac{\omega}{q v_F} + \frac{i}{ql} \right) \sqrt{2\pi r \gamma^2 y} \right] dy. \quad (7.47)$$

This function can be expressed in terms of Fresnel integrals. This term X' does not exhibit the oscillating dependence of the ultrasound velocity and describes oscillations similar to the oscillations of the electron density of states.

We can simplify the expressions for the functions X and X' under the following condition:

$$\frac{\omega}{q v_F} \gamma \gg 1. \quad (7.48)$$

Under this condition Re X has the form

$$\text{Re } X = \pi \frac{\omega}{q v_F} \sum_{r=1} (-1)^r \psi_r(\theta) \cos\left(\pi r \frac{m_\perp}{m} \right)$$

$$\times \exp\left[-2\pi r \gamma^2 \frac{\omega}{q^2 v_F l} \right] \sin\left(\pi r \gamma^2 \left[1 - \left(\frac{\omega}{q v_F} \right)^2 \right] \right). \quad (7.49)$$

Using asymptotic expressions for the Fresnel integrals we can show that under condition (7.48), $G_r^\pm \approx 1$. Therefore, in this case, X' coincides with the function Δ, which describes oscillations of the density of states of the electron system in a quantizing magnetic field. In the opposite limit case, when

$$\frac{\omega}{q v_F} \gamma \ll 1, \quad (7.50)$$

we obtain the following expressions for Re X and X':

$$\text{Re } X = \frac{1}{\lambda q \gamma} \sum_{r=1} \frac{(-1)^r}{r} \psi_r(\theta) \exp\left[-\frac{\pi r}{\Omega \tau} \right] \cos\left(\pi r \frac{m_\perp}{m} \right) \cos\left(\pi r \left[\gamma^2 - \left(\frac{\lambda q}{2} \right)^2 \right] \right), \quad (7.51)$$

$$X' = \frac{1}{\gamma(\lambda q)^2} \sum_{r=1} \frac{(-1)^r}{r^{3/2}} \psi_r(\theta) \cos\left(\pi r \frac{m_\perp}{m} \right) \cos\left(\pi r \gamma^2 + \frac{\pi}{4} \right), \quad (7.52)$$

here $\lambda = (\hbar c / |e| B)^{1/2}$ is the magnetic length.

Thus the quantum correction to the velocity of sound (7.42) includes two oscillating functions (X and X') which differ in period, phase and field dependence of the magnitude. For moderately low temperatures ($\theta \simeq 1$) both oscillating contributions are of the same order of magnitude. For low temperatures and in pure metals when the inequalities

$$\frac{ql}{\gamma^2} > 1, \qquad \frac{\hbar\omega}{T} \frac{\omega}{q v_F} > 1, \quad (7.53)$$

are satisfied, the oscillations of the function X are of the order of unity in magnitude and critically exceed the oscillations described by the function X'. These inequalities (7.53) define the conditions of occurrence of the quantum waves in metals which are analyzed in [163]–[165]. The interaction with these waves mainly determines the character of the quantum oscillations of the ultrasound velocity in the low-temperature range. The coupling of the ultrasound to quantum waves is analyzed in detal in [162], therefore we do not discuss this problem here.

The principal qualitative features of the sound velocity oscillations in the non-local propagation regime obtained for an isotropic metal are also retained in the case of a nonspherical Fermi surface. We now consider the quantum oscillations of the longitudinal sound velocity as the sound propagates in a direction normal to the external magnetic field. The Fermi surface is assumed to be closed, and its curvature to be finite and nonzero everywhere. When the number of Landau levels below the Fermi surface is sufficiently large, we can represent matrix elements $n_{\nu\nu'}(-\mathbf{q})$ and $\Lambda_{\nu\nu'}(-\mathbf{q})$ in the form

$$
\begin{aligned}
n_{\nu\nu'}(-\mathbf{q}) &\approx \delta_{\sigma\sigma'}\delta_{p_z p_z'}\delta_{x_0;x_0'+\lambda^2 q}\, I_{Nlp_z}(-\mathbf{q}), \\
\Lambda_{\nu\nu'}(-\mathbf{q}) &\approx \delta_{\sigma\sigma'}\delta_{p_z p_z'}\delta_{x_0;x_0'+\lambda^2 q}\, \Lambda_{Nlp_z}(-\mathbf{q}),
\end{aligned}
\tag{7.54}
$$

where $N = n + n'$ and $l = n - n'$:

$$
\begin{aligned}
I_{Nlp_z}(-\mathbf{q}) &= \frac{1}{2\pi}\int_0^{2\pi} d\psi \exp\left[-\frac{i}{\Omega}\int_0^{\psi}(qv_y - l\Omega)d\psi'\right], \\
\Lambda_{Nlp_z}(-\mathbf{q}) &= \frac{1}{2\pi}\int_0^{2\pi} d\psi\, \Lambda_{Np_z}(\psi)\exp\left[-\frac{i}{\Omega}\int_0^{\psi}(qv_y - l\Omega)d\psi'\right].
\end{aligned}
\tag{7.55}
$$

Assume that the condition $qv_\perp/\Omega \gg 1$ is met for the cyclotron orbits corresponding to the extreme cross-sections of the Fermi surface. In this case, the integrals may be evaluated by means of the stationary phase method, and expression (7.55) may be considerably simplified. Assume that there are only two stationary points on the cyclotron orbit, A and B, and

$$
\begin{aligned}
\psi_B - \psi_A &= \pi, \\
\Lambda_{Np_z}(\psi_B) = \Lambda_{Np_z}(\psi_A) &\equiv \Lambda_{Np_z}^0, \\
q\frac{\partial v_y}{\partial \psi}\Big|_{\psi=\psi_A} = -q\frac{\partial v_y}{\partial \psi}\Big|_{\psi=\psi_B} &\equiv qU_0.
\end{aligned}
$$

If so, keeping only the lowest-order terms in the expansions of $I_{Nlp_z}(-\mathbf{q})$, and $\Lambda_{Nlp_z}(-\mathbf{q})$, in powers of the small parameter $\Omega/qv_\perp$, leads to the expressions

$$
I_{Nlp_z}(-\mathbf{q}) \approx \sqrt{\frac{2\Omega}{\pi qU_0}}\cos\left(qR_0 + \pi\frac{l}{2} - \frac{\pi}{4}\right),
\tag{7.56}
$$

$$
\Lambda_{Nlp_z}(-\mathbf{q}) \approx \Lambda_{Nlp_z}^0 \sqrt{\frac{2\Omega}{\pi qU_0}}\cos\left(qR_0 + \pi\frac{l}{2} - \frac{\pi}{4}\right),
\tag{7.57}
$$

where, R_0 denotes $|v_y(\psi_A)|/\Omega$.

In calculating the renormalization of the elastic moduli by electrons, the last term in (7.2) may be neglected, because the magnetostriction does not markedly influence the results at $q R_0 \gg 1$. By using approximations like (7.56) and (7.57) for the matrix elements, it is simple to obtain the asymptotic expansions of the averages (A, B) in powers of γ^{-1}. This leads finally to the following result for the oscillating part of λ_{yyyy}:

$$\tilde{\lambda}_{yyyy} = -\frac{2}{\pi q} \sum_i \frac{(L_i^{yy})^2}{U_{0i}} \cos^2 \left(q R_{0i} - \frac{\pi}{4} \right) \Delta_i, \qquad (7.58)$$

where $L_i^{yy} = \Lambda_i^* - \langle \Lambda^* \rangle / g + N/g$, and Δ_i is a function describing the oscillations of the density of states in a magnetic field in a metal with a nonspherical Fermi surface.

The summation is carried out over all extreme cross-sections. The obtained expression, as the analogous result for an isotropic metal (7.40), describes the quantum oscillations akin to the oscillations of the density of states with their amplitude modulated by the geometrical oscillations. The ratio between the periods of the quantum and geometrical oscillations is a small value of the order of $\hbar q / p_0$, as well as in the isotropic model.

Quantum oscillations of the velocity of longitudinal ultrasound waves propagating along the magnetic field **B** in a metal with a nonspherical Fermi surface, are described by formula (7.42). Suppose that the Fermi surface is axisymmetric and that the inequality $\omega/q v_m \gg \hbar q / p_m$ (v_m, p_m are the maximum values of the longitudinal components of the electron velocity and kinematic momentum) is satisfied. Then we can obtain the following expression for the principal term in the expansion of the function X in powers of the small parameter γ^{-1}:

$$\mathrm{Re}\, X = a \frac{\pi \omega}{q v_m} \sum_{r=1} \psi_r(\theta) \cos \left(\pi r \frac{m_\perp}{m} \right) \exp \left[-r \lambda^2 \frac{(2\pi m_\perp \omega)^2}{\omega \tau q^2 |\partial^2 S/\partial p_z^2|} \right]$$

$$\times \sin \left[r \frac{\lambda^2}{\hbar^2} \left(S_{ex} + \pi \frac{\hbar^2}{\lambda^2} \delta - \frac{(2\pi m_\perp \omega)^2}{q^2 |\partial^2 S/\partial p_z^2|} \right) \right], \qquad (7.59)$$

where a is a dimensionless constant of the order of unity.

This result (7.59) is valid when the Fermi surface has a unique extreme cross-section. When there are several extreme cross-sections, each gives the contribution to the real part of the function X. Besides, each extreme cross-section gives an additional contribution to the oscillating part of the function F_q. The real parts of these extra contributions nearly coincide with the functions Δ_i. These oscillating corrections describe the conventional oscillations of the density of the electron states.

Some changes of qualitative character will take place when the Fermi surface has extreme cross-sections of zero curvature. Assume that the magnetic field is directed along the symmetry axis of the axisymmetric Fermi surface. When the curvature tends to zero at the points of the extreme cross-section, the oscillating function Δ_i corresponding to this cross-section, has the form (7.9) and describes oscillations which are larger in magnitude than oscillations corresponding to the remaining conventional extreme cross-sections. As a result, contributions from these nearly cylindrical cross-sections are dominant in expression (7.58) describing the quantum oscillations of the velocity of sound propagating perpendicularly to the magnetic field **B**.

As previously, the quantum correction to the velocity of longitudinal sound propagating along the magnetic field has the form (7.42). However, when the Fermi surface has an extreme cross-section with zero curvature, it leads to changes in the expressions for the functions X and X'. In particular, we obtain the following expression for the real part of the function X:

$$\mathrm{Re}\,X = a\pi \left(\frac{\omega}{q v_0}\right)^{1/(2l-1)} \sum_{r=1} \psi_r(\theta) \cos\left(\pi r \frac{m_\perp}{m}\right)$$

$$\times \exp\left[-\frac{2\pi r}{2l-1}\frac{p_0}{\hbar q}\frac{1}{\Omega \tau}\left(\frac{\omega}{q v_0}\right)^{1/(2l-1)}\right]$$

$$\times \sin\left\{r\frac{\lambda^2}{\hbar^2}S_{\mathrm{ex}}\left[1-\left(\frac{\omega}{q v_0}\right)^{2l/(2l-1)}\right]+\pi r\delta\right\}. \qquad (7.60)$$

Here $v_0 = l S_{\mathrm{ex}}/\pi m_\perp p_0$ is a constant of the dimensions of velocity, and $p_0 = \left[(2l)!\,S_{\mathrm{ex}}/|\partial^{2l} S/\partial p_z^{2l}|_0\right]^{1/2l}$ is a constant of the dimensions of momentum. This result (7.60) corresponds to the contribution from the center cross-section of the Fermi surface. It is assumed that the area of this cross-section $S(\zeta, p_z)$ near $p_z = 0$ is described by formula (7.8). The second oscillating term X' under these assumptions has the form

$$X' = \frac{a}{l}\Gamma\left(\frac{1}{2l}\right)\left(\frac{\hbar|e|B}{c S_{\mathrm{ex}}}\right)^{1/2l} \sum_{r=1}^{\infty} \frac{\psi_r(\theta)}{r^{2l}} \cos\left(\frac{r c S_{\mathrm{ex}}}{\hbar|e|B} - \frac{\pi}{4l}\right) \cos\left(\pi r \frac{m_\perp}{m}\right).$$

$$\qquad (7.61)$$

Thus the presence of the extreme cross-section of zero curvature significantly changes the shape of the oscillations and increases their magnitude. It also leads to a change in ratio of the amplitudes of the oscillating terms X and X'. In a metal with the spherical Fermi surface, the ratio Re X'/Re X is of the order of $q v_F \sqrt{\theta}/\omega\gamma$. In a metal with the Fermi surface nearly cylindrical in shape in the vicinity of the center extreme cross-section we have that Re X'/Re X is of the order of $\theta^{1/2l}(q v_0/\omega)^{1/(2l-1)}\gamma^{-1/l}$. At last, the inequalities defining the conditions of occurrence of the quantum waves now have the form

$$\frac{\hbar\Omega}{T}\left(\frac{\omega}{q v_0}\right)^{1/(2l-1)} > 1, \qquad \frac{q l^*}{\gamma^2}\left(\frac{q v_0}{\omega}\right)^{(l-1)/(2l-1)} > 1, \qquad (7.62)$$

where $l^* = v_0 \tau$. A comparison of (7.62) and (7.52) shows that these quantum waves are more available for experimental observations in metals whose Fermi surfaces include nearly cylindrical segments.

It was shown before that a characteristic feature of the effects arising in the presence of the external magnetic field, due to the local geometry of the Fermi surface, is the dependence of their manifestations on the direction of the magnetic field. We can expect that a similar dependence will exhibit itself for amplitudes of quantum oscillations of the velocity and absorption coefficient of ultrasound waves in a high-frequency range.

Local anomalies of curvature at extreme sections of the Fermi surface affect the amplitude of giant quantum oscillations of ultrasound absorption, because the absorption involves electrons whose velocity component in the direction of the sound-wave propagation coincides with the velocity of sound. When sound propagates along the direction of the applied magnetic field **B**, such electrons are concentrated in the neighborhood of the extremum cross-sections of the Fermi surface by planes perpendicular to the magnetic field. Should any of the extremum cross-sections coincide with a line of parabolic points, the relative number of electrons concentrated in the neighborhood of that section at the Fermi surface will be markedly greater than in the ordinary case where the Fermi surface has a finite, nonzero curvature. Accordingly, their contribution to the absorption of sound-wave energy must be greater.

When the direction of the field is changed, the extreme cross-section on the Fermi surface by a plane perpendicular to **B** will move out of coincidence with the line of parabolic points. Accordingly, this must lead to a reduction in the relative number of electrons contributing to the absorption of sound-wave energy. Hence, the amplitude of giant quantum oscillations in ultrasonic absorption, associated with a quasi-cylindrical segment on the Fermi surface, must strongly decrease.

The principal qualitative features, in the way the amplitude of giant oscillations in ultrasound absorption varies with the direction of the magnetic field relative to the crystallographic axes of a singe-crystal sample, can be analyzed within a fairly simple model of a metal with a single-connected axially symmetric Fermi surface that has only one extremal section for any direction of the applied magnetic field.

Suppose that the extreme cross-sectional area is described by expression (7.19) and $k > 1$. The condition $k > 1$ ensures that the curvature at the central section of the Fermi surface goes to zero when **B** is directed along the axis of symmetry of the Fermi surface.

In reality, the shape and height of absorption peaks are to a great extent determined by the character and intensity of the electron collisions. However, the behavior of the oscillation amplitude with the angle (Φ) considered here must manifest itself regardless of the broadening of the absorption peaks caused by electron scattering. For this reason, in the case at hand, one may limit oneself to the collision-free limit, and this will significantly simplify the calculations.

The coefficient of absorption for a longitudinal ultrasonic wave with a wave vector q and a frequency $\omega = qs$, propagating along the applied magnetic field, may be written as

$$\Gamma = \frac{\pi L_{zz}^2 q}{2\rho_m s^2} Y, \tag{7.63}$$

where ρ_m is the mass density of the lattice and Y equals the imaginary part of the quantity F_q defined by (7.43), which has to be calculated in the collisionless limit

$\tau \to \infty$ (see [27]):

$$Y = \frac{2}{(2\pi\hbar)^2} \frac{|e|B}{cg} \sum_{n,\sigma} \int dp_z (f^\sigma_{np_z - \hbar q} - f^\sigma_{np_z}) \delta(E^\sigma_{np_z} - E^\sigma_{np_z - \hbar q} - \hbar\omega). \quad (7.64)$$

In the long-wave limit, when the wavelength of sound is considerably greater than the radius of the cyclotron orbit of electrons related to an extremal section of the Fermi surface, the quantity Y may be written as

$$Y = C(\Phi)\Delta'. \quad (7.65)$$

The function Δ' in (7.65) describes the giant quantum oscillations of ultrasonic attenuation

$$\Delta' = 1 + 2\sum_{r=1}^{\infty}(-1)^r \psi_r(\theta) \cos\left(\frac{rcS_{ex}}{\hbar|e|B}\right) \cos\left(\frac{\pi r\Omega_0}{\Omega}\right). \quad (7.66)$$

The dimensionless amplitude factor $C(\Phi)$ is

$$C(\Phi) = \frac{s/v_m}{a(\Phi) + w(s/v_m)^{1-\beta}}, \quad (7.67)$$

where $w = (\beta + 1)/[2\beta^2(akb)^\beta]$, $a = S_{ex}/\pi p_m v_m m_\perp$ are positive dimensionless constants of the order of unity, v_m is the maximum value of the longitudinal conponent of velocity, and $\beta = 1/(2k - 1)$.

Precisely this factor describes how the amplitude of oscillations of the ultrasonic attenuation varies with the angle that the magnetic field makes with the axis of symmetry of the Fermi surface. For $\Phi = 0$, when the absorption is due to the electrons from the neighborhood of a quasi-cylindrical extremal section of the Fermi surface, $C(\Phi)$ takes on a maximum value

$$C(0) = \frac{1}{w}\left(\frac{s}{v_m}\right)^\beta, \qquad (0 < \beta < 1). \quad (7.68)$$

An increase in Φ is accompanied by an increase in $a(\Phi)$, and this leads to a decrease in the amplitude factor $C(\Phi)$. When the angle Φ becomes equal to or greater than Φ_0 the second term in the denominator of (7.67) becomes negligible in comparison with the first term, and $C(\Phi)$ takes on a value close to (s/v_m), which characterizes the amplitude of giant quantum oscillations of ultrasonic absorption related to an extremal section of the Fermi surface with a finite and nonzero curvature.

The ratio of the amplitude of oscillations of the sound attenuation rate which involves electrons from the neighborhood of a quasi-cylindrical extremal section to the amplitude of "normal" giant quantum oscillations is equal to about $(s/v_m)^{\beta-1}$. In typical metals, the velocity of electrons at the Fermi surface is two or three orders of magnitude greater than the velocity of sound, which is why this ratio must be significantly greater than unity. Thus, as the angle Φ changes, the amplitude of oscillations also varies over a fairly broad range, a thing that can readily be detected experimentally.

When the behavior of the amplitude of giant quantum oscillations in ultrasonic absorption is observed experimentally, the value of Φ_0, which characterizes the shape of the Fermi surface in the neighborhood of its central section, can be determined from experimental data.

7.3   Quantum Oscillations of the Velocity and Absorption Coefficient of Ultrasound as it Interacts with a Helical Wave in a Compensated Metal

The experimental results of [167], [168], [57] on the quantum oscillations of the ellipticity and the angle of rotation of the polarization plane of transverse ultra-sound, propagating in the direction of the magnetic field in tungsten, revealed substantial anomalies in these oscillations. It turned out that in strong fields, when the magnitude of B is considerably higher than the threshold values for Doppler-shifted cyclotron resonances for all groups of charge carriers, the amplitudes of the oscillations of the polarization parameters do not decrease with growth of the field for this range, in accordance with simple formulas, but have nonmonotonic dependences on B and are quite large. Nor can the phases of the oscillations be explained by the asymptotic formulas for strong fields. These anomalies might be associated with the influence of the interaction of ultrasound with the helical wave predicted in [166] for a compensated metal.

Here we give a full theoretical description of the quantum oscillations of the parameters of circularly polarized transverse ultrasounds in a compensated metal under conditions of the occurrence of helical waves. The calculation starts from the equations for the amplitudes $u_\pm$ of the circularly polarized components of the lattice displacement vector which behave as $\exp(-i\omega t + iqz)$ in a magnetic field applied in the direction of the high-order symmetry axis which here coincides with the z-axis of the coordinate system

$$- \omega^2 \rho_m u_\pm = -\lambda q^2 u_\pm \mp i \frac{B}{c} \overline{J}_\pm + iq \overline{\Lambda}_\pm. \tag{7.69}$$

Here ρ_m and λ are the density and corresponding elastic modulus of the lattice, and $\overline{J}_\pm$ and $\overline{\Lambda}_\pm = \overline{\Lambda}_{xz} \pm \overline{\Lambda}_{yz}$ are the amplitudes of the components of the current density and a mean value of the deformation potential tensor. The values $\overline{J}_\pm$ and $\overline{\Lambda}_\pm$ are expressed in terms of the amplitudes of the components of the displacements $u_\pm$ and the strength of the electrical field $E_\pm$ by the relations

$$\overline{J}_\pm = \sigma_\pm \left(E_\pm \mp \omega \frac{B}{c} u_\pm \right) + \omega q \beta_\pm u_\pm, \tag{7.70}$$

$$\overline{\Lambda}_\pm = \beta_\pm \left(E_\pm \mp \omega \frac{B}{c} u_\pm \right) + \omega q \alpha_\pm u_\pm, \tag{7.71}$$

which contain the electronic kinetic coefficients $\alpha_\pm$, $\beta_\pm$, and $\sigma_\pm$. Eliminating $E_\pm$ from (7.70), (7.71) and using the Maxwell equations

$$E_\pm = \frac{4\pi i \omega}{c^2 q^2} \overline{J}_\pm, \tag{7.72}$$

we obtain the dispersion equation for the wave vectors $q_\pm$ of the elastic waves with frequency ω. Since the electronic system has a fairly weak influence on sound, $q_\pm$ can be put in the form of a basic term $\omega\sqrt{\rho_m/\lambda} \equiv \omega/s$, and small additions $\delta q_\pm$, for which we can write

$$\Delta q_\pm = \frac{q}{2\lambda} i\omega\alpha_\pm + \frac{B^2 q}{8\pi\lambda}\left[\frac{\left(1 \mp 4\pi i\omega\beta_\pm/cqB\right)^2}{1 - 4\pi i\omega\sigma_\pm/c^2 q^2} - 1\right], \qquad (7.73)$$

where q is assumed to be equal to ω/s.

To calculate the kinetic coefficients, we use the kinetic equation derived within a framework of the quantum theory of the propagation of ultrasound in a metal omitting terms which describe the Fermi-liquid interaction of electrons (which is insignificant in this case). In calculating the values $J_\pm$ and $\overline{\Lambda}_\pm$, we perform averaging with the perturbed matrix of the electronic density ρ. When considering quantum effects it is important that ρ contains a term to describe the local equilibrium of the electronic system in the field of the sonic wave. So the expressions for the kinetic coefficients include locally equilibrium and dynamic parts

$$\alpha_\pm = \alpha^L + \ll \Lambda_{xz}, \Lambda_{xz} \pm i\Lambda_{yz} \gg, \qquad (7.74)$$

$$\beta_\pm = \beta_\pm^L + \ll \Lambda_{xz}, j_x \pm ij_y \gg, \qquad (7.75)$$

$$\sigma_\pm = \sigma^L + \ll j_x, j_x \pm ij_y \gg, \qquad (7.76)$$

where, for example,

$$\ll \Lambda_{xz}, \Lambda_{xz} \gg = -\sum_{\nu\nu'} \frac{f_\nu - f_{\nu'}}{E_\nu - E_{\nu'}} \frac{\Lambda_{\nu'\nu}^{xz}(-\mathbf{q})\Lambda_{\nu'\nu}^{xz}(\mathbf{q})}{\gamma + (i/\hbar)(E_\nu - E_\nu')}. \qquad (7.77)$$

Expressions for $\ll \Lambda_{xz}, \Lambda_{xz} \pm \Lambda_{yz} \gg$ and $\ll \Lambda_{xz}, j_x, \pm ij_y \gg$ can be obtained from (7.77) as a result of replacing the matrix elements $\Lambda_{\nu'\nu}^{xz}(\mathbf{q})$ by $\Lambda_{\nu'\nu}^{xz}(\mathbf{q}) \pm i\Lambda_{\nu'\nu}^{yz}(\mathbf{q})$ or by $j_{\nu'\nu}^x(\mathbf{q}) \pm ij_{\nu'\nu}^y(\mathbf{q})$, respectively. To arrive at the expression for $\ll j_x, j_x \pm ij_y \gg$ we have to replace the matrix elements $\Lambda_{\nu\nu'}^{xz}(-\mathbf{q})$ by $j_{\nu\nu'}(-\mathbf{q})$ in the expression for $\ll \Lambda_{xz}, j_x \pm ij_y \gg$. The local equilibrium terms in (7.74)–(7.76) can be written as follows:

$$\alpha^L = \frac{1}{i\omega}(\Lambda_{xz}, \Lambda_{xz}), \qquad (7.78)$$

$$\beta_\pm^L = \pm\frac{1}{\omega}(\Lambda_{xz}, j_y), \qquad (7.79)$$

$$\sigma^L = c^2 q^2 \chi / i\omega, \qquad (7.80)$$

where χ is the static transverse magnetic susceptibility of the electrons. The variables σ_L and $\beta_\pm^L$ are purely quantum terms and are associated, respectively, with the magnetization current and the influence of the magnetic field on the elastic moduli, α^L describes the contributon of electrons to the static elastic moduli. If quantization of the electronic energy is insignificant, the terms σ^L and $\beta_\pm^L$ become zero and the value of α^L is independent of $\mathbf{B}$. Thus, within the framework of the semiclassical approach, the locally equilibrium electron system can influence the lattice only implicitly by renormalization of the static elastic moduli and this renormalization does not depend on the external magnetic field.

We consider the strong field range, in which the characteristic values of the electronic displacement d in the field direction in the cyclotronic period are small compared to the wavelength of the ultrasound ($qd \ll 1$). We shall also assume that the characteristic free path length of electrons l is large compared to the wavelength of the ultrasound ($ql \gg 1$). We do the calculations for a compensated metal with a closed Fermi surface, allowing as usual that the number of occupied Landau levels is large and $\hbar\Omega \ll \zeta$.

The kinetic coefficients can be written in the form of sums of the semiclassical variables and oscillating quantum additions. Consider at first the semiclassical parts of the asymptotic expressions for $\alpha_\pm$, $\sigma_\pm$, and $\beta_\pm$. In calculations we can omit the local equilibrium terms in (7.74), (7.75) and transform expression (7.77) to the form

$$\ll \Lambda_{xz}, \Lambda_{xz} \gg = \frac{1}{2\pi^2\hbar^3} \int dp_z m_\perp \sum_n \frac{\Lambda_{-n}^{xz}(p_z, -q)\Lambda_n^{xz}(p_z, q)}{\gamma + iq\langle v_z\rangle + in\Omega}. \tag{7.81}$$

To obtain the desired asymptotic expressions we can expand the terms with $n \neq 0$ included in (7.81) in powers of the parameter $(\gamma + iq\langle v_z\rangle)/\Omega$ and the term with $n = 0$ in powers of $\gamma/q\langle v_z\rangle$. As a result, we arrive at the well-known formula for the conductivity

$$\sigma_\pm = i\frac{c^2 q^2}{4\pi}\left(\pm\frac{1}{\omega_s} - i\tau_s\right), \tag{7.82}$$

where

$$\pm\frac{1}{\omega_s} - i\tau_s = \frac{4\pi}{c^2}\frac{e^2}{4\pi^2\hbar^2} \int dp_z \sum_{n\neq 0} \frac{|m_\perp|}{in\Omega}v_{-n}^x\left[\pm\left(\frac{\langle v_z\rangle}{n\Omega}\right)^2 v_n^y + \frac{1}{q^2}\frac{\gamma}{n\Omega}v_n^x\right]. \tag{7.83}$$

Here v_{-n}^x, v_n^y are the Fourier transforms of the corresponding velocity components in their expansions in the variable ψ, ω_s is the frequency of the helical wave, and the real part of the complex parameter τ_s characterizes attenuation. The first and second terms in (7.83) respectively, determine ω_s and τ_s, of the wave. However, expression (7.83) does not contain contributions from the collisionless part of the conductivity. Such contribution can appear due to pulsations of the longitudinal velocity in the cyclotron orbit. Because of these pulsations (dependence v_z on ψ) the Fourier components of $v_0^x(p_z, q)$ and $v_0^y(p_z, q)$ do not tend to zero. For small q we have

$$v_0^x(p_z, q) \approx -iq\frac{c}{eB}\Pi_{yz}(p_z),$$

$$v_0^y(p_z, q) \approx iq\frac{c}{eB}\Pi_{xz}(p_z). \tag{7.84}$$

Here $v_0^x(p_z, q)$ and $v_0^y(p_z, q)$ are expressed in terms of the components of the momentum flux tensor averaged over the cyclotron orbit, $\Pi_{xz} = \langle p_x \Delta v_z\rangle$,

$\Pi_{yz} = \langle p_y \Delta v_z \rangle$, and the difference Δv_z characterizes the pulsation of the longitudinal velocity on the cyclotron orbit. For some orbits these averages are nonzero and this is the condition of appearance of the effects being considered here. The collisionless part of the parameter τ_s originates from the term with $n = 0$ in the sum over n in (7.83). In the limit, $(\gamma/(q\langle v_z \rangle)) \to 0$, this part has the form

$$\tau_s^0 = \frac{4\pi}{B^2} \frac{|m_\perp| m_{||}}{2\pi\hbar^3 q} (\Pi_{xz})^2 |_{\langle v_z \rangle = 0}, \tag{7.85}$$

where $m_{||} = (\partial\langle v_z \rangle / \partial p_z)^{-1}$. In (7.85) we imply summation over all cavities of the Fermi surface. If the collision-free contribution τ_s^0 satisfies the condition $\omega_s \tau_s^0 \ll 1$, then a weakly attenuating helical wave can exist (provided the collision frequency of the electrons τ_s^{-1} is small enough) [166]. This condition can only be satisfied when the Fermi surface of the metal has definite geometric properties. The analysis of the experimental results on dopplerons in tungsten and molybdenum made in [169] showed that in these metals the collision-free part of the conductivity is negligibly small, that is, the basic condition for the appearance of helical waves can be satisfied.

Under the same approximations as (7.82) the semiclassical expression for $\alpha_\pm$ can be written in the form

$$\alpha_\pm = \alpha' \pm i\alpha'' = \frac{|m_\perp| m_{||}}{2\pi\hbar^3 q} (\Lambda_0^{xz}) |_{\langle v_z \rangle = 0}$$

$$\pm \frac{1}{2\pi\hbar^3} \int dp_z \sum_{n \neq 0} \frac{|m_\perp|}{in\Omega} \Lambda_{-n}^{xz}(p_z, 0) \Lambda_n^{yz}(p_z, 0). \tag{7.86}$$

The semiclassical parts of the kinetic coefficients $\beta_\pm$ include four terms:

$$\frac{\mp 4\pi i\omega\beta_\pm}{cqB} = i(\eta_0'' + \eta_1'') \mp (\eta_0' + \eta_1'). \tag{7.87}$$

Here two terms originate from the pulsations of the longitudinal velocity

$$\mp \eta_0' + i\eta_0'' = \frac{4\pi}{B^2} \times \left\{ \mp \frac{1}{2\pi^2\hbar^3} \int dp_z \sum_{n,n' \neq 0} \frac{m_\perp^2}{nn'\Omega} \Lambda_{-n}^{xz}(p_z, 0) \Delta v_z^{n-n'} v_{n'}^x \right.$$

$$\left. + i\frac{|m_\perp| m_{||}}{2\pi\hbar^3 q} \Lambda_0^{xz} \Pi_{xz} |_{\langle v_z \rangle = 0} \right\}, \tag{7.88}$$

and the remaining terms are not connected with these pulsations

$$\mp \eta_1' + i\eta_1'' = \frac{4\pi\omega}{B^2} \frac{1}{2\pi^2\hbar^3} \int dp_z \sum_{n \neq 0} \frac{m_\perp^2 \langle v_z \rangle}{n^2\Omega} \Lambda_{-n}^{xz}(p_z, 0) \left(\mp v_n^x + i\frac{2\gamma v_n^y}{in\Omega} \right). \tag{7.89}$$

According to the obtained asymptotics all the terms included into the kinetic coefficients besides α' decrease with the growth of the magnetic field: $\alpha'' \sim 1/B$; $\tau_s, \eta_0'' \sim 1/B^2$; $1/\omega_s, \eta_0', \eta_1' \sim 1/B^3$; $\eta_1'' \sim 1/B^4$. Correspondingly, the electron contributions to the wave vectors of circularly polarized waves $\Delta q_\pm$

decrease when the magnetic field increases. For strong magnetic fields we have

$$\operatorname{Re} \Delta q_{\pm} = q \frac{\omega}{2\lambda} \left\{ \frac{B^2}{4\pi\omega} \eta_1' \left(\eta_1'' + 2\frac{\omega}{\omega_s} \right) \mp \left[\alpha'' + \frac{B^2}{4\pi\omega} \left(-\frac{\omega}{\omega_s} + 2\eta_0' + 2\eta_1' \right) \right] \right\},$$
(7.90)

$$\operatorname{Im} \Delta q_{\pm} = q \frac{\omega}{2\lambda} \left\{ \alpha' + \frac{B^2}{4\pi\omega} (\tau_s + 2\eta_0'' + 2\eta_1' \mp 2\eta_1'\tau_s) \right\}.$$
(7.91)

These formulas can be applied for $B \gg B_s$ where B_s is the resonance value of the magnetic field defined by the relation $\omega/\omega_s = (B_s/B)^3$. When B is of the order of B_s and larger than the threshold values for Doppler-shifted cyclotron resonances, one of the circularly polarized ultrasound waves interacts with the helical wave. It is supposed that $\omega_s \tau_s' \ll 1$, so that the helical wave can exist. For $B \sim B_s$ we can neglect small collisionless contributions to $\sigma_{\pm}$ and $\beta_{\pm}$ and the correction proportional to the small parameter η_1''. Also we can omit the term $\alpha_{\pm}$ which does not depend on the interaction between the ultrasound and helical waves. As a result, we can simplify the expression for the resonance part of the correction $\Delta q_{\pm}$ and represent the latter in the form

$$\Delta q_{\pm}^s = \frac{B^2 q}{8\pi\lambda} \frac{(1 \mp \omega D/\omega_s)^2}{1 \mp \omega/\omega_s - i(\omega\tau_s)},$$
(7.92)

where D is the dimensionless parameter, characterizing the value of the component Λ_{xz} of the deformation potential tensor:

$$D = -\frac{\int dp_z \sum_{n\neq0} \frac{|m_\perp|^3}{n^2} \langle v_z \rangle \Lambda_{-n}^{xz}(p_z, 0) v_n^x}{\int dp_z \sum_{n\neq0} \frac{|m_\perp|^3 m_\|}{in^3} \langle v_z \rangle^2 v_{-n}^x v_n^y}.$$
(7.93)

When $\omega_s \tau_s' \ll 1$ the magnetic field or frequency dependence of $\Delta q_{\pm}^s$ is of resonance character (see Figure 7.3).

However, when $\omega_s \tau_s'$ is of the order of unity the coupling of ultrasound to the damping helical wave also produces noticeable polarization effects, as is clear from Figure 7.4.

Now let us turn to a consideration of quantum oscillating additions to the kinetic coefficients, restricting ourselves in the calculations to the principal terms of expansions in powers of the small parameters given above. Under this restriction we have to take into account only those contributions which originate from those parts of the sums in (7.77)–(7.80) which are diagonal with respect to the oscillator quantum number "n".

The experiments [167], [168], [58] were done on tungsten, for which we have fairly full information on the Fermi surface. The magnetic field was in the direction of the [001] axes so that the x- and y-axes of the laboratory system of coordinates are best taken along [100] and [010].

The Fermi surface of tungsten consists of an electron jack, centered at point Γ of the Brillouin zone, a hole octahedron at point H, and small hole ellipsoid at points N. The extremal sections correspond to the central orbits τ and ν on the

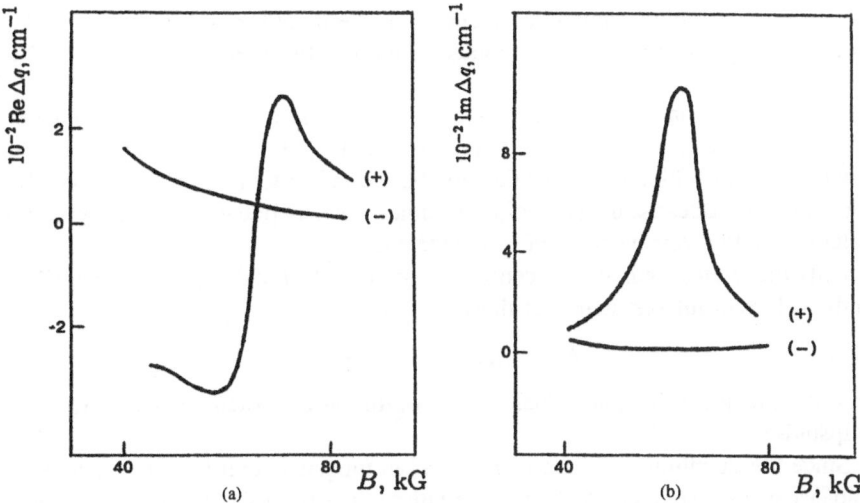

FIGURE 7.3. The dynamical corrections to the wave vectors of circularly polarized ultrasonic waves $\Delta q_{\pm}$ versus a magnetic field in the region of their coupling to a weakly damping helical wave.

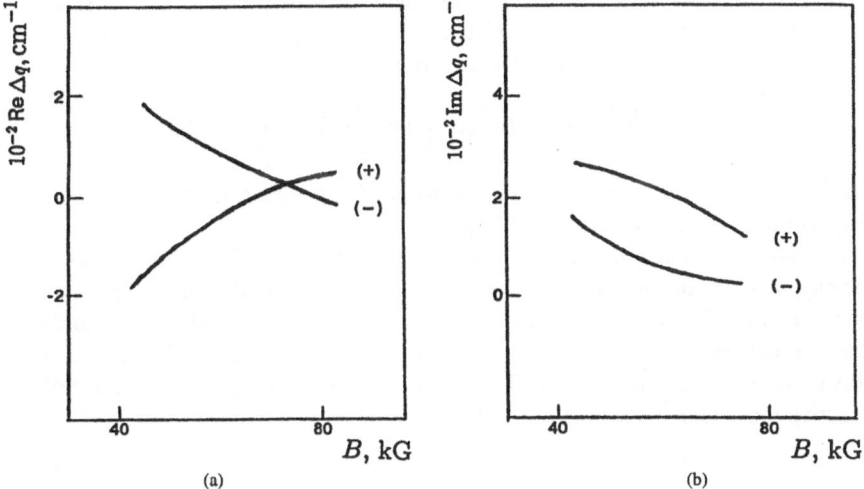

FIGURE 7.4. The dynamical corrections to the wave vectors of circularly polarized ultrasonic waves versus a magnetic field in the region of their coupling to a strongly damping helical wave.

jack and octahedron, orbits σ and π on the neck and spheroid of the jack, orbit ρ_1 on the ellipsoids, the centers of which lie on a plane $p_z = 0$, and orbits ρ_2 on the ellipsoids with centres which do not lie on this plane. Orbits τ, ν, and ρ_1 lie on the plane of symmetry perpendicular to the z-axis and, because of that, the component

Λ_{xz} of the deformation potential is zero. For the σ and π -orbits $\Lambda_{xz} \neq 0$, but since the z-axis is an axis of symmetry of the fourth order, $(\Lambda_{xz})_{extr}$ is equal to zero.

The contribution $\Delta q_{\pm}$ can only give ρ_2 orbits. These belong to the ellipsoids, whose centers can be taken to lie on the axes [101], [011], [$\bar{1}$01], [0$\bar{1}$1]. The principal axes of the ellipsoids coincide with the second-order axes of symmetry NP, $N\Gamma$, and NH. Because the ellipsoids are small, the dependence of the deformation potential on the wave vector can be neglected.

Only those ellipsoids whose centers lie on the [101] and [$\bar{1}$01] axes contribute to the values of interest here. For these

$$\Lambda_{xz} = (\Lambda_2 - \Lambda_3)/2 \equiv \Lambda_e$$

(Λ_2, Λ_3 are the principal values of the deformation potential tensor for hole ellipsoids).

Since the quantum oscillations observed in tungsten stem from the hole ellipsoids we need only calculate their contributions. The basic oscillatory correction to the local equilibrium term α^L equals $\lambda'_e \Delta / i\omega$ where $\lambda'_e = 2g\Lambda_e^2$ (g is the density of states on a unique ellipsoid), and the function Δ describes the quantum oscillations of the density of states on the Fermi surface. Oscillations of the dynamic part of the kinetic coefficient $\alpha_{\pm}$ are of a more complicated character (see the preceding sections of this chapter). However, under the condition $qv_m/\omega \gg \sqrt{\zeta/\hbar\Omega}$ which is satisfied in the experiments, the principal oscillating term in the expression for $\ll \Lambda_{xz}, \Lambda_{\pm} \gg$ is proportional to the function

$$\Delta' = 2\sum_{r=1}(-1)^r \psi_r(\theta) \cos\left(2\pi r \frac{\zeta}{\hbar\Omega}\right) \cos\left(\pi r \frac{m_\perp}{m}\right), \qquad (7.94)$$

which describes giant quantum oscillations of the absorption coefficient. The proportionality factor equals $\pi\lambda'_e/2qv_m$.

The basic oscillatory correction to the coefficient $\beta_{\pm}$ is also proportional to λ_e. Besides, it is proportional to the average value of an off-diagonal component of the momentum flux tensor. In the geometry of experiments, two hole ellipsoids are characterized nonzero and independent of the p_z average value of $\Pi_{xz}(\Pi_{xz} \equiv \Pi_e)$. This constant Π_e can be expressed in terms of ζ and in the principal values of the effective mass tensor for hole ellipsoids

$$\Pi_e = \zeta \frac{m_3 - m_2}{m_3 + m_2}. \qquad (7.95)$$

The quantity Π_e is nonzero by virtue of pulsations in the z-component of the carrier velocity in the cyclotron orbit.

Thus, the kinetic coefficients $\beta_{\pm}$ and $\sigma_{\pm}$ include oscillating corrections similar to (7.20) and (7.94), which arise due to the geometry of the Fermi surface of tungsten. The result calculated for the quantum correction to the kinetic coefficient $\beta_{\pm}$ can be written as follows:

$$\Delta\beta_{\pm} = \pm i \frac{cq}{B} \lambda'_e \mu \left(\Delta + i \frac{\pi}{2} \frac{\omega}{qv_m} \Delta'\right), \qquad (7.96)$$

where $\mu = \Pi_e / \Lambda_e$.

The quantity μ is small ($\sim 10^{-2}$) because the dimensions of the hole ellipsoids are relatively small. Accordingly, if we retain in $\Delta q_{\pm}$ only those corrections which are linear in μ, we can ignore the quantum increments in the conductivity $\sigma_{\pm}$. The latter are proportional to Π_e^2 and describe the carrier component giant oscillations in the absorption of energy of the electromagnetic field.

In a description of the quantum oscillations, a matter of fundamental importance is to take into account the oscillating increment in the kinetic coefficient $\beta_{\pm}$. It follows from (7.92) that this term provides manifestations of the effects connecting with the interaction of ultrasound with helical waves, which cause the observed anomalies in quantum oscillations of the ultrasound velocity and absorption coefficient. The principal term in the semiclassical expression for $\beta_{\pm}$ can be written in the form

$$\beta_{\pm} = -i \frac{cq}{B} \frac{1}{2\pi^2 \hbar^3} \int dp_z \sum_{n \neq 0} \frac{m_{\perp}^2 \langle v_z \rangle^2}{n^2 \Omega} \Lambda_{-n}(p_z, 0) v_n^x. \tag{7.97}$$

As a result, we find the following expressions for the oscillating real and imaginary parts of the quantum increments in the wave vectors of the circularly polarized waves:

$$\text{Re } \Delta \tilde{q}_{\pm} = \frac{\lambda_e' q}{2\lambda} \left\{ (1 + 2\mu \Phi_{\pm}') \Delta - \frac{\pi}{2} \frac{\omega}{q v_m} 2\mu \Phi_{\pm}'' \Delta' \right\}, \tag{7.98}$$

$$\text{Im } \Delta \tilde{q}_{\pm} = \frac{\lambda_e' q}{2\lambda} \left\{ 2\mu \Phi_{\pm}'' \Delta + \frac{\pi}{2} \frac{\omega}{q v_m} (1 + 2\mu \Phi_{\pm}') \Delta' \right\}. \tag{7.99}$$

Here $\Phi_{\pm}'$ and $\Phi_{\pm}''$ are the real and imaginary parts of the function

$$\Phi_{\pm} = \frac{1 \mp D\omega/\omega_s}{1 \mp \omega/\omega_s - i(\omega\tau_s)}, \tag{7.100}$$

and the dimensionless parameter D is defined by (7.93).

According to formulas (7.98), (7.99), the difference between the amplitudes and phases of the oscillations of the velocities or, the coefficients of absorption of waves of different polarizations is due to the contribution of the term containing the function $\Phi_{\pm}$. The differences $\text{Im } \Delta q_+ - \text{Im } \Delta q_-$ and $\text{Re } \Delta q_+ - \text{Re } \Delta q_-$, owing to the smallness of the parameter μ have an appreciable value, which are consistent with the experimental data provided that $\omega/\omega_s \sim 1$, which corresponds to the interaction between sound and the helical wave. This explains both the increase of the amplitudes of the oscillations and their nonmonotonic dependence on B (due to the corresponding dependence $\Phi_{\pm}$), as well as the phase shift of the oscillations of the parameters (especially the absorption coefficients) of different modes which arise due to the superposition of the functions Δ and Δ' with different coefficients.

The difference between Φ_+ and Φ_-, which appears due to the coupling of the ultrasonic wave with the polarization "+", to the helical wave reaches its maximum at the crossover point of the dispersion curves of these waves, when $\omega = \omega_s$. If the condition $\omega\tau_s \ll 1$ is held, the helical wave would have been slightly damped and the interaction would have been resonant. The interaction would have occurred in a narrow field interval near the resonant field. In this interval $|\Phi_+|$ would exceed $|\Phi_-|$ by several orders in magnitude. Even for $\omega_s \tau_s' \sim 1$ the interaction between the ultrasound and damped helical wave would give considerable polarization effects.

We can use the obtained results to estimate the difference in amplitudes and phases for the quantum oscillations of the velocity and absorption coefficients of ultrasonic waves of different polarizations. We will first give numerical values for the quantities which appear in the amplitude factors in (7.98), (7.99). Among these quantities are the effective masses and the Fermi energy for hole ellipsoids: $m_1 = 0.54m$, $m_2 = 0.29m$, $m_3 = 0.22m$, $\zeta = 4.26 \cdot 10^{-13}$ erg, the deformation constant $\Lambda_e = 8 \cdot 10^{-12}$ erg, and the elastic modulus $\lambda = 1.5 \cdot 10^{12}$dyn/cm^2 [168]. For the ultrasound frequency 196 MHz, T=2.5 K, and the magnetic field $\sim$ 60–70 kG, (7.98) gives us the amplitude of the quantum oscillations in Re $\Delta q_\pm$ of the order of 2 cm^{-1}. Taking into account that the amplitude of the quantum oscillations described by the function Δ' is approximately one-fourth of the amplitude of the oscillations described by the function Δ, we can expect the quantum oscillations in Im $\Delta_\pm$ to be smaller in amplitude than the oscillations in Re $\Delta q_\pm$. However, the difference in amplitudes between Re $\Delta q_\pm$ and Im $\Delta q_\pm$ will be smaller than that between Δ and Δ', because of the term proportional to Δ which is included into expression (7.99) for the imaginary part of $\Delta q_\pm$. These estimates agree with the experimental data of [168].

This proportional to the function Δ term in (7.99) can cause a significant difference in the phases of oscillations of Im Δq_+ and Im Δq_-. In the magnetic field interval corresponding to the interaction between the ultrasound and helical wave, this quantity $|\Phi_+''|$ can be large enough to provide the first term in (7.99) for Im Δq_+ to be of the same order as the second term, or to exceed it instead of a small value of the parameter $\mu(\mu \sim 0.02)$. At the same time the corresponding term in the expression for ImΔq_- remains small. For $|D| > 1$, the functions Φ_+'' and Φ_-'' near the interaction take values of different signs, therefore the phase shift can be very significant.

A concrete form of the magnetic field dependences of the oscillating contributions to $\Delta q_\pm$ is determined by the values of the dimensionless parameter D and the constants C_1 and C_2. These parameters contain new information about the Fermi surface shape, electron scattering and the deformation potential. They can be determined as a result of comparison of the observed oscillations and the theory. This comparison was performed in [58] and gave: $C_1 = 3.6 \cdot 10^{-6}$s^{-1}(G)$^{-3}$, $C_2 = 2.7$ s(G)^2, and $D = 12.7$. Substituting these values for the parameters C_1, C_2, and D into (7.98), (7.99) we can obtain a reasonable agreement between the theoretical $\Delta q_\pm(B)$ dependence and the experimental results, as is shown in Figure 7.5.

Thus, the anomalies in the quantum oscillations of the polarization parameters in tungsten arise due to the interaction between one of the circular by polarized ultrasonic waves and the helical wave. This interaction influences the quantum oscillations due to some special features in the geometry of the Fermi surface of tungsten.

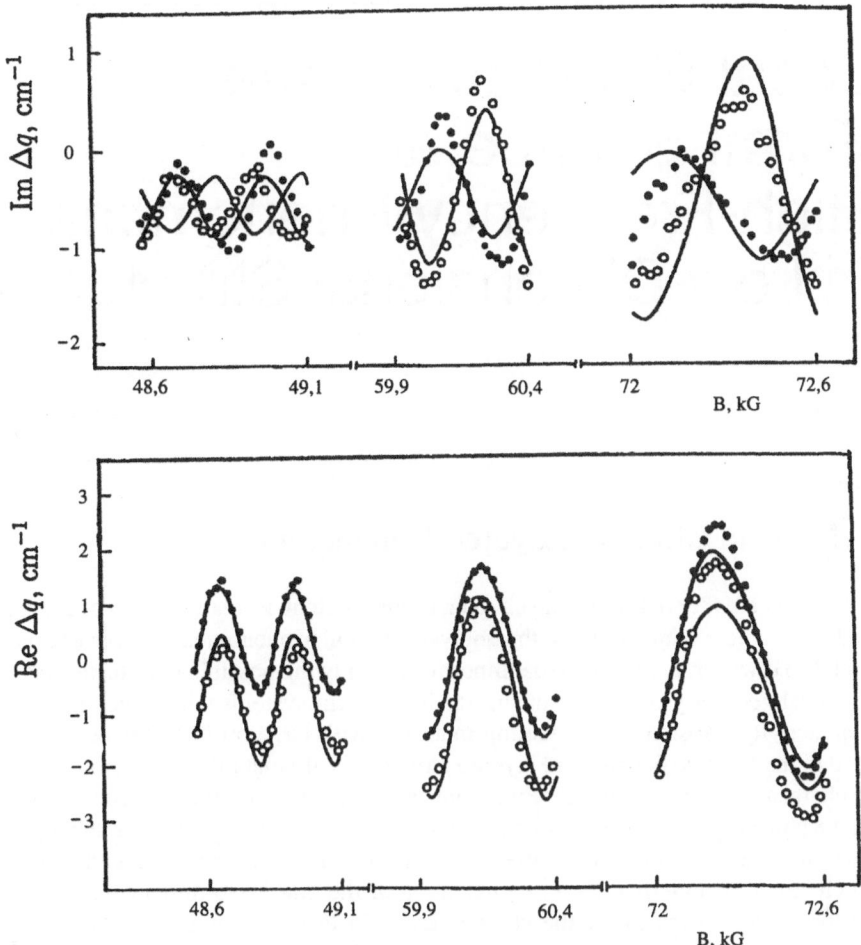

FIGURE 7.5. Magnetic field dependence of the (a) imaginary and (b) real parts of the oscillating contributions to the wave vector of the ultrasound from the experiments of [58] for the "−" and "+" polarization; solid lines—theoretical predictions from (7.98), (7.99).

CHAPTER 8

Local Geometry of the Fermi Surface and High-Frequency Phenomena in Low-Dimensional Structures

8.1   Skin Effect in Layered Conductors

The subject matter of the present chapter, to some extent, is related to that of the preceding chapters which contain the analysis of similar phenomena in conventional metals. However, it appears to be important to study the possible manifestations of the local geometry of the Fermi surface in different systems with a metallic-type conductivity. Most superconducting materials with large critical parameters created in the last two decades are layered structures with metallic-type conductivity. A characteristic feature of these materials is strong anisotropy of the conductivity in the nonsuperconducting state: the conductivity in the layer plane is much higher than in the direction normal to the layers. It is common to assume that anisotropy in electrical conductivity is a manifestation of the quasi two-dimensional nature of the energy spectrum of the charge carries in layered conductors. The Fermi surface of such conductors is a system of weakly rippled cylinders (isolated or connected by links) whose axes are perpendicular to the layers. The experimental data (see, e.g., [171]–[179]) support this assumption. However, study of the Fermi surfaces of layered conductors is far from completion. Many aspects of the profiles of periodically pinched cylinders have yet to be investigated. The local geometrical characteristics of the Fermi surface strongly affect the high-frequency properties of layered conductors, just as they do in ordinary metals. Here we study the effect of the local geometry of the Fermi surface on the skin effect in layered conductors. The exposure of the features of the skin effect, related to the specific geometrical characteristics of the Fermi surface, should create additional possibilities for reconstructing the Fermi surfaces of such materials from the experimental data.

The Fermi surface of a conductor with a quasi two-dimensional energy spectrum can be described by the following equation:

$$E(\mathbf{p}) = \sum_{k=0}^{\infty} E_k(p_x, p_y) \cos \frac{akp_z}{\hbar}, \tag{8.1}$$

where $\mathbf{p}$ is the electron quasi momentum, $E_k(p_x, p_y)$ are coefficients with dimensions of energy, p_z is the projection of the quasi momentum on the direction normal to the layers, and a is the distance between the layers. If we ignore the anisotropy of the energy spectrum in the layer plane, we can write the simpler equation, instead of (8.1):

$$E_{\mathbf{p}} = \frac{p_{\perp}^2}{2m_{\perp}} + \sum_{k=1}^{\infty} E_k(p_{\perp}) \cos \frac{akp_z}{\hbar}, \tag{8.2}$$

where $p_{\perp}$ is the projection of the quasi momentum on the layer plane, and $m_{\perp}$ is the effective mass corresponding to the motion of the quasi particles in that plane. Equation (8.2) describes an axisymmetric open Fermi surface with the axis directed along a normal to the layers.

The usual approach, in theoretical studies of the electron properties of layered conductors, is to keep only the first few terms in the sum over k in (8.2). As a rule, only the first term is taken into account, which corresponds to results obtained in the tight-binding approximation. Here we use a different approach to describe the electron energy spectrum of the charge carriers in layered conductors, whose Fermi surface is defined by the equation

$$E(\mathbf{p}) = \frac{p_{\perp}^2}{2m_{\perp}} + \eta v_0 p_0 E\left(\frac{p_z}{p_0}\right), \tag{8.3}$$

where $v_0 = (2\zeta/m_{\perp})^{1/2}$, $p_0 = \pi\hbar/a$, $E(p_z/p_0)$ is an even function periodic in its argument p_z/p_0 with a period equal to 2, and η is a dimensionless parameter characterizing the rate of rippling of the Fermi surface. The quantity $-\eta v_0 p_0 E(p_z/p_0)$ is the sum of the trigonometric series in (8.2). By selecting this type of function we can obtain Fermi surfaces shaped as pinched cylinders with different profiles. This approach provides broad possibilities in analyzing the effect of the shape of the Fermi surface on the observed characteristics of layered conductors.

Let us assume that the function $E(p_z/p_0)$ in the interval $-p_0 \le p_z \le p_0$ is described by the expression

$$E\left(\frac{p_z}{p_0}\right) = \frac{1}{rl}\left[1 - \left|\frac{p_z}{p_0}\right|^l\right]^r, \tag{8.4}$$

where the parameters l and r take values greater than unity. The model specified by (8.3) and (8.4) makes it possible to describe a broad class of Fermi surfaces.

The Gaussian curvature of the Fermi surface is described by (3.7). At $l = r = 2$ the curvature of the Fermi surface at its sections by the planes $p_z = 0$ and $p_z = \pm p_0$, equals

$$K(0) = \frac{\delta S}{S_{\max}} \frac{1}{p_0^2}, \tag{8.5}$$

$$K(\pm p_0) = -\frac{2\delta S}{S_{\min}} \frac{1}{p_0^2}, \tag{8.6}$$

where $S_{\max}$ and $S_{\min}$ are the maximum and minimum sectional areas of the Fermi surface: $S_{\max} = S(0)$, $S_{\min} = S(\pm p_0)$, and $\delta S = S_{\max} - S_{\min} = (\pi/2)m_\perp \eta v_0 p_0$. Thus, if the Fermi surface remains a pinched cylinder ($\eta \neq 0$), its curvature at all points of the sections with extremal diameters is finite. Similar results are obtained if the tight-binding approximation is used to describe the electron energy spectrum.

For $r \neq 2$ and $l = 2$, the curvature of the Fermi surface near $p_z = \pm p_0$ remains finite and $K(0)$ is still described by (8.6). However, the asymptotic behavior of the curvature of the Fermi surface near $p_z = \pm p_0$ is different:

$$K(p_z) = -2(r-1)\frac{\delta S}{S_{\min}} \frac{1}{p_0^2} \left[1 - \left(\frac{p_z}{p_0} \right)^2 \right]^{r-2}. \tag{8.7}$$

Thus, for $1 < r < 2$, the curvature of the Fermi surface has singularities in these sections. When $r > 2$ the curvature $K(p_z)$ vanishes at $p_z = \pm p_0$. The corresponding sections of the Fermi surface are lines of parabolic points. The larger the value of the parameter r, the more the Fermi surface near these sections resembles a cylinder.

The anomalies in the curvature of the Fermi surface near $p_z = 0$ can be described by the model (8.3) and (8.4) with $r = 2$ and $l \neq 2$. Here the curvature of the Fermi surface near $p_z = 0$ is described by the asymptotic expression

$$K(p_z) = (l-1)\frac{\delta S}{S_{\max}} \frac{1}{p_0^2} \left| \frac{p_z}{p_0} \right|^{l-2}. \tag{8.8}$$

For $1 < l < 2$ the curvature of the Fermi surface has a singularity at $p_z = 0$; if $l > 2$ the Fermi surface near $p_z = 0$ resembles a cylinder, and the larger the value of l the closer the resemblance. Finally, if $r \neq 2$ and $l \neq 2$, we have a surface in the form of a pinched cylinder with curvature singularities in all the sections with extremal diameters. The profiles of the Fermi surface described by (8.3) and (8.4) are depicted schematically in Figure 8.1

Thus, the proposed model makes it possible to analyze the effect of local curvature anomalies of the Fermi surface on the observed characteristics of layered conductors. This makes it preferable to the tight-binding approximation, which is commonly used to conduct specific calculations (see, e.g., [125], [180]–[183]).

We assume that the conductor fills a half-space $z < 0$ and its surface is parallel to the layer planes. We also assume that a plane electromagnetic wave is incident normally on the surface. Since in layered organic metals the ratio v_z/v_0 is small, we can limit ourselves to a mirror reflection of the electrons from the boundary. In this case, the surface impedance tensor is diagonal and can be described with a formula similar to (3.30):

$$Z_{\alpha\alpha} = 8i\omega \int_0^\infty \frac{dq}{4\pi i\omega\sigma_{\alpha\alpha}(\omega, q) - c^2 q^2}, \tag{8.9}$$

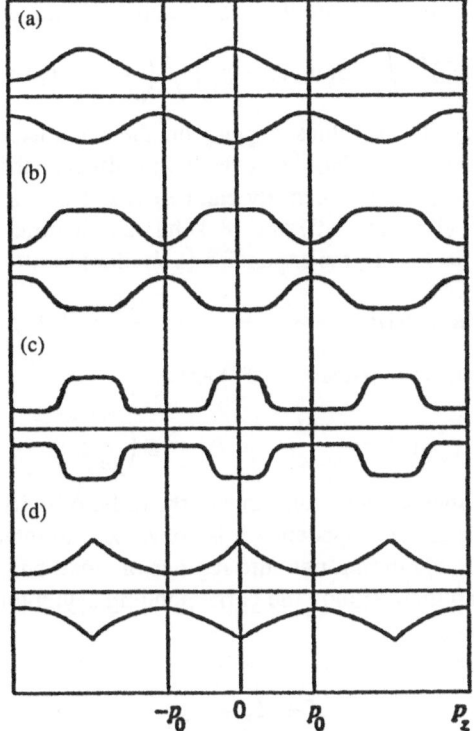

FIGURE 8.1. Profiles of corrugated cylinders described by (8.3), (8.4) for different values of the parameters r and l: (a) $l = r = 2$, (b) $r = 2$, $l > 2$; (c) $r > 2$, $l > 2$; (d) $r = 2$, $1 < l < 2$.

where ω and $\mathbf{q}$ are the frequency and the wave vector of the wave ($\mathbf{q} = (0, 0, q)$) and $\sigma_{\alpha\alpha}(\omega, q)$ are the diagonal components of the electrical conductivity tensor ($\alpha = x, y$).

In this geometry the components σ_{xx} and σ_{yy}, of the conductivity tensor for an axisymmetric Fermi surface and, correspondingly, the components of the impedance tensor, are equal to each other:

$$\sigma_{xx} = \sigma_{yy} = \sigma = \frac{2ie^2}{(2\pi\hbar)^3 m_\perp} \int_{-p_0}^{p_0} \frac{S(p_z)\, dp_z}{\tilde{\omega} - qv_z}, \qquad (8.10)$$

where $\tilde{\omega} = \omega + i/\tau$, and τ is the quasi particle effective relaxation time. The maximum value of the longitudinal component of the velocity, v_z, is of the order of ηv_0. For small values of q, where the parameter u ($u = \omega/\eta q v_0$) assumes values much larger than unity, we can expand σ in a series in inverse powers of u:

$$\sigma = \sigma_0(1 + Q_2 u^{-2} + Q_4 u^{-4} + \cdots) = \sigma_0 [1 + \Phi(1/u)]. \qquad (8.11)$$

The leading approximation for σ_0 equals $iNe^2/m_\perp\tilde{\omega}$ where N is the charge carrier number density and is independent of the specific features of the local geometry of the Fermi surface. The dimensionless coefficients Q_{2n} are specified by

$$Q_{2n} = \left((\omega/\tilde{\omega})^{2n} \int_0^{p_0} S(p_z)\bar{v}_z^{2n}\, dp_z\right) \Big/ \left(\int_0^{p_0} S(p_z)\, dp_z\right) \qquad (8.12)$$

where $\bar{v}_z = v_z/\eta v_0$. Their values depend on the parameters r and l, which determine the function $v_z(p_z)$. This, however, has no effect on the expansion (8.11).

For q large ($u \ll 1$), the conductivity can be expanded in a power series in u. If the Fermi surface given by (8.3) and (8.4) has no curvature anomalies in the extremal sections ($r = l = 2$), the expansion is similar to (3.9):

$$\sigma = \frac{\pi}{i}\frac{\tilde{\omega}}{\omega}\sigma_0(u + \Lambda_1 u^2 + \Lambda_2 u^3 + \cdots) = \frac{\pi}{i}\frac{\tilde{\omega}}{\omega}\sigma_0(1 + f(u)). \qquad (8.13)$$

The first coefficients in expansion (8.13) are

$$\Lambda_1 = -\frac{ib}{\pi}\frac{\tilde{\omega}}{\omega}, \qquad \Lambda_2 = g\left(\frac{\tilde{\omega}}{\omega}\right)^2, \qquad (8.14)$$

where b and g are dimensionless constants of the order of unity.

In calculating the surface impedance it is convenient to integrate with respect to u and divide the integration range into regions of small and large values of u. In each region we can then employ the corresponding asymptotic behavior of the conductivity.

Thus ($Z_{xx} = Z_{yy} = Z$):

$$Z + Z_1 + Z_2, \qquad (8.15)$$

where

$$Z_1 = \frac{8\eta v_0}{c^2}\int_0^1 \frac{du}{\pi(\tilde{\omega}/\omega)\xi^2 u^3(1 + f(u)) + i}, \qquad (8.16)$$

$$Z_2 = \frac{8i\eta v_0}{c^2}\int_1^\infty \frac{du}{\xi^2 u^2(1 + \Phi(1/u)) + 1}. \qquad (8.17)$$

Integration with respect to u in (8.16) and (8.17) can easily be done in the limits of large and small absolute values of the anomaly parameter ξ:

$$\xi = \eta\frac{\omega_p}{\sqrt{\omega\tilde{\omega}}}\frac{v_0}{c},$$

where $\omega_p = \sqrt{4\pi Ne^2/m^*}$ is the plasma frequency. For $|\xi| \ll 1$, the main contribution to the surface impedance is provided by the region of large values of u and is

$$Z \approx \frac{4\pi}{ic}\frac{\sqrt{\omega\tilde{\omega}}}{\omega_p}\left(1 + \frac{\xi^2}{2\pi}Q_2\right). \qquad (8.18)$$

In the opposite limit $|\xi| \gg 1$, the principal part of the impedance is determined by the region of small values of u and is

$$Z \approx \frac{8\pi}{3\sqrt{3}c}\left(\frac{\omega^2}{\omega_p^2}\frac{\eta v_0}{\pi c}\right)^{1/3}\left[1 - i\sqrt{3} - \frac{2\Lambda_1}{3\xi^{2/3}}\left(\frac{\omega}{\pi\tilde{\omega}}\right)^{1/3}(\sqrt{3} - i)\right]. \qquad (8.19)$$

The leading approximation for the conductivity in the region of small values of q (large u) is independent of q. Thus, for $|\xi| \ll 1$, the link between the electric field and current is local, which is characteristic of the skin effect. The skin depth δ is given by the following expression:

$$\frac{1}{\delta} = \frac{\omega}{\eta v_0} \xi',$$

where $\xi = \xi' + i\xi''$. At low frequencies ($\omega\tau \ll 1$):

$$\xi' = \xi'' = \eta \frac{\omega_p}{\omega} \frac{v_0}{c} \sqrt{\frac{\omega\tau}{2}},$$

i.e., $|\xi| = \sqrt{2}l/\delta\omega\tau$. Here $l = \eta v_0 \tau$ is the mean free path of the charge carriers along the normal to the layer plane. The inequality $|\xi| \ll 1$ is valid under the normal skin effect condition ($l \ll \delta$). At high frequencies ($\omega\tau \gg 1$) we have $|\xi| = \xi' = 1/\delta\omega\tau$. Due to the presence of the large factor $\omega\tau$ in the denominator of the expression for $|\xi|$, the inequality $|\xi| \ll 1$ is valid if $l < \delta$, a condition that can easily be met in layered conductors.

The leading term in the asymptotic expression for the impedance with $|\xi| \gg 1$ corresponds to an anomalous skin effect with a skin depth

$$\delta = \frac{2\eta v_0}{\sqrt{3}\omega} \left(\frac{\omega}{\pi\tilde{\omega}\xi^2} \right)^{1/3} = \frac{2}{\sqrt{3}} \left(\frac{c^2\eta v_0}{\pi\omega\omega_p^2} \right)^{1/3}.$$

When $\omega\tau \ll 1$ holds, $|\xi|$ takes a value of the order of $(l/\delta)^{3/2}/\omega\tau$, while in the opposite limit $\omega\tau \gg 1$ this parameter is of the order of $(l/\delta\omega\tau)^{3/2}$. Since the mean free path in the direction perpendicular to the conducting layers is small, for layered organic metals it is essentially impossible to meet the condition $l \gg \delta$. This means that, in contrast to ordinary metals, it is impossible to observe an anomalous skin effect in such substances in the high-frequency range ($\omega\tau \gg 1$).

In the limit $\omega\tau \ll 1$, the skin effect is anomalous ($|\xi| \gg 1$) for $l > \delta$. This condition can easily be met at moderate frequencies, since the skin depth increases with decreasing frequency. The intermediate frequency range $\omega\tau \sim 1$ is optimal for realizing an anomalous skin effect in layered conductors. The maximum value of $|x|$ is reached at $\omega\tau = 1/\sqrt{2}$ and is of the order of $\eta\omega_p\tau v_0/c$. The ratio $\omega_p\tau v_0/c$ in a pure ($\tau \sim 10^{-8}$ s) layered conductor is of the order of 10^3–10^4. Thus, at moderate values of η ($\eta \sim 10^{-2}$) the maximum value of $|\xi|$ may reach 10^2. This means that in layered conductors both the normal skin effect ($|\xi| \ll 1$) and the anomalous skin effect ($|\xi| \gg 1$) can be present, although the latter is observed in a frequency range narrower than in ordinary metals. On the other hand, at very small values of η, when the Fermi surface is for all practical purposes a cylinder, the condition $|\xi| \ll 1$ is met over the entire frequency range. Accordingly, we can use the asymptotic formula (8.18) for the surface impedance corresponding to a normal skin effect at all frequencies.

The leading terms in the asymptotic expressions for the surface impedance in both limits, the normal skin effect ($|\xi| \ll 1$) and the anomalous skin effect

$(|\xi|) \gg 1$, are independent of the specific characteristics of the Fermi surface and coincide with the results obtained by Gokhfel'd and Peschanskii [183], who used a model of the Fermi surface based on the tight-binding approximation for electrons.

Let us now examine a conductor whose Fermi surface has anomalies of the Gaussian curvature in the effective sections. For definiteness, we assume that in the model (8.3) and (8.4) we have $l = 2$ and $r \neq 2$. This corresponds to curvature anomalies at $p_z = \pm p_0$. In this case, the asymptotic expansion of the conductivity in the region where the parameter u is small contains an additional term σ_a which now equals

$$\sigma_a = \frac{\pi}{2i}\sigma_0\mu_\beta \left(u\frac{\tilde{\omega}}{\omega}\right)^{\beta+1}, \qquad (8.20)$$

where

$$\mu_\beta = (\beta + 1)\left(1 - i\tan\frac{\pi\beta}{2}\right), \qquad -\beta = \frac{r-2}{r-1}.$$

If the sections of the Fermi surface at $p_z \pm p_0$ are lines of parabolic points ($r > 2$), the parameter β takes negative values ($-1 < \beta < 0$). The shape of the Fermi surface in the vicinity of these sections is close to cylindrical, and as $\beta \to -1$, the Fermi surface resembles a cylinder more closely. For $1 < r < 2$, the parameter β assumes positive values. In the given case, the curvature of the Fermi surface in the vicinity of the sections corresponding to $p_z = \pm p_0$ becomes anomalously large. In the low-frequency limit, (8.20) can be written as const./$(ql)^\gamma$, where $\gamma = 1/(r-1)$. The same asymptotic behavior of the contribution to the conductivity of the quasi-cylindrical section of the Fermi surface of a three-dimensional metal was obtained by Kaganov and Kontreras [184] (see also [54]).

A comparison of (8.20) and (8.13) suggests that if the Fermi surface of a layered conductor near the sections with an extremal (in our case minimum) diameter closely resembles a cylinder, σ_a exceeds all the other terms in the expansion of the conductivity in powers of u. Accordingly, the leading approximation for the impedance, in the event of an anomalous skin effect determined by the contribution of the quasi-cylindrical section of the Fermi surface, and practically coincides with (3.33):

$$Z = \frac{8\pi}{c}W(\beta)\frac{\sqrt{\omega\tilde{\omega}}}{i\omega_p}\left(\xi\frac{\omega}{\tilde{\omega}}\right)^{(\beta+1)/(\beta+3)}, \qquad (8.21)$$

where

$$W(\beta) = \left[\frac{2\cos(\pi\beta/2)}{\pi(\beta+1)}\right]^{1/(\beta+3)}\left(1 + i\cot\frac{\pi}{\beta+3}\right)\frac{1}{\beta+3}$$

$$\equiv w(\beta)\left(1 + i\cot\frac{\pi}{\beta+3}\right). \qquad (8.22)$$

In the limit $\beta \to -1$, the complex-valued function $W(\beta)$ tends to $\frac{1}{2}$ and (8.21) becomes the leading term in (8.18). Here the dependence on the parameter η, which characterizes the extent to which the Fermi surface is rippled, disappears. Thus, the presence of wide cylindrical belts on a highly pinched Fermi surface leads to the same result for the surface impedance as in the case of a weakly pinched Fermi surface.

Equation (8.21) implies that, to within a complex-valued constant ζ ($|\zeta| \sim 1$):

$$Z = \zeta |Z_0|(1 - i\omega\tau)^{-\beta/(\beta+3)}(\delta/l)^{-\beta/(\beta+3)}. \tag{8.23}$$

Here Z_0 is the leading approximation of the impedance for an anomalous skin effect in the case where the Fermi surface of the conductor has no curvature anomalies in the effective sections (the first term in (8.19)), and δ is the skin depth for the anomalous skin effect. We see that the impedance depends on the mean free path of the charge carriers. Kaganov and Kontreras [184] found that such a dependence exists only if there are quasi-cylindrical sections on the Fermi surface. If there are no such sections, the leading approximation Z_0 of the impedance for an anomalous skin effect in independent of l.

In the limit $\beta \to -1$, the exponent of δ/l in (8.23) assumes values close to $\frac{1}{2}$. Thus, if the Fermi surface in the vicinity of the effective sections at $p_z = \pm p_0$ resembles a cylinder, the impedance for an anomalous skin effect is proportional to $1/\sqrt{l}$, in the same way as it is for a normal skin effect. For $\omega\tau \ll 1$ and $\beta = -\frac{2}{3}$ ($r = 4$), (8.23) coincides with the result of [184].

At sufficiently high frequencies ($\omega\tau > 1$), the real part of the surface impedance (8.23) is

$$R(0) = \frac{8\pi}{c}w(\beta)\left(\eta\frac{\omega_p}{\omega}\frac{v_0}{c}\right)^{(\beta+1)/(\beta+3)}\frac{\omega}{\omega_p}\left[\cot\frac{\pi}{\beta+3} - \frac{\beta}{\beta+3}\frac{1}{\omega\tau}\right]. \tag{8.24}$$

For negative values of β not too close to -1 we have $R \sim \omega^{2/(\beta+3)}$. The real part of the impedance increases with frequency faster than in the case of a conductor for which the curvature of the Fermi surface in the effective section is finite. In the limit $\beta \to -1$ the frequency dependence of R disappears, as it does for a purely cylindrical surface.

Let us assume that the parameter β, which characterizes the shape of the Fermi surface, takes positive values. This means that the curvature of the Fermi surface at $p_z = \pm p_0$ becomes infinite. The term σ_a ceases to be the leading term in the asymptotic expression for the conductivity for small u. However, for $0 < \beta < 1$, when the singularity in the curvature is not pronounced, σ_a is the first correction to the leading approximation. Here the asymptotic behavior for an anomalous skin effect is given by an expression similar to (3.34):

$$Z = \frac{8\pi}{3\sqrt{3}}\left(\frac{\omega^2}{\omega_p^2}\frac{\eta v_0}{\pi c}\right)^{1/3}\left\{1 - i\sqrt{3} - \frac{U(\beta)}{2\sqrt{3}}\frac{1}{(\pi\xi^2)^{\beta/3}}\left(\frac{\tilde{\omega}}{\omega}\right)^{2\beta/3}\right\}, \tag{8.25}$$

where

$$U(\beta) = \frac{(1+\beta)^2}{\cos(\pi\beta/2)}\left[\cot\frac{\pi(1+\beta)}{3} - i\right]. \tag{8.26}$$

The first correction to the leading approximation for the impedance is now larger than in the absence of curvature anomalies in the effective belts on the Fermi surface. The frequency dependence of this correction also changes. The additional term is proportional to $\omega^{2(1+\beta)/3}$ rather than to $\omega^{4/3}$ (the latter case corresponds to a Fermi surface without curvature anomalies).

Note that the surface impedance for an anomalous skin effect is also described by expressions of the form (8.25) in the case where narrow neighborhoods of some (but not all) extremal sections of the Fermi surface resemble narrow cylindrical bands. Here the anomalous additional term in the conductivity contains a small positive factor ρ describing the relative number of effective charge carriers related to the cylindrical section of the Fermi surface. The parameter β takes negative values in the interval $-1 < \beta < 0$. For small values of ρ and moderate values of β, not too close to -1, σ_a is smaller than the leading term in expansion (8.13) for σ and must be taken into account as the first-order correction. Here, for the component of the surface impedance tensor, we arrive at a result that differs from (8.25) in that the term describing the first correction contains an additional factor ρ and $\beta < 0$ in this term.

The above analysis can be repeated for a Fermi surface that has anomalies in the Gaussian curvature at $p_z = 0$. Such a surface is described by (8.3) and (8.4) with $r = 2$ and $l \neq 2$. As a result we arrive at expressions that coincide with (8.19)–(8.25) with a parameter β, which characterizes the shape of the Fermi surface near the effective section $p_z = 0$, expressed in terms of $l\left(-\beta = (l-2)/(l-1)\right)$.

Finally, let us consider the case where all the effective sections of the Fermi surface have curvature anomalies. If $r > 2$ and $l > 2$ hold, the neighborhood of each section of the Fermi surface with an extremal diameter resembles a cylinder. Here the asymptotic behavior of the surface impedance for an anomalous skin effect retains its form (8.21). The same expression (8.21) describes the asymptotic behavior of the surface impedance of a layered conductor for an anomalous skin effect, in the case where the curvature of the Fermi surface of the conductor is anomalously large in all sections with minimum and maximum diameters ($1 < r < 2$ and $1 < l < 2$). In both cases, the value of β is expressed in terms of the larger of the two parameters, r and l.

8.2 Cyclotron Resonance in Layered Conductors in a Normal Magnetic Field

The local features of the geometry of the Fermi surface of layered conductors lead to specific singularities in their observed properties in an external magnetic field. This is true, in particular, for cyclotron resonance, which in recent years has been repeatedly observed in organic metals [185]–[190].

Suppose that an external magnetic field is directed along a normal to the surface of a semi-infinite conductor and that this surface is parallel to the planes of the conducting layers. When the charge carriers undergo a specular reflection from the boundary, the surface impedance tensor becomes diagonal in terms of the circular components and is described by formula (4.1).

The asymptotic expressions for the circular components of the transverse conductivity $\sigma_\pm$ for large and small u coincide with the expansions (8.11) and (8.13) in which $\tilde{\omega}$ is replaced by $\tilde{\omega}_\pm$ ($\tilde{\omega}_\pm = \omega \mp \Omega + i/\tau$, where Ω is the cyclotron frequency of the charge carries). The same is true of the expression for $\sigma_a^\pm$. Below we examine the case where the polarization corresponds to cyclotron resonance.

If the Fermi surface of the layered conductor has wide sections resembling cylinders, the impedance for an anomalous skin effect is described by an expression obtained from (8.21) by substituting $\tilde{\omega}_+$ for $\tilde{\omega}$. This substitution must be carried out everywhere, including the expression for the parameter ξ characterizing the extent to which the skin effect is anomalous. The real part of the impedance in such conditions is

$$\frac{R(B)}{R(0)} = \left(\sqrt{\left(1 - \frac{\Omega}{\omega}\right)^2 + \frac{1}{(\omega\tau)^2}} \right)^{-\beta/(\beta+3)} \frac{\cos(Y_\beta(\Omega, \omega, \tau))}{\cos(\pi/(\beta+3))}, \tag{8.27}$$

where

$$Y_\beta(\Omega, \omega, \tau) = \frac{1}{\beta+3} \left\{ \pi + \pi\beta\theta(\Omega - \omega) + \beta \operatorname{arccot}\left[\omega\tau \left(1 - \frac{\Omega}{\omega}\right) \right] \right\},$$

and

$$\theta(x) = \begin{cases} 1, & x \geq 0, \\ 0, & x < 0, \end{cases} \qquad -1 < \beta < 0. \tag{8.28}$$

At a fixed frequency ω the value of $R(B)$ rapidly increases with the magnetic field strength in fields near cyclotron resonance. The discontinuity in $R(B)$ at $B = B_r$ is no smaller in order of magnitude than $R(0)$ at the same frequency. In strong magnetic fields ($B \gg B_r$), where $\omega \ll \Omega$, the value of $R(B)$ increases in proportion to $(\Omega/\omega)^{-\beta/(\beta+3)}$. For moderate (in absolute value) values of β, the increase in $R(B)$ in a strong magnetic field is slow as $\beta \to -1$, when the shape of the effective belts on the Fermi surface in essentially cylindrical, and for $B \gg B_r$ we have $R(B) \sim \sqrt{\Omega/\omega}$. The field dependence of the ratio $R(B)/R(0)$ near cyclotron resonance is depicted in Figure 8.2 where the curves are described by (8.27).

If the cylindrical belts near the extremal sections of the Fermi surface are narrow, the related additional term in the conductivity determines the first correction to the leading approximation of impedance for an anomalous skin effect. The leading term in the surface impedance in this case is independent of the magnetic field. Substituting $\tilde{\omega}_+$ for $\tilde{\omega}$ in (8.25) and allowing for the small factor ρ, which describes the width of the cylindrical belts on the Fermi surface, we get

$$\frac{R(B)}{R(0)} = 1 + \tilde{\rho}\tilde{\xi}^{-2\beta/3} \left(\sqrt{\left(1 - \frac{\Omega}{\omega}\right)^2 + \frac{1}{(\omega\tau)^2}} \right)^\beta \frac{\cos(Y'(\Omega, \omega, \tau))}{\sin(\pi(1+\beta)/3)}. \tag{8.29}$$

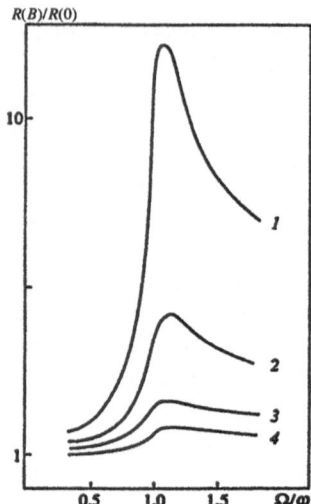

FIGURE 8.2. Dependence of the active part of the surface impedance of a layered organic metal whose Fermi surface has nearly cylindrical wide bands on the magnetic field near cyclotron resonance. The curves are described by (8.27) and are plotted for $\omega\tau = 10$ with: $\beta = -0.2$ (curve 1), $\beta = -0.4$ (curve 2), $\beta = -0.6$ (curve 3); $\beta = -0.8$ (curve 4).

Here

$$Y'_\beta(\Omega, \omega, \tau) = \frac{\pi}{3}(2 - \beta) + \pi\beta\theta(\Omega - \omega) + \beta \operatorname{arccot}\left[\omega\tau\left(1 - \frac{\Omega}{\omega}\right)\right],$$

$$\tilde{\rho} = \frac{(1 + \beta)^2}{2\sqrt{3}\cos(\pi\beta/2)}\rho, \qquad \tilde{\xi} = \sqrt{\pi}\frac{\omega_p}{\omega}\frac{\eta v_0}{c}. \tag{8.30}$$

This result is valid for $\tilde{\rho}\tilde{\xi}^{-2\beta/3} \ll 1$. At high frequencies, the parameter $\tilde{\xi}$ assumes values close to those of the anomaly parameter ξ in the absence of an external magnetic field. Taking for $\tilde{\xi}$ a value of the order of the maximum of ξ ($\tilde{\xi} \sim 10^{-2}$), we see that for a moderate curvature anomaly ($-0.5 < \beta < 0$) the above inequality holds if we have $\rho \sim 0.1$.

For $\beta < 0$ the second term in (8.29) describes the peak in the field dependence of $R(B)$ related to the cyclotron resonance (Figure 8.3). The height of this peak depends on ρ. For moderate values of ρ ($\rho < 0.1$) the height may amount to 10% of the leading approximation of the real part of the impedance. For moderate values of $\omega\tau$ the top of the peak (it corresponds to the minimum of the resonance term in (8.29)) is appreciably shifted in relations to B_r:

$$\frac{B_r - B}{B_r} \equiv \frac{\Delta B}{B_r} = \frac{\cot \Phi}{\omega\tau}. \tag{8.31}$$

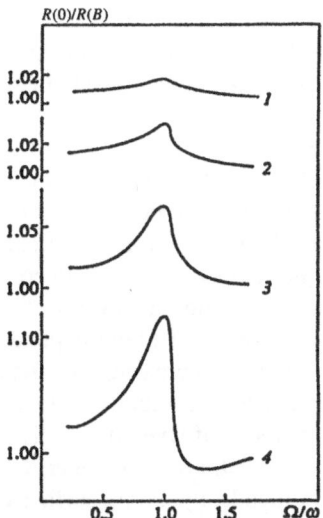

FIGURE 8.3. Dependence of the active part of the surface impedance of a layered organic metal on the magnetic field near cyclotron resonance for the case where one of the extremal sections of the Fermi surface coincides with the line of parabolic points. The curves are described by (8.29) and are plotted for $\omega\tau = 10$ and $\tilde{\rho} = 0.05$ with: $\beta = -0.2$ (curve 1), $\beta = -0.3$ (curve 2), $\beta = -0.4$ (curve 3), $\beta = -0.5$ (curve 4).

The value of Φ is determined by the shape of the effective section of the Fermi surface

$$\Phi = \frac{\pi}{6}\left(1 - \frac{\beta}{1-\beta}\right). \tag{8.32}$$

Such field behavior near the cyclotron resonance has been observed in some layered organic metals of the α-(BEDT-TTF)$_2$MHg(SCN)$_4$ group [188] and in the organic conductor (BEDO-TTF)$_2$ReO$_4$(H$_2$O) [189]. The experiments of [188], [189] were conducted in the 30–100 GHz frequency range in magnetic fields of about 50 kG with $\omega\tau \sim 10$. The magnitude of the singularities related to cyclotron resonance amounted to several percentage points of the leading impedance value [188], which agrees with estimates that follow from (8.29). These results can serve as proof of the presence of cylindrical sections on the Fermi surface of the layered conductors used in these experiments.

The data of [190] allow us to estimate the parameter β, which characterizes the shape of the quasi-cylindrical section of the Fermi surface of (BEDO-TTF)$_2$ReO$_4$(H$_2$O), via (8.31) and (8.32). The experiment was conducted at frequencies corresponding to variations in $\omega\tau$ in the interval from 10 to 20 and to variations in $\Delta B/B_r$ in the interval from 0.11 to 0.06. This yields $\beta \approx -(0.25\text{--}0.35)$.

The resonance peak in the field dependence of the active part of the impedance must be accompanied by a stronger singularity in the dependence of dR/dB. If the curvature of the Fermi surface rapidly increases as we move away from the line of parabolic points, the height of the peak in the field dependence of $R(B)$ corresponding to cyclotron resonance may prove to be too small to be observable. However, the resonance singularity in the derivative of the impedance under these conditions may clearly manifest itself. The resonant singularities in the field and frequency dependence of dR/dB can also be observed if $\beta > 0$. This corresponds to an anomalously large curvature of the Fermi surface in the vicinity of sections with minimum or maximum diameter. Cyclotron resonance can be observed at moderate values of the parameter β ($0 < \beta < 1$). In contrast to the above case for positive β, cyclotron resonance does not manifest itself in the field dependence of the real part of the surface impedance. The field-dependent term in the expression (8.29) decreases monotonically with increasing magnetic field.

It is too early to draw any conclusions about the local features of the geometry of the Fermi surface of the majority of layered organic metals, since there is a lot to study in the electron energy spectra of such materials. It can be assumed, however, that here, as in ordinary metals, the Fermi surface contains quasi-cylindrical bands or sections with an anomalously large curvature. These features of the local geometry of the Fermi surface can be created (if they are absent) or enhanced by applying an agent that changes the shape of the constant-energy surfaces, e.g., by applying external pressure along the normal to the conducting planes.

The above analysis shows that the special features in the profile of the corrugated cylinder, which is the main part of the Fermi surface of layerd organic metals, can substantially change the high-frequency properties of these materials. The model developed here makes it possible to study in detail the observable manifestations of the local geometry of the Fermi surface of layered conductors. It resolves some of the difficulties that emerge when one used the model of tightly bound electrons. For instance, the characteristic features of the observable properties of layered conductors, for which the strong anisotropy of the electrical conductivity is responsible, can be described and analyzed without passing to the limit $\eta \rightarrow 0$, which corresponds to a conductor with a two-dimensional energy spectrum of charge carriers.

The model specified by (8.3) and (8.4) makes it possible to do a detailed study of the frequency dependence of the surface impedance of layered conductors with different Fermi-surface profiles; with it we can analyze on a unified basis all possible manifestations of cyclotron resonance in a magnetic field that is perpendicular to the surface of the conductor. By using this model, we also can investigate the quantum oscillations of the thermodynamic characteristics of layered conductors.

8.3   Quantum Oscillatory Phenomena in Layered Conductors

Experimental data, with respect to quantum oscillations in the various characteristics of metals, were widely used as the instruments of study of the electron energy spectra in the usual metals. Quantum oscillatory phenomena were repeatedly observed in the organic layered conductors [171], [173], [174], [190]. Quantum oscillations in thermodynamic observables in the layered conductors with the quasi two-dimensional energy spectrum of charge carriers can exhibit some characteristic properties which have to be taken into account in the interpretation of experimental data.

Quantum oscillations in thermodynamic quantities are due to the oscillations in the electron density of states on a Fermi surface in a strong magnetic field:

$$N_\zeta = -\sum_\nu \frac{\partial f_\nu}{\partial E_\nu}. \tag{8.33}$$

Here f_ν is the Fermi distribution function for the quasi particles with energies E_ν.

We shall use the model (8.3), (8.4) to describe the charge carriers energy spectrum and consider oscillations in their density of states in a magnetic field $\mathbf{B}$ directed along the cylinder symmetry axis. Under semiclassical conditions, when the cyclotron quantum energy $\hbar\Omega$ is small compared to the Fermi energy ζ, we can transform expression (8.33) to the form

$$N_\zeta = g_0(1 + \Delta). \tag{8.34}$$

The function Δ represents the oscillating contribution to the density of states.

The form of the function Δ depends importantly on the characteristic properties of the electron energy spectrum. Assuming that the Fermi surface curvature on the effective strips is nonzero ($r = l = 2$) we can obtain

$$
\begin{aligned}
\Delta &= \Delta(0) + \Delta(p_0) \\
&= \frac{1}{\gamma}\sqrt{\frac{S_0}{\delta S}} \left\{ \sum_{n=1}^\infty \frac{(-1)^n}{\sqrt{n}} \psi_n(\theta) \cos\left(\frac{ncS_{\max}}{\hbar|e|B} - \frac{\pi}{4}\right) \cos\left(\pi n \frac{\Omega_0}{\Omega}\right) \right. \\
&\quad \left. + \sum_{n=1}^\infty \frac{(-1)^n}{\sqrt{n}} \psi_n(\theta) \cos\left(\frac{ncS_{\min}}{\hbar|e|B} + \frac{\pi}{4}\right) \cos\left(\pi n \frac{\Omega_0}{\Omega}\right) \right\}.
\end{aligned}
\tag{8.35}
$$

Here $S_0 = 2\pi m_\perp \zeta$ is the area of the Fermi surface cross-section in the limit $\eta \to 0$, when the Fermi surface becomes a pure cylinder, and $\hbar\Omega_0$ is the spin splitting energy.

In the case when there is an anomaly in the Fermi surface curvature at $p_z = \pm p_0$ ($l = 2, r \neq 2$) the contribution to the oscillating function Δ from the vicinities of the corresponding cross-sections assumes the form

$$
\begin{aligned}
\Delta(p_0) &= \left(\frac{rS_0}{2\gamma^2\delta S}\right)^{1/r} \frac{\Gamma(1/r)}{r} \sum_{n=1}^\infty \frac{(-1)^n \psi_n(\theta)}{n^{1/r}} \\
&\quad \times \cos\left(\frac{ncS_{\min}}{\hbar|e|B} + \frac{\pi}{2r}\right) \cos\left(\pi n \frac{\Omega_0}{\Omega}\right),
\end{aligned}
\tag{8.36}
$$

where Γ is the gamma function. Under this condition the first term in (8.35) retains its form. Correspondingly, when the local geometry of the Fermi surface is characterized with the anomalous curvature at $p_z = 0$ ($r = 2, l \neq 2$) the oscillating function $\Delta(0)$ is described by the expression

$$\Delta(p_0) = \left(\frac{lS_0}{2\gamma^2\delta S}\right)^{1/l} \frac{\Gamma(1/l)}{l} \sum_{n=1}^{\infty} \frac{(-1)^n}{n^{1/l}} \cos\left(\frac{ncS_{\max}}{\hbar|e|B} - \frac{\pi}{2l}\right) \cos\left(\pi n\frac{\Omega_0}{\Omega}\right). \tag{8.37}$$

At the same time, expression (8.35) for $\Delta(p_0)$ remains valid.

When the effective strip on the Fermi surface, centered at $p_z = 0$ or $p_z = \pm p_0$, is close to a cylinder in shape, its contribution to the function Δ has the same form as for the conductor with a two-dimensional energy spectrum. For a instance, if $l \gg 1$, expression (8.37) can be transformed to the form

$$\Delta(0) = \sum_{n=1}^{\infty} (-1)^n \cos\left(\frac{ncS_{\max}}{\hbar|e|B}\right) \cos\left(\pi n\frac{\Omega_0}{\Omega}\right). \tag{8.38}$$

In the case when both parameters (r and l) are large, in comparison with unity, the expression for Δ contains two terms of the form (8.38). Both terms are the contributions from the quasi-cylindrical strips on the Fermi surface. These oscillating terms have to differ in period. The difference in their periods arises due to the distinction in the extremal cross-sectional areas. It may be noticeable if the magnetic field is not too strong and the inequality $\gamma^2\delta S/S_0 \gg 1$ is satisfied.

In a very strong magnetic field the inequality $\gamma^2\delta S/S_0 \ll 1$ needs to be satisfied. Under this condition the difference in form between the Fermi surface and the cylinder does not affect the oscillating function Δ:

$$\Delta = 2\sum_{n=1}^{\infty} (-1)^n \psi_n(\theta) \cos(\pi n\gamma^2) \cos\left(\pi n\frac{\Omega_0}{\Omega}\right). \tag{8.39}$$

When a temperature is not too low ($\theta \sim 1$) one can retain only the first term in the sum over "n" in (8.35)–(8.39). Thus, under $\theta \sim 1$, the dependence of the function Δ on the inverse magnetic field has to be of a harmonic type. Because of the factor $\exp(-\theta)$, the amplitude of the oscillations of the function Δ under $\theta \sim 1$ is small compared to unity, even in strong magnetic fields.

When $\theta \ll 1$, the form of the oscillations becomes much more complicated and their amplitude increases. To evaluate the amplitude of the oscillations under this condition, one can use the asymptotic formulas following from the Euler–Maclaurin summation formula. As a result under this condition, we obtain the following estimation for the amplitude of the oscillations

$$\Delta_{\max}(p_0) \approx \left(\frac{S_0\theta}{\gamma^2\delta S}\right)^{1/r} \frac{1}{\theta}, \tag{8.40}$$

$$\Delta_{\max}(0) \approx \left(\frac{S_0\theta}{\gamma^2\delta S}\right)^{1/l} \frac{1}{\theta}. \tag{8.41}$$

In the case when the Fermi surface curvature remains finite and nonzero at $p_z = 0$ and $p_z = \pm p_0$, we have

$$\Delta_{\max}(0) = \Delta_{\max}(p_0) = \frac{1}{\gamma}\sqrt{\frac{S_0}{\delta S}} \sum_{n=1}^{\infty} \frac{1}{2\sqrt{2n}} \psi_n(\theta) \approx \frac{3}{2}\frac{1}{\gamma\sqrt{\theta}}\sqrt{\frac{S_0}{\delta S}}. \tag{8.42}$$

For the same values of the parameters γ, $\delta S/S_0$, and θ, the ratio of the amplitudes of oscillations described by formulas (8.37) and (8.35) is of the order of $(\gamma^2 \delta S/S_0 \theta)^{(l-2)/2l}$. Thus, under typical experimental conditions when $\gamma^2 \delta S/S_0 \gg 1$, the oscillations associated with a quasi-cylindrical extremal cross-section of the Fermi surface have an amplitude much larger than the amplitude of ordinary oscillations described by formula (8.35). If $1 < l < 2$, one of the principal radii of curvature vanishes at all points of the extremal cross-section of the Fermi surface at $p_z = 0$. In this case, the amplitude of oscillations described by formula (8.37) has to be much smaller than the amplitude of the oscillating parts of the contributions from the Fermi surface ordinary extremal cross-sections.

When $\theta \ll 1$, the oscillation amplitude value may be of the order of unity. In this case, the oscillating contribution to the charge carriers density of states on the Fermi surface becomes comparable to its classical value g_0. Analysis of low-temperature ($\theta \ll 1$) quantum oscillations in thermodynamic observables in metals shows that the oscillations must have some characteristic features. The charge carriers concentration in layered organic metals is far below that in ordinary metals. It generates much more favorable conditions for observation of the salient features in low-temperature quantum oscillations in thermodynamic quantities. If we take, for $m_\perp$ and S_0, the values obtained in experimental study for the Fermi surface of the organic metal $\beta - (ET)_2 I Br_2$ [174] ($m_\perp \approx 4, 5m_0$; $S_0 \approx 1, 26 \times 10^{-39}$ g^2 cm^2/s^2; m_0, free electron mass) the condition

$$\frac{3}{2} \frac{1}{\gamma\sqrt{\theta}} \sqrt{\frac{S_0}{\delta S}} \sim 1,$$

in magnetic fields $B \sim 200$ kG and for $\delta S/S_0 \approx 0.05$, will be satisfied at temperatures of the order of 1 K.

The results obtained for metals show that the interactions among electrons affect the quantum oscillations in various thermodynamic characteristics of metals to a considerable extent. It forces us to take the possible Fermi-liquid effects into account in the study of low-temperature quantum oscillatory phenomena in layered conductors.

If the Fermi-liquid interaction is taken into account, the charge carrier's density of states on the Fermi surface has to be changed by the quantity

$$N_\zeta^* = -\sum_{\nu\nu'} \frac{f_\nu - f_{\nu'}}{E_\nu - E_{\nu'}} n_{\nu\nu'}^*(-\mathbf{q}) n_{\nu'\nu}(\mathbf{q})\bigg|_{q=0}, \qquad (8.43)$$

in the expressions for the various the thermodynamic susceptibilities.

Here $n_{\nu'\nu}(\mathbf{q})$ are the Fourier components in the expansion of matrix elements of the number density operator for particles in spatial variables: $n_{\nu'\nu}^*(\mathbf{q})$ is the corresponding effective operator renormalized due to the Fermi-liquid interaction. The Fermi surface under consideration is axisymmetric. Therefore we can write matrix elements of the Fermi-liquid functions $\varphi_{\alpha\alpha'}^{\alpha_1\alpha_2}$ and $\psi_{\alpha\alpha'}^{\alpha_1\alpha_2}$ in the form (1.82). We can use (1.82) to calculate the renormalized density of states N_ζ^*. Keeping only the main terms in the asymptotic expansions of N_ζ^* in powers of the small parameter γ^{-1}, we can derive the following expression:

$$N_\zeta^* = g_0(1 - \alpha_0 + K). \tag{8.44}$$

Here the oscillating function K has the form

$$K = \frac{[1 - \bar{\alpha}_0(0)]^2 \Delta(0) + [1 - \bar{\alpha}_0(p_0)]^2 \Delta(p_0)}{1 + [\alpha_0(0) + \beta_0(0)]\Delta(0) + [\alpha_0(p_0) + \beta_0(p_0)]\Delta(p_0)}. \tag{8.45}$$

The functions $\Delta(0)$ and $\Delta(p_0)$ are described by (8.35)–(8.39). Their concrete form is determined by the Fermi surface local geometry near $p_z = 0$ and $p_z = \pm p_0$. Parameters α_0, $\alpha_0(p_z)$, $\bar{\alpha}_0(p_z)$, and $\beta_0(p_z)$ describe the Fermi-liquid interaction among the charge carriers

$$\alpha_0 = \frac{A_0}{1 + \tilde{A}_0}, \qquad\qquad \bar{\alpha}_0(p_z) = \frac{\overline{A}_0(p_z)}{1 + \tilde{A}_0},$$

$$\alpha_0(p_z) = \frac{[P_{00}^0(p_z)]^2}{1 + \tilde{A}_0}, \qquad\qquad \beta_0(p_z) = \frac{[Q_{00}^0(p_z)]^2}{1 + \tilde{B}_0},$$

$$A_0 = \frac{m_\perp}{2\pi^2 \hbar^3 g_0} \left(\int_0^{p_0} P_{00}^0(p_z)\, dp_z \right)^2, \qquad \tilde{A}_0 = \frac{m_\perp}{2\pi^2 \hbar^3 g_0} \int_0^{p_0} [P_{00}^0(p_z)]^2\, dp_z,$$

$$\tilde{B}_0 = \frac{m_\perp}{2\pi^2 \hbar^3 g_0} \int_0^{p_0} [Q_{00}^0(p_z)]^2\, dp_z, \qquad \overline{A}_0 = \frac{m_\perp}{2\pi^2 \hbar^3 g_0} P_{00}^0(p_z) \int_0^{p_0} P_{00}^0(p_z)\, dp_z.$$

Comparison of (8.44) with (8.34) shows that there are pronounced distinctions between the charge carrier density of states on the Fermi surface N_ζ and the renormalized due to the interaction among the quasi-paticles N_ζ^* in the low-temperature range. Under $\theta \ll 1$ the amplitude and the form of the oscillations described by the function K can differ significantly from the corresponding characteristics of the function Δ. In particular, the denominator in (8.44) can turn to zero at the peaks of the oscillations. Thus, the function K can appear to have singularity caused by the Fermi-liquid interaction.

The specific features of the low-temperature quantum oscillations in the renormalized density of states have to be manifested in the oscillations of the thermodynamic observables. We consider, for example, the oscillating contribution to the magnetic susceptibility for a uniform field $\mathbf{B}$. To simplify the following calculations we shall assume that the effective strip on the Fermi surface near $p_z = 0$ is close to a cylindrical shape, and near $p_z = \pm p_0$ the Fermi surface curvature is nonzero. In this case, we can neglect in the expression (8.45) all the terms proportional to $\Delta(p_0)$ in comparison with the terms proportional to $\Delta(0)$. Generalizing the corresponding result of [160], we can show that in the case of the axisymmetric Fermi surface, $\chi_\parallel$ is proportional to the oscillating function

$$K' = \frac{K}{1 + (1 - \bar{\alpha}_0(0))K} = \frac{\Delta(0)}{1 + (1 + \alpha_0(0) - \bar{\alpha}_0(0) + \beta_0(0))\Delta(0)}. \tag{8.46}$$

In the low-temperature range, the amplitude of the oscillations in $\Delta(0)$ might increase to such an extent that it reaches values of the order of unity. Under these conditions the form of the de Haaz–van Alphen oscillations is defined to a great

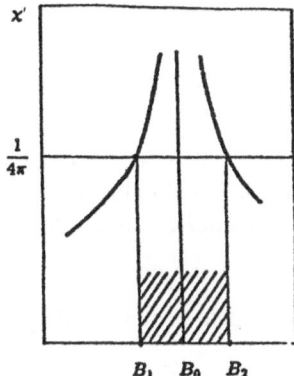

FIGURE 8.4. Dependence of the magnetic susceptibility $\chi_{||}$ on the magnetic field at $T = 0$. The domain of the values of the magnetic field where $4\pi\chi > 1$ is section lined.

extent by the denominator in (8.46) and depends critically on the Fermi-liquid interaction. Under $(1 + \alpha_0(0) - \overline{\alpha}_0(0) + \beta_0(0)) < 0$ the denominator in (8.46) can go to zero at the peaks of the oscillations and the susceptibility $\chi_{||}$ would increase beyond all bounds. This would lead to disturbing the stability ($\chi_{||} < 1/4\pi$) at the peaks of the oscillations (See Figure 8.4). Thus, the low-temperature analog to the Shoenberg effect can be observed. The basic opportunity for the appearance of such an effect was predicted in [160] for the case of the isotropic electron liquid (see also [191]). However, as was mentioned above, the conditions for observing this effect in conventional metals are severely confined.

It was established by experiment that the superconducting transition temperature for some layered organic conductors changes considerably under pressure [192]. In this connection it is of interest to study various aspects of the response of the electron system of such conductors for the deformation of its crystal lattice. That is why it is expedient to consider the features of the low-temperature quantum oscillations in the layered conductor elastic moduli in more detail. To simplify the derivation of the formulas, describing the quantum oscillations in the elastic moduli, one may fail to consider the deformation interaction between the electrons and the lattice. A more careful treatment carried out here for the isotropic model of metal (see the first section of Chapter 7) does not cause a qualitative change of results.

Under the accepted assumption the effect of the electrons on a lattice arises due to a self-consistent electrical field, which occurs at deformation. Besides, the deformation of the lattice causes the occurrence of an additional nonuniform magnetic field $\mathbf{b}(\mathbf{r})$. The presence of these fields leads to a redistribution of the electron density. The additive $\delta N(r)$ equals

$$\delta N(\mathbf{r}) = -\frac{\partial N}{\partial \zeta} e\varphi(\mathbf{r}) + \frac{\partial N}{\partial \mathbf{B}} \mathbf{b}(\mathbf{r}) \equiv -N_\zeta^* \left(e\varphi(\mathbf{r}) + \frac{\partial \zeta}{\partial \mathbf{B}} \mathbf{b}(\mathbf{r}) \right). \qquad (8.47)$$

The magnetic field $\mathbf{b}(\mathbf{r})$ satisfies the equation

$$\text{curl } \mathbf{b}(\mathbf{r}) = 4\pi \text{ curl } \delta\mathbf{M}(\mathbf{r}) = 4\pi \text{ curl } \left(\frac{\partial \mathbf{M}}{\partial \zeta} e\varphi(\mathbf{r}) + \frac{\partial \mathbf{M}}{\partial \mathbf{B}} \mathbf{b}(\mathbf{r}) \right) \text{ div } \mathbf{b}(\mathbf{r}) = 0.$$

(8.48)

Here $\mathbf{M}$ is the magnetization vector, ζ is the chemical potential of charge carriers, and $\varphi(\mathbf{r})$ is the potential of the electrical field, arising at deformation.

We should add the electrical neutrality condition to the relations (8.47), (8.48):

$$\delta N(\mathbf{r}) + eN \text{ div } \mathbf{u}(\mathbf{r}) = 0,$$

(8.49)

where $\mathbf{u}(\mathbf{r})$ is the lattice displacement vector. It follows from (8.47)–(8.49) that

$$\text{curl } \{(1 - 4\pi\chi)\mathbf{b}(\mathbf{r})\} = -4\pi \text{ curl } \left\{ \frac{\partial \mathbf{M}}{\partial \zeta} \frac{N}{N_\zeta^*} \text{ div } \mathbf{u}(\mathbf{r}) \right\}.$$

(8.50)

The set of simultaneous equations (8.47)–(8.49) was stated in [50]. It allows us to exclude $\mathbf{b}(\mathbf{r})$ and to express the potential $\varphi(\mathbf{r})$ in terms of the lattice displacement vector. As a result, the expression for the electron force $\mathbf{F}(\mathbf{r})$, acting upon the lattice under its displacement by the vector $\mathbf{u}(\mathbf{r})$, can be derived. For the conductor with an axisymmetric Fermi surface in the magnetic field $\mathbf{B}$ directed along the symmetry axis, the force $\mathbf{F}(\mathbf{r})$ equals

$$\mathbf{F}(\mathbf{r}) = \lambda_0 \mathbf{b}_0 (\mathbf{b}_0 \nabla(\nabla \mathbf{u}(\mathbf{r}))) + \lambda [\mathbf{b}_0 \times [\nabla(\nabla \mathbf{u}(\mathbf{r})) \times \mathbf{b}_0]].$$

(8.51)

Here $\mathbf{b}_0$ is the unit vector directed along $\mathbf{B}$. In the considered case $\mathbf{b}_0$ coincides with the normal vector to the layers $\mathbf{n}$. It follows from (8.51), that under longitudinal in the $\mathbf{b}_0$ (perpendicular to the layers) direction of gradients of the deformation tensor, the electron contribution to the corresponding elastic modulus is equal to λ_0, and under their direction across $\mathbf{b}_0$ (in a plane of layers) it equals λ. The elastic constants λ_0 and λ are described by the expressions

$$\lambda_0 = \frac{N^2}{N_\zeta^*} = \frac{N^2}{(1 - \alpha_0)g_0} [1 - (1 - \bar{\alpha}_0(0))K'],$$

(8.52)

$$\lambda = \lambda_0 \left(1 + \frac{4\pi\chi_\zeta}{1 - 4\pi\chi_\parallel} \right) = \frac{N^2}{(1 - \alpha_0)g_0} [1 - (1 - \bar{\alpha}_0(0))K''].$$

(8.53)

Here the oscillating function K'' has the form,

$$K'' = \frac{\Delta}{1 + (1 + \alpha_0(0) - \bar{\alpha}_0(0) + \beta_0(0) - 4\pi\chi_0\gamma^4)\Delta(0)},$$

(8.54)

where the longitudinal part of the magnetic susceptibility $\chi_\parallel$ is described by expression (8.46), $\chi_\zeta = (\partial M_z/\partial \zeta)(\partial \zeta/\partial B)$, and the quantity χ_0 defines the Landau diamagnetic susceptibility, which equals $-\chi_0/3$.

As follows from (8.52), when the deformation interaction is neglected, the quantity λ_0 coincides with the compression modulus of the electron liquid. The structure of the quantity λ is more complicated. Besides the contribution, connected with the electron compression, λ also contains the contribution arising due to magnetostriction. Comparison the oscillating function K' with the function K'' shows that under low temperatures there is a significant distinction between the oscillating corrections to the elastic moduli λ_0 and λ arising due to magnetostriction. Beyond the bounds of the low-temperature region the contribution, connected with the magnetostriction, becomes negligible and the distinction between λ_0 and λ is smoothed.

The more sharply the difference between λ and λ_0 is displayed, at those values of the magnetic field **B**, at which the quantity $1 - 4\pi\chi_{\parallel}$ goes to zero and the denominator in (8.54) turns to zero at the peaks of the quantum oscillations. It makes the elastic constant λ vanish near the oscillations maxima. Thus, there can arise lattice instability. It is connected with the instability in the magnetization and arises because of the magnetostriction. Under these conditions the Fermi-liquid correlation plays the major part. When $4\pi\chi_0\gamma^4 < 1$ the instability can arise only under $\alpha_0(0) - \overline{\alpha}_0(0) + \beta_0(0) < 0$.

In the considered case $(\mathbf{B}||\mathbf{n})$ the velocity of the longitudinal sound, propagating along the normal to the layers, is proportional to the root square of λ_0, and when it propagates in the plane of layers, to the root square of λ. Therefore, the distinction between λ_0 and λ, due to the magnetostriction, will be displayed in the quantum oscillations of the velocity of sound.

The characterictic properties of the abnormal skin effect, the cyclotron resonance, the acoustic–electron phenomena, and the quantum oscillations of thermodynamic observables in layered conductors are caused by the quasi two-dimensional character of the energy spectra of charge carriers. It is acceptable to suppose that the most favorable conditions for the manifestation of these specific effects are created when the Fermi surface rippling is too weak (the parameter η is too small in comparison with unity). Under these conditions one can consider the Fermi surface as a pure cylinder. The model of the energy spectrum of the charge carriers assumed here allows us to show that the specific features in the electron properties of the layered conductors are caused rather by the local geometry of their Fermi surfaces in the vicinities of their extremal cross-sections than by the weakness of their surfaces rippling by itself.

The small values of the charge carrier's velocity component along the normal to the conducting planes can take place when the Fermi surface is strongly rippled $(\eta \sim 1)$. For instance, when the parameters r and l in (8.4) take on values large in comparison with unity, it corresponds to the Fermi surface which is a steplike cylinder. The longitudinal velocity of the quasi particles on such Fermi surfaces equals zero at any value of the parameter η. The model (8.3), (8.4) presented here allows us to treat in detail the manifestations of the layered conductor's Fermi surface local geometry. It allows us to describe all the types of singularities arising due to the cyclotron resonance in the normal magnetic field, and to analyze the correspondense between the resonance curve's profiles and the Fermi surface's local geometry. It also allows us to analyze the specific low-temperature oscillatory effects in the organic layered metals in quantizing the magnetic fields. The experimental observation of these effects can provide additional information about the parameters of the electronic energy spectra and the interelectron correlation in such substances.

8.4   Deformed Fermi Surfaces and the Magnetoacoustic Response of the Modulated Quantum Hall System Near a Half-Filling of the Lowest Landau Level

The integer and fractional quantum Hall effect still remains at the cutting edge of condensed matter physics, nearly two decades after the initial discovery. New experiments continue to give us unexpected results, and theoretical work on the strongly correlated two-dimensional electron gas, at very low temperatures and under a high magnetic field and with possibly added external "modulating" perturbation, continues at a very intense level.

Surface acoustic waves, propagating above a two-dimensional electron gas in GaAS/AlGaAs heterostructures in a strong magnetic field, change their properties due to the conductivity of the adjacent electron system. Hence the surface acoustic waves' experiments provide an efficient tool for experimental studies of the conductivity of two-dimensional electron systems in quantum Hall states at even denominator filling factors ν [193]–[198]. It was observed repeatedly that the dependence of the surface acoustic waves' velocity shift on the magnetic field B exhibits a pronounced minimum at B corresponding to $\nu = \frac{1}{2}$, implying a maximum in the conductivity. Exactly at $\nu = \frac{1}{2}$ the conductivity appeared to be linear in the sound-wave vector q when the wavelength is small enough. These results are in good agreement with the theoretical calculations based on the Chern–Simons approach of Halperin, Lee, and Read (HLR) [199], [200], describing the quantum Hall state at and near $\nu = \frac{1}{2}$, thus providing strong support for this theory. This theory corresponds to the physical picture of the electrons decorated by attached quantum flux tubes which are the relevant fermionic quasi particles of the system (the so called composite fermions). Near half-filling the composite fermions move in the reduced effective magnetic field, $B_{\text{eff}} = B - 4\pi \hbar c n / e$ (n is the electron density). Just at $\nu = \frac{1}{2}$, the composite fermions form a Fermi sea and exhibit a Fermi surface. The composite fermion Fermi surface is a circle in a two-dimensional quasi-momenta space and its radius p_F equals $\sqrt{4\pi n \hbar^2}$.

Due to the piezoelectric properties of GaAS, the velocity shift ($\Delta s / s$) and the attenuation rate (Γ) for the surface acoustic waves, propagating along the x-axis across the surface of a heterostructure, containing a two-dimensional electron gas, take the form [201], [202]:

$$\frac{\Delta s}{s} = \frac{\alpha^2}{2} \operatorname{Re} \left(1 + \frac{i \sigma_{xx}}{\sigma_m} \right)^{-1}, \tag{8.55}$$

$$\Gamma = -q \frac{\alpha^2}{2} \operatorname{Im} \left(1 + \frac{i \sigma_{xx}}{\sigma_m} \right)^{-1}. \tag{8.56}$$

Here $\mathbf{q}$, $\omega = sq$ is the surface acoustic wave wave vector and frequency, α is the piezoelectric coupling constant, $\sigma_m = \varepsilon s / 2\pi$, ε is an effective dielectric constant of the background, and σ_{xx} is the component of the electron conductivity tensor.

According to the semiclassical composite fermion theory, the electron resistivity tensor ρ (at finite q and ω) is given by [199]:

$$\rho = \sigma^{-1} = \rho^{\mathrm{cf}} + \rho^{\mathrm{cs}}, \tag{8.57}$$

where ρ^{cf} is the composite fermion resistivity tensor which has to be calculated by means of a Boltzmann transport equation; the contribution ρ^{cs} arises due to a fictitious magnetic field which originates from the Chern–Simons formulation of the theory. This tensor contains only off-diagonal elements $\rho^{\mathrm{cs}}_{xy} = -\rho^{\mathrm{cs}}_{yx} = 4\pi \hbar/e^2$.

Recently, new results were obtained near $\nu = \frac{1}{2}$ on samples whose electron density was periodically modulated in one direction by applying an additional static electric field [203], [204]. It was observed in the experiments [203] that the periodic density modulation in the two-dimensional electron gas produces dramatic and highly anisotropic response to the surface acoustic waves. When the density modulation wave vector $\mathbf{g}$ is orthogonal to the surface acoustic waves' wave vector $\mathbf{q}$, the minimum in the magnetic field dependence of the sound velocity shift at $\nu = \frac{1}{2}$ was converted to a large maximum. The maximum was observed only for sufficiently large modulation wave vectors and for sufficiently high magnitudes of the electric field producing the electron density modulation. In some experiments reported in [203], the peak disappeared on further increasing the electron density modulation but was replaced by a minimum in the surface acoustic waves' velocity shift again. In the case when the surface acoustic waves propagated along the electron density modulation direction, no anomaly in the electron system response was observed.

In the framework of the approach proposed in [199] we can see that the grating will influence the composite fermions' system in two ways: through the direct effect of the modulating potential and through the effect of the magnetic field $\Delta B(\mathbf{r})$ proportional to the local density modulation $\Delta n(\mathbf{r})$ ($\Delta B(\mathbf{r}) = -4\pi \hbar c \Delta n(\mathbf{r})/e$).

It was shown [205], [206] that, in the local limit $ql \ll 1$, the direct effect of the external modulating electric field on the two-dimensional electron gas conductivity at $ql \ll 1$ is small compared to the effect due to the modulation of the effective magnetic field B_{eff}. But under the experimental conditions $ql > 1$ we cannot neglect the direct effect of the external modulating potential on the conductivity of the two-dimensional electron gas. The simplest physical implementation of this result is to invoke a new concrete model whereby the periodic modulating electric field deforms the originally circular composite fermions' Fermi surface. We can have in mind the analogous situation of the action of an extra crystalline field on the Fermi surface of a conventional metal. In our case, the modulating wave vector $\mathbf{g}$ replaces the reciprocal lattice vector of the usual metal.

In the local limit ($ql \ll 1$) all parts of the composite fermions' Fermi surface contribute to the composite fermion conductivity essentially equally, and the difference in shape between the deformed and undeformed Fermi surfaces does not lead to a significant change in the composite fermions' conductivity. In contrast, for large q ($ql > 1$), the most important mechanism of the absorption of the energy of the electric field connected with the surface acoustic waves is the Landau damping, therefore the main contribution to the conductivity originates from those small effective parts of the Fermi surface where the quasi-particle velocity vector $\mathbf{v}$ and the surface acoustic waves' wave vector $\mathbf{q}$ are nearly transverse ($\mathbf{q} \cdot \mathbf{v} \approx 0$). Hence the scalar potential modulation deforming the composite fermions' Fermi surface can be very important at $ql > 1$. Moreover, in this case, it may be the most important factor determining the magnetic field dependence of the linear response functions of the composite fermions near one half-filling.

We will show below that a modulation-induced deformation of the composite fermions' Fermi surface can be at the origin of the observed transport anomalies. Assume that electronic density is modulated in the y-direction and that the modulation period is small enough ($\hbar g > 2p_F$). Using the nearly free electron model we can obtain the energy-momentum relation for the composite fermions in the form

$$E(\mathbf{p}) = \frac{p_x^2}{2m^*} + \frac{p_y^{*2}}{2m^*} + \frac{(\hbar g)^2}{8m^*} - \sqrt{\left(\frac{\hbar g p_y^*}{2m^*}\right)^2 + V_g^2}. \tag{8.58}$$

Here $p_y^* = p_y - \hbar g/2$, m^* is the composite fermions' effective mass, and V_g is the magnitude of the quasi-particle potential energy in the periodic electric field. V_g is assumed to be small in comparison with the quasi-particles' Fermi energy E_F. Calculating the Fermi surface curvature, we can find it tending to zero when p_x tends to $\pm p_F (V_g/E_F)^{1/2}$ (see Figure 8.5).

In the vicinities of the corresponding points on the Fermi surface the quasi particles' velocities are nearly parallel to the y-direction. So these parts of the Fermi surface contribute strongly to the attenuation and velocity shift of the surface

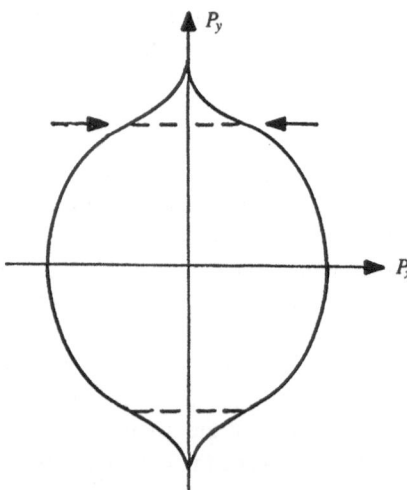

FIGURE 8.5. The shape of the composite fermions Fermi surface deformed due to an external periodic modulation in the "y"-direction in the nearly free electron approximation (solid line) at $2p_F < \hbar g$ and the composite fermions' Fermi surface corresponding to (8.61) (dashed line).

acoustic waves, propagating in the x-direction. Near these zero-curvature points we will use asymptotic expressions for (8.58). Determining (p_{x_0}, p_{y_0}) by $p_{x_0} = \zeta p_F$, $p_{y_0} = p_F\left(1 - (1/\sqrt{2})\zeta^2\right)$, where $\zeta = \sqrt{V_g/E_F}$, $E_F = p_F^2/2m^*$, we can expand the variable p_y in powers of $(p_x - p_{x_0})$, and keep the lowest-order terms in the expansion. We obtain

$$p_y - p_{y_0} = -\zeta(p_x - p_{x_0}) - \frac{2}{\zeta^4}\frac{(p_x - p_{x_0})^3}{p_F^2}. \tag{8.59}$$

Near p_{x_0}, where $(|p_x - p_{x_0}| < \zeta^2 p_F)$, the first term on the right-hand side of (8.59) is small compared to the second one and can be omitted.

Hence, near p_{x_0} we have

$$E(\mathbf{p}) = \frac{4}{\zeta^4}\frac{p_F^2}{2m^*}\left(\frac{p_x - p_{x_0}}{p_F}\right)^3 + \frac{p_y^2}{2m^*}. \tag{8.60}$$

The "nearly free" particle model can be used when ζ^2 is very small. For larger V_g, corresponding to $\Delta n/n$ of the order of a few percent (as in the experiments [230]), the local flattening of the composite fermions' Fermi surface can be more significant. To analyze the contribution to the conductivity from these flattened parts, we generalize the expression for $E(\mathbf{p})$ and define our dispersion as

$$E(\mathbf{p}) = \frac{p_0^2}{2m_1}\left|\frac{p_x}{p_0}\right|^\gamma + \frac{p_y^2}{2m_2}, \tag{8.61}$$

where p_0 is a constant with the dimension of momentum, the m_i are effective masses, and γ is a dimensionless parameter which will determine the shape of the composite fermions' Fermi surface. When $\gamma > 2$, the Fermi surface looks like an ellipse flattened near the vertices $(0, \pm\sqrt{m_2/m_1}\,p_0)$. Near these points the curvature is

$$K = -\frac{\gamma(\gamma-1)}{2p_0\sqrt{m_1/m_2}}\left|\frac{p_x}{p_0}\right|^{\gamma-2} \tag{8.62}$$

and $K \to 0$ at $p_x \to 0$. The surface will be the flatter at $(0, \pm\sqrt{m_2/m_1}\,p_0)$, the larger is the parameter γ. A separate investigation is required to establish how γ depends on V_g. Here we postulate (8.61) as a natural generalization of (8.62) and we then derive the resulting surface acoustic waves' response. Thus (8.61) is assumed here as a concrete model for the deformed composite fermions' Fermi surface.

When $p_F > \hbar g$ we have to consider the composite fermions' Fermi surface as consisting of several branches belonging to several "bands" or Brillouin zones, as is shown in Figure 8.6. The modulating potential wave vector $\mathbf{g}$ in this case determines the size of the "unit cell." (If the modulating potential is applied in one direction, as in experiments [203], the "unit cell" is a strip extended transverse to $\mathbf{g}$.) However, with this condition we may expect some branches of the composite fermions' Fermi surface to be flattened.

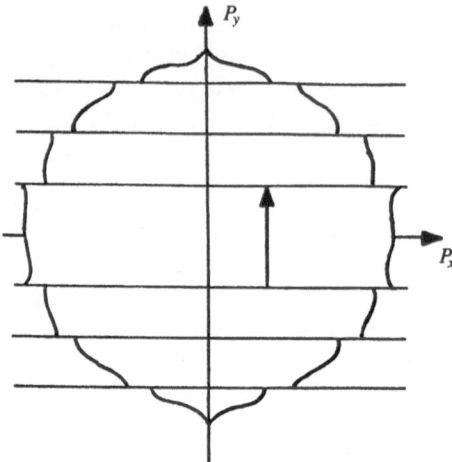

FIGURE 8.6. The deformation of the composite fermions' Fermi surface in a modulated two-dimensional system in the nearly free electron approximation under the condition $2p_F > \hbar g$. The modulation wave vector $\mathbf{g}$ is directed along the "y"-axis.

Representing the composite fermions as semiclassical particles with simultaneously defined values of position and momentum, we can introduce their distribution function in a phase space $n(\mathbf{r}, \mathbf{p}, t)$. For the equilibrium state of composite fermions the distribution function is the Fermi–Dirac function f_p, depending on the composite fermions' energy E_p alone. In the presence of a small disturbance, the function $n(\mathbf{r}, \mathbf{p}, t)$ contains a small correction $g(\mathbf{r}, \mathbf{p}, t)$ describing its deviation from the equilibrium value. The correction can be written in the form $g(\mathbf{r}, \mathbf{p}, t) = -(\partial f_\mathbf{p}/\partial E_\mathbf{p})\chi_\mathbf{p}(\mathbf{r}, t)$.

To calculate the surface acoustic wave response for the modulated two-dimensional electron gas, we have to solve the transport problem in both the local and non-local regimes starting from the Boltzmann equation. Using linear response theory we have

$$\left(\frac{\partial}{\partial t} + \mathbf{v} \cdot \nabla_\mathbf{r} + \frac{\mathbf{e}}{\mathbf{c}}[\mathbf{v} \times \mathbf{B}(\mathbf{r})]\nabla_\mathbf{p}\right)\chi_\mathbf{p}(\mathbf{r}, t) - e\mathbf{E}(\mathbf{r}, t)\mathbf{v} = I[g]. \qquad (8.63)$$

Here $I[g]$ is the collision integral describing scattering with a relaxation time τ. The field $\mathbf{E}(\mathbf{r}, t)$ is the effective electric field including both externally (connected with the surface acoustic waves) and internally induced contributions.

The effect of the density modulation is taken into account through the composite fermions' Fermi surface deformation and through the inhomogeneity of the magnetic field $B(\mathbf{r})\big(B(\mathbf{r}) = B_{\text{eff}} + \Delta B(\mathbf{r}) \equiv B_{\text{eff}} - 4\pi \hbar c \Delta n(\mathbf{r})/e\big)$. A systematic analysis of the problem for longwave external disturbances was performed in [205], [206]. However, in the opposite case, when $ql > 1$ the problem is too complicated to be solved straightforwardly, therefore we will proceed as follows. As a first step, we will omit the correction $\Delta B(\mathbf{r})$ in order to consider the surface acoustic waves' response for the composite fermions' system, whose Fermi surface is deformed due to the density modulation, in a uniform magnetic field B_{eff}. Afterward, we will analyze semiquantitatively the influence of the weak inhomogeneous magnetic field $\Delta B(\mathbf{r})$ on the response functions and we will modify our results to include the corresponding corrections.

We assume that $\mathbf{E}(\mathbf{r}, t) = \mathbf{E} \exp(i\mathbf{qr} - i\omega t)$. Exactly at $\nu = \frac{1}{2}$ the q-and ω-dependent composite fermions' conductivity tensor is given by

$$\sigma_{\alpha\beta}^{cf}\left(\nu = \tfrac{1}{2}\right) = \frac{e^2}{(2\pi\hbar)^2} \int \frac{v_\alpha(\mathbf{p})v_\beta(\mathbf{p})}{iqv_x - i\omega + 1/\tau} \frac{d\lambda}{|v|}, \tag{8.64}$$

where $d\lambda = \sqrt{dp_x^2 + dp_y^2}$. The integration has to be performed over the composite fermions' Fermi surface. When the composite fermions' Fermi surface consists of several branches the summation over these branches has to be done in (8.64).

At first we shall consider the contribution to the conductivity from the flattened part of the composite fermion Fermi surface ($\sigma_{(1)}^{cf}$) which corresponds to the model (8.61). We can parametrize the dispersion equation of the model as follows:

$$p_x = \pm p_0 |\cos u|^{2/\gamma}, \qquad p_y = p_0 \sqrt{m_2/m_1} \sin u, \tag{8.65}$$

where the parameter u takes values belonging to the interval $0 \le u < 2\pi$, and the "+" and "−" signs are chosen corresponding to the normal domains of the positive and negative values of cosine.

Taking into account the Fermi surface symmetry, we can transform the expression (8.64) for $\sigma_{(1)yy}^{cf}$ ($\nu = \frac{1}{2}$) to the form

$$\sigma_{(1)yy}^{cf}\left(\nu = \tfrac{1}{2}\right) = \frac{8e^2}{(2\pi\hbar)^2} p_0 l \sqrt{\frac{m_1}{m_2}} \frac{2\mu^2}{(\mu + 1)^2} \int_0^{\pi/2} \frac{\sin^2 u (\cos u)^{(\mu-1)/(\mu+1)}}{1 + (ql)^2 \cos u^{4/(\mu+1)}} \, du. \tag{8.66}$$

Here $\mu = 1/(\gamma - 1)$, which defines a new dimensionless parameter ($0 < \mu \le 1$). The value $\mu = 1$ corresponds to the case $\gamma = 2$ (the Fermi surface, shaped as an ellipse). The composite fermions' mean free path l equals

$$l = \frac{\mu + 1}{2\mu} \frac{p_0}{m_1} \tau. \tag{8.67}$$

For typical experimental conditions [203], [204] $\omega\tau \ll 1$, therefore $\omega\tau$ is neglected in (8.66). Making the change of variables

$$y = (ql)^2 (\cos u)^{4/(\mu+1)} \tag{8.68}$$

we obtain the following asymptotic expression

$$\sigma_{(1)yy}^{cf}\left(\nu = \tfrac{1}{2}\right) = \frac{4e^2\mu^2}{(\mu + 1)(2\pi\hbar)^2} \frac{p_0 l}{(ql)^\mu} \int_0^{(ql)^2} \frac{y^{\mu/2-1}}{1 + y} \, dy. \tag{8.69}$$

Replacing the upper limit in the integrand (8.69) by infinity, we have

$$\sigma_{(1)yy}^{cf}\left(\nu = \tfrac{1}{2}\right) = b \frac{e^2}{4\pi\hbar^2} p_0 \frac{l}{(ql)^\mu}. \tag{8.70}$$

Here

$$b = \sqrt{\frac{m_1}{m_2} \frac{4\mu^2}{\mu + 1}} \frac{1}{\sin(\pi\mu/2)}.$$

is a dimensionless constant of the order of unity. When the Fermi surface is a circle $(m_1 = m_2, p_0 = p_F, \mu = 1)$ b equals 2 and our result (8.70) coincides with the corresponding [199], [200] formula for zero flattening:

$$\sigma_{(1)yy}^{cf}\left(\nu = \tfrac{1}{2}\right) = \frac{e^2 p_F}{4\pi \hbar^2 q}. \tag{8.71}$$

Comparing (8.70) and (8.71) we can conclude that the flattening of the effective segments of the composite fermions' Fermi surface critically changes both the magnitude and wave vector dependence of the conductivity. When the flattening is very pronounced ($\gamma \gg 1$) the parameter μ takes values close to zero. In this case, the conductivity is larger in magnitude than in the case when the Fermi surface is undeformed and nearly independent of q. This result is in agreement with the experimental data [203]. Thus we can conclude that the deformed Fermi surface model is supported by experiment.

When $2p_F < \hbar g$ the entire composite fermions' Fermi surface is described by the model (8.61), and (8.70) gives the approximation for the composite fermions' conductivity component $\sigma_{yy}^{cf}(\nu = \tfrac{1}{2})$ at $ql \gg 1$. When $2p_F > \hbar g$, as in experiment [203], we have to evaluate the contributions to the conductivity from the composite fermions' Fermi surface branches (presumably nonflattened) belonging to other "Brillouin zones." We shall consider them to be constructed from the pieces of the undeformed circular Fermi surface. Some of them have no effective parts where $\mathbf{q} \cdot \mathbf{v} \approx 0$. Contributions from these parts can be evaluated as follows:

$$\sigma_{(2)yy}^{cf}\left(\nu = \tfrac{1}{2}\right) \approx \frac{2e^2}{(2\pi\hbar)^2} p_F l \sum_i \int_{u_{i0}}^{u_{i1}} \frac{\sin^2 u \, du}{1 + (ql)^2 \cos^2 u}. \tag{8.72}$$

The summation in (8.72) has to be done over all branches of the composite fermions' Fermi surface which have no effective parts. The limits in the integrals over u are determined by a range of possible values of the composite fermions' velocity component v_x on these branches. The upper limits (u_{i1}) take values significantly less than $\pi/2$, therefore we can use the approximation $1 + (ql)^2 \cos^2 u \approx (ql)^2 \cos^2 u$. It gives us

$$\sigma_{(2)yy}^{cf}\left(\nu = \tfrac{1}{2}\right) \approx \frac{e^2}{\pi\hbar^2} \frac{p_F}{q} \frac{W}{\pi ql}. \tag{8.73}$$

Those parts of the Fermi surface which have nonflattened effective parts contribute to the conductivity as follows (summation is done over these parts of the composite fermions' Fermi surface):

$$\sigma_{(3)yy}^{cf}\left(\nu = \tfrac{1}{2}\right) \approx \frac{4e^2}{(2\pi\hbar)^2} p_F l \sum_k \int_{u_{k0}}^{\pi/2} \frac{\sin^2 u \, du}{1 + (ql)^2 \cos^2 u}. \tag{8.74}$$

Evaluating the integrals we have

$$\sigma_{(3)yy}^{cf}\left(\nu = \tfrac{1}{2}\right) \approx \frac{Ue^2}{2\pi\,\hbar^2}\frac{p_F}{q}, \tag{8.75}$$

where U is a constant of the order of unity. Comparing σ_{3yy}^{cf} $(\nu = \tfrac{1}{2})$ and σ_{1yy}^{cf} $(\nu = \tfrac{1}{2})$ we can see that, even in the case when the composite fermions' Fermi surface has additional effective parts besides the flattened segments their contribution to the composite fermions' conductivity for large q is smaller in magnitude than the contribution from the flattened segments. The ratio of the magnitudes is of the order of $(ql)^{\mu-1}$. Thus the contribution from the effective segments with anomalously small curvature is the principal term of the conductivity in the limit of large q for the strong composite fermions' Fermi surface flattening ($\gamma \gg 1$, $\mu \ll 1$).

The contribution from the flattened part of the Fermi surface determines the main approximation of the composite fermions' conductivity and, consequently, the electron conductivity in the modulated system at $ql \gg 1$ when $\mathbf{q} \perp \mathbf{g}$. The component σ_{xx} of the electron conductivity tensor which appears in the expressions (8.55), (8.56) for the ultrasound velocity shift and attenuation, equals

$$\sigma_{xx}\left(\nu = \tfrac{1}{2}\right) = \frac{\rho_{yy}}{\rho_{xy^2}} \approx \frac{e^2(ql)^\mu}{4b\pi p_0 l}. \tag{8.76}$$

It follows from (8.76) that the electron conductivity in a modulated quantum Hall system at $\nu = \tfrac{1}{7}2$ can be smaller in magnitude than in unmodulated systems, and is almost independent of q when $ql \gg 1$ and flattening of the composite fermions' Fermi surface is strong ($\mu \ll 1$).

When $\mathbf{g}\|\mathbf{q}$ (the modulation applied in the "x"-direction) the flattened segments are not effective parts of the Fermi surface (see Figures 8.5 and 8.6). Then the main approximation to the composite fermions' conductivity component $\sigma_{yy}^{cf}(\nu = \tfrac{1}{2})$ is determined by the contribution from the nonflattened effective parts and described by (8.75). Correspondingly, the electron conductivity σ_{xx} is

$$\sigma_{xx}\left(\nu = \tfrac{1}{2}\right) = \frac{e^2q}{8\pi U p_F}. \tag{8.77}$$

Under the condition $2p_F < \hbar g$ (the composite fermions' Fermi surface wholly is in the first Brillouin zone) $U = 1$, and the result coincides with that for the unmodulated system. Hence the effect of the density modulation on the response of the two-dimensional electron system is very anisotropic.

To analyze the composite fermions' conductivity dependence on the magnetic field B_{eff} at ν near, but not equal $\tfrac{1}{2}$, we can start from the Boltzmann transport equation for the composite fermions' distribution function in a uniform magnetic field $\mathbf{B}_{\text{eff}}$. Following the standard methods, we obtain for the conductivity tensor component

$$\sigma_{\alpha\beta}^{cf} = \frac{e^2 m_c}{(2\pi\,\hbar)^2}\frac{1}{\Omega}\int_0^{2\pi} d\psi \exp\left[-\frac{iq}{\Omega}\int_0^\psi v_x(\psi'')\,d\psi''\right] v_\alpha(\psi)$$

$$\times \int_{-\infty}^{\psi} v_\beta(\psi') \exp\left[\frac{iq}{\Omega} \int_0^\psi v_x(\psi'') \, d\psi'' + \frac{1}{\Omega\tau}(\psi' - \psi)\right] d\psi'. \quad (8.78)$$

Here, m_c, Ω are the cyclotron mass and the cyclotron frequency at the field B_{eff}, and ψ is the angular coordinate of the composite fermions' cyclotron orbit ($\psi = \Omega t$, t is the time of the composite fermions' motion along the cyclotron orbit). We have taken $\omega\tau \ll 1$.

Introducing a new variable η, which is defined by the expression,

$$\eta \equiv \left(-i\omega + \frac{1}{\tau} + ik\Omega + iqv_x(\psi)\right)\theta + iq \int_0^\theta [v_x(\psi + \Omega\theta') - v_x(\psi)] \, d\theta', \quad (8.79)$$

we arrive at the result $\theta = (\psi' - \psi)/\Omega$ which is similar to formula (4.61):

$$\sigma_{\alpha\beta}^{\text{cf}} = \frac{ie^2}{(2\pi\hbar)^2} m_c \sum_k v_{k\beta} \int_{-\infty}^0 e^\eta \, d\eta \int_0^{2\pi} \frac{v_\alpha(\psi) \exp(ik\psi)}{\omega + i/\tau - k\Omega - qv_x(\psi + \Omega\theta)} \, d\psi. \quad (8.80)$$

When the composite fermions' Fermi surface consists of several branches the summation over the branches has to be done in (8.80).

It was shown in the previous section that for $\mathbf{q} \perp \mathbf{g}$ the main contribution to the conductivity arises from the flattened parts of the composite fermions' Fermi surface. To evaluate this contribution we transform the integration over ψ to an integration over the composite fermions' Fermi surface

$$m_c \int_0^{2\pi} d\psi = \int \frac{d\lambda}{|v|}. \quad (8.81)$$

Under the conditions $ql \gg 1$, $\omega\tau \ll 1$, $\Omega\tau < 1$, the variable θ is approximately equal to $\eta\tau(1 + iqv_x\tau)^{-1}$. Expanding the last term in the denominator of (8.80) in powers of $\Omega\theta$, and keeping the first terms in the expansion, one obtains:

$$qv_x(\psi + \Omega\theta) = qv_x(\psi) + \eta\Omega\tau q(dv_x/d\psi)(1 + iqv_x\tau)^{-1}$$
$$+ \frac{\eta^2}{2}(\Omega\tau)^2 q(d^2v_x/d\psi^2)(1 + iqv_x\tau)^{-2}$$
$$\equiv qv_x(\psi) - \eta\delta_1(q, \psi) + i\eta^2\delta_2(q, \psi). \quad (8.82)$$

We can estimate the averages of the functions $\delta_1(q, \psi)$ and $\delta_2(q, \psi)$ over the composite fermions' Fermi surface. These are

$$\langle\delta_1(q, \psi)\rangle = \frac{1}{2\pi} \int_0^{2\pi} \delta_1(q, \psi) \, d\psi = 0,$$
$$\langle\delta_2(q, \psi)\rangle = \frac{1}{2\pi} \int_0^{2\pi} \delta_2(q, \psi) \, d\psi \approx (\Omega\tau)^2 \bar\delta(q)/\tau, \quad (8.83)$$

this defines $\bar\delta(q)$.

Using the model (8.61) we can prove that for moderately flattened composite fermions' Fermi surface the function $\bar\delta(q)$ takes positive values of the order of $(ql)^{-(1+\mu)/2}$ (for an underformed Fermi surface it is of the order of $(ql)^{-1}$). For strong flattening of the composite fermions' Fermi surface $\bar{\delta}(q)$ is practically independent of q and takes negative values of the order of unity. Using this estimate we obtain the approximation

$$\sigma_{1yy}^{cf} = \tfrac{1}{2}\sigma_{1yy}^{cf}(\nu = \tfrac{1}{2})[S_\mu^+(\Omega\tau) + S_\mu^-(\Omega\tau)]. \tag{8.84}$$

Here, $\sigma_{1yy}^{cf}(\nu = \tfrac{1}{2})$ is the conductivity at one half-filling (8.70), and the functions $S_\pm(\Omega\tau)$ are

$$S_\mu^\pm(\Omega\tau) = \int_{-\infty}^0 e^\eta (\kappa_\pm(\eta))^{\mu-1}\, d\eta, \tag{8.85}$$

where $\kappa_\pm(\eta) = 1 \mp i\Omega\tau(1 \mp i\Omega\tau\bar{\delta}\eta^2)$.

The approximation (8.84), (8.85) represents the main term in the expansion of the conductivity in inverse powers of the large parameter ql. When $\mu < 1$ the next term of the expansion equals

$$\tfrac{1}{2}\sigma_{1yy}^{cf}(\nu = \tfrac{1}{2})\frac{i}{ql}\tan\frac{\pi\mu}{2}\left(S_{\mu+1}^-(\Omega\tau) - S_{\mu+1}^+(\Omega\tau)\right). \tag{8.86}$$

For $\mu = 1$ the first correction is of the order of $\ln(ql)/ql$.

For small $\bar{\delta}\Omega\tau$ we can expand the function $S_\mu^\pm(\Omega\tau)$ in powers of $\Omega\tau$. Keeping terms larger than $(\Omega\tau)^3$ we have

$$\sigma_{1yy}^{cf} = \sigma_{1yy}^{cf}(\nu = \tfrac{1}{2})[1 - a^2(\Omega\tau)^2 + \xi(\Omega\tau)^2], \tag{8.87}$$

where $a^2 = (1 - \mu)(2 - \mu)/2$ and $\xi = 4(1 - \mu)\bar{\delta}$.

For moderate flattening of the effective parts of the composite fermions' Fermi surface the constant ξ (positive in this case) is small compared to a^2 because of the small factor $\bar{\delta}$. For strong flattening, this constant ξ takes negative values of the order of a^2. In both cases we can omit the last term in (8.87) neglecting it for $\mu \sim 1$, or combining it with the previous term (for $\mu \ll 1$). The contributions from the other (nonflattened) parts of the Fermi surface to the composite fermions' conductivity component σ_{yy}^{cf} are small compared to σ_{1yy}^{cf}. Hence, (8.87) gives the main approximation for σ_{yy}^{cf} in a modulated system at ν near $\tfrac{1}{2}$. Other components of the composite fermions' conductivity tensor can be evaluated similarly. For $ql \gg 1$ we obtain $\sigma_{xx}^{cf} \sim \sigma_0/(ql)^2$, $\sigma_{yx}^{cf} = -\sigma_{xy}^{cf} \sim \Omega\tau\sigma_0/(ql)^{1+\mu}$, where $\sigma_0 = \sigma_{1yy}^{cf}(\nu = \tfrac{1}{2})(ql)^\mu = b(e^2/4\pi\hbar^2)p_0l$. Hence under conditions of the experiment [203] $(\Omega\tau < 1)$ $(\sigma_{xy}^{cf})^2/\sigma_{xx}^{cf}\sigma_{yy}^{cf} \sim (\Omega\tau)^2/(ql)^\mu \ll 1$ and we can calculate the electron conductivity using the formula

$$\sigma_{xx}(q) = \frac{e^4}{(4\pi\hbar)^2}\rho_{yy}^{cf} = \frac{e^4}{(4\pi\hbar)^2}\frac{\sigma_{xx}^{cf}}{\sigma_{xx}^{cf}\sigma_{yy}^{cf} + (\sigma_{xy}^{cf})^2} \approx \frac{e^4}{(4\pi\hbar)^2}(\sigma_{yy}^{cf})^{-1}. \tag{8.88}$$

Substituting this result into formulas (8.55) and (8.56) we obtain the following approximation ($\omega\tau \ll 1$, $ql \gg 1$, $\Omega\tau < 1$):

$$\frac{\Delta s}{s} = \frac{\alpha^2}{2}\frac{1}{1+\tilde{\sigma}^2}\left(1 - \frac{2a^2\tilde{\sigma}^2}{1+\tilde{\sigma}^2}(\Omega\tau^2)\right), \tag{8.89}$$

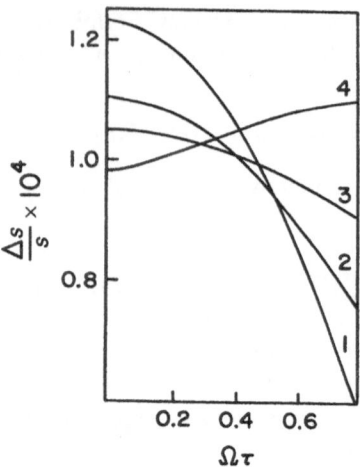

FIGURE 8.7. The surface acoustic wave velocity shift $\Delta s/s$ as a function of $\Omega\tau$ calculated for the following values of the parameters: $n = 0.7 \cdot 10^{11}$ cm^{-2}, $\tau = 2 \cdot 10^{-11}$ s, $q = 9 \cdot 10^4$ cm^{-1}, $\alpha^2/2 = 3.2 \cdot 10^{-4}$, $\sigma_m = 0.6 \cdot 10^6$ cm/s [218]; $\mu = \frac{3}{4}$ (curve 1), $\frac{1}{2}$ (curve 2), $\frac{1}{4}$ (curve 3). Curve 4 represents $\Delta s/s$ as a function of $\Omega\tau$ for the circular (undeformed) Fermi surface.

$$\Gamma = q \frac{\alpha^2}{2} \frac{\tilde{\sigma}}{1+\tilde{\sigma}^2} \left(1 - \frac{a^2\tilde{\sigma}^2}{1+\tilde{\sigma}^2}(\Omega\tau^2) \right). \qquad (8.90)$$

Here

$$\tilde{\sigma} = \frac{\sigma_{xx}(\nu = \frac{1}{2})}{\sigma_m} \equiv \left(\frac{e^2}{4b\pi} \frac{(ql)^\mu}{p_0 l} \right) \frac{1}{\sigma_m}.$$

For a strong flattening of the composite fermions' Fermi surface ($\mu \ll 1$) we have to replace a^2 by $a^2 + |\xi|$.

Expressions (8.89), (8.90) are the new results in our theory. They predict peaks both in the surface acoustic waves' attenuation and velocity shift at one half-filling. This follows from the fact that $\Omega = 0$ at $\nu = \frac{1}{2}$ when $B_{\text{eff}} = 0$. The peaks arise due to the distortion of the composite fermions' Fermi surface in the presence of the electron density modulation. These peaks, in $\Delta s/s$ for several values of the parameter μ, are shown in Figure 8.7. For $\alpha^2/2 = 3, 2 \cdot 10^{-4}$ in GaAs, $n = 0, 7 \cdot 10^{11}$ cm^{-2}, $\tau \sim 10^{-11}$ s, $q \sim 10^5$ cm^{-1}, which agree with the conditions of the experiments [203], amplitudes of the peaks can reach the order of 10^{-4} when the parameter μ characterizing the rate of the distortion of the CF–FS is small enough. This order of magnitude of the peaks in $\Delta s/s$ agrees with the experimental data of [203]. When the composite fermions' Fermi surface flattening is strong ($\mu \ll 1$), the magnitude of the peak of the velocity shift is practically independent of the surface acoustic waves' wave vector q. Also these anomalies are not sensitive to any relation between q and the density modulation wave vector g. The effect is strongly anisotropic. The peaks can arise only in the geometry when $\mathbf{g} \perp \mathbf{q}$ and the composite fermions' Fermi surface effective parts coincide with its segments flattened due to the density modulation.

To analyze the effect of the inhomogeneity of the effective magnetic field arising in the modulated composite fermions' systems due to the density modulation we shall start from the Lorentz force equations describing the composite fermions' motion along the cyclotron orbit

$$\frac{dp_x}{dt} = -\frac{|e|}{c} B(\mathbf{r})v_y, \qquad \frac{dp_y}{dt} = \pm\frac{|e|}{c} B(\mathbf{r})v_x, \tag{8.91}$$

where $B(\mathbf{r}) = B_{\text{eff}} + \Delta B(\mathbf{r}) \equiv B_{\text{eff}} - 4\pi \hbar c \Delta n(\mathbf{r})/e$. For a single-harmonic density modulation along the "y"-direction we can write $\Delta n(\mathbf{r}) \equiv \Delta n(y) = \Delta n \cos(gy)$. In the following, the correction term $\Delta B(\mathbf{r})$ is assumed to be small compared to B_{eff}. Under this condition we can write the composite fermions' velocity $\mathbf{v}$ in the form $\mathbf{v} = \mathbf{v_0} + \delta\mathbf{v}$, where $\mathbf{v_0}$ is the uniform-field velocity and $\delta\mathbf{v}$ is a small correction term arising in the presence of the inhomogeneous magnetic field $\Delta B(\mathbf{r})$.

At first we will suppose that the composite fermions' Fermi surface is undeformed (a circle). In this case we have $v_{0x} = v_F \cos \Omega t$, $v_{0y} = v_F \sin \Omega t$, and $\Delta n(y) \approx \Delta n \cos(gY - gR \cos \Omega t)$, where v_F is the composite fermions' Fermi velocity, $R = v_F/\Omega$ is the radius of the cyclotron orbit, and Y is the "y"-coordinate of the guiding center. Substituting the expressions for $\mathbf{v}$ and $\Delta n(y)$ into (8.91) and keeping only the first-order terms we obtain

$$\frac{d(\delta v_x)}{dt} = -\Omega \delta v_y - \frac{\Delta B}{B_{\text{eff}}} v_F \sin \Omega t \cos(gY - gR \cos(\Omega t)),$$
$$\tag{8.92}$$
$$\frac{d(\delta v_y)}{dt} = -\Omega \delta v_x - \frac{\Delta B}{B_{\text{eff}}} v_F \cos \Omega t \cos(gY - gR \cos(\Omega t)).$$

It is natural to assume that to the first order in the modulating field the corrections δv_x and δv_y are periodic over the unperturbed cyclotron orbit. This assumption is equivalent to that used in [207].

Using these equations we calculate the averages of the corrections δv_x and δv_y over the cyclotron orbit, and obtain the expressions for the components of the velocity of the guiding center V_x and V_y defined below. Expanding the functions $\cos(gR \cos \psi)$ and $\sin(gR \cos \psi)$ in Bessel functions and substituting these expansions into (8.92) we arrive at the formulas

$$V_x \equiv \langle \delta v_x \rangle = -\frac{v_F}{2\pi} \frac{\Delta B}{B_{\text{eff}}} \int_0^{2\pi} \cos \psi \cos(gY - gR \cos \psi)\, d\psi$$
$$= -v_F \frac{\Delta B}{B} \sin gY\, J_1(gR), \tag{8.93}$$
$$V_y \equiv \langle \delta v_y \rangle = -\frac{v_F}{2\pi} \frac{\Delta B}{B_{\text{eff}}} \int_0^{2\pi} \sin \psi \cos(gY - gR \cos \psi)\, d\psi = 0,$$

here $\psi = \Omega t$.

Formulas (8.93) describe the drift of the guiding center along the direction perpendicular to both the magnetic field and to the electric field producing the density modulation in the electron system. Similar effects in the periodically modulated electron system in low-magnetic fields were considered before [207]–[208].

It is necessary to remark here that the general solution of the system (8.92) is

$$\delta v_x = C_1 \cos \psi + C_2 \sin \psi - v_F \frac{\Delta B}{B_{\text{eff}}} \int_0^\psi \cos(gY - gR \cos \psi')\, d\psi',$$

$$\delta v_y = C_1 \sin \psi - C_2 \cos \psi + v_F \frac{\Delta B}{B_{\text{eff}}} \int_0^\psi \cos(gY - gR \cos \psi')\, d\psi', \quad (8.94)$$

where C_1 and C_2 are arbitrary constants. For the average of δv_x over the unperturbed cyclotron orbit this gives the result which differs from our (8.93):

$$\langle \delta v_x \rangle = v_F \frac{\Delta B}{B_{\text{eff}}} \{\cos gY\, J_0(gR) - \sin gY\, J_1(gR)\}.$$

This extra term proportional to $J_0(gR)$ arises because the general solution of the system (8.92) is not periodic in ψ. The origin of this aperiodicity lies in the system (8.92) which is valid to first order in the modulation field. When we replaced the co-ordinate "y" in the argument of cosine in the expression for $\Delta n(y)$ by $Y - R \cos \Omega t$, we arrived at the system of differential equations which can be reduced to a linear inhomogeneous differential equation with the right-hand side having the same period as the general solution of the corresponding linear homogeneous differential equation. Such equations describe a resonance which occurs when the frequency of an external perturbation coincides with a natural frequency of free oscillations. Correspondingly, they have to have an aperiodic general solution describing the unbounded increase in the magnitude of oscillations. For our situation it means that strictly we cannot use expansion in the powers of a small parameter $\Delta B / B_{\text{eff}}$ to obtain the system (8.92), because the solution of this system increases in magnitude in time, and we cannot assume it to be small in ΔB all the time.

To evaluate this contribution semiquantitatively for the extended range of values of the perturbing wave vector q including the nonlocal region ($ql > 1$), we will make the following assumption. We assume that in the presence of a weak inhomogeneous magnetic field the components of the composite fermions' velocity can be written $v_x = v_{x0} + V_x$ and $v_y = v_{y0} + V_y$. Using these expressions we can evaluate the composite fermions' conductivity as follows:

$$\sigma_{\alpha\beta}^{\text{cf}} \approx \frac{g}{2\pi} \int_{-\pi/g}^{\pi/g} \sigma_{\alpha\beta}^{\text{cf}}(Y)\, dY, \quad (8.95)$$

where

$$\sigma_{\alpha\beta}^{\text{cf}}(Y) = \frac{ie^2 m_c}{(2\pi \hbar)^2} \sum_k (v_{k\beta0} + V_\beta(Y)\delta_{k0})$$

$$\times \int_{-\infty}^0 e^\eta\, d\eta \int_0^{2\pi} \frac{[v_{\alpha0}(\psi) + V_\alpha(Y)]\exp(ik\psi)}{\omega + i/\tau - k\Omega - q v_{x0}(\psi + \Omega\theta) - q V_x(Y)}\, d\psi. \quad (8.96)$$

The correction V_x produces additional terms in the expressions for the $\sigma_{\alpha\beta}^{\text{cf}}$. When the composite fermions' Fermi surface is a circle, a nonzero correction arises only in the expression for σ_{xx}^{cf}. For small q $(ql \ll 1)$ we have

$$\sigma_{xx}^{cf} = \tfrac{1}{2}\sigma_0 \left(1 + 2\frac{\delta\sigma}{\sigma_0}\right) \equiv \tfrac{1}{2}\sigma_0 \left\{[1+(\Omega\tau)^2]^{-1} + 2\left(\frac{\Delta B}{B_{\text{eff}}}\right)^2 J_1^2(gR)\right\}. \quad (8.97)$$

The remaining components of the composite fermions' conductivity tensor are the same as in the absence of the density modulation and equal, correspondingly, $\sigma_{yy}^{cf} = \tfrac{1}{2}\sigma_0[1+(\Omega\tau)^2]^{-1}$, and $\sigma_{xy}^{cf} = -\sigma_{yx}^{cf} = \sigma_{yy}^{cf}\Omega\tau$. These results are valid for arbitrary values of the parameter $\Omega\tau$.

It should be noted that the expression (8.97) may be obtained without using (8.95), (8.96). Following [208] we can consider the guiding center drift to lead to a diffusion along the "x"-direction with diffusion coefficient δD which equals

$$\delta D = \tau \frac{g}{2\pi} \int_{-\pi/g}^{\pi/g} V_x^2(Y)\,dY. \quad (8.98)$$

The term δD is the additional contribution to the xx-component of the diffusion tensor. The latter is connected with the conductivity through the Einstein relation $\sigma_{\alpha\beta} = Ne^2 D_{\alpha\beta}$ ($N = m^*/(2\pi\hbar^2)$ is the composite fermions' density of states). Substituting (8.98) into this relation we obtain for $\delta\sigma$ the expression which coincides with the second term in (8.97). This coincidence confirms that we may use the formulas (8.95), (8.96) for the evaluation of the order of magnitude of the corrections arising in the conductivity due to the inhomogeneity of the magnetic field.

The electron conductivity σ_{xx} can be evaluated as follows:

$$\left(\frac{\Delta B}{B_{\text{eff}}}\Omega\tau = \frac{\Delta n}{n}\frac{p_F l}{\hbar} = \frac{\Delta n}{n}k_F l\right):$$

$$\sigma_{xx} = \frac{e^4}{(4\pi\hbar)^2}\rho_{yy}^{cf} = \frac{e^4}{(4\pi\hbar)^2}\frac{2}{\sigma_0}\left\{1 + 2\left(\frac{\Delta B}{B_{\text{eff}}}\Omega\tau\right)^2 \frac{J_1^2(gR)}{1+2(\Delta B/B_{\text{eff}})^2 J_1^2(gR)}\right\}$$

$$\approx \frac{e^4}{(4\pi\hbar)^2\sigma_0}\left[1 + 2\left(\frac{\Delta n}{n}k_F l\right)^2 J_1^2(gR)\right], \quad (8.99)$$

$$\sigma_{yy} = \frac{e^4}{(4\pi\hbar)^2}\rho_{xx}^{cf} = \frac{e^4}{(4\pi\hbar)^2}\frac{2}{\sigma_0}\left(\frac{1}{1+2(\Delta B/B_{\text{eff}})^2 J_1^2(gR)}\right) \approx \frac{e^4}{(4\pi\hbar)^2}\frac{2}{\sigma_0}. \quad (8.100)$$

The correction to the conductivity component σ_{xx} is of the same order as the corresponding result of [205]. It confirms once more that the simplified expression assumed here may be used to evaluate the composite fermions conductivity.

When the density modulation is weak $((\Delta n/n)k_F l \ll 1)$ the corresponding correction to the conductivity σ_{xx} is small and we can neglect it. In this case, the inhomogeneity of the effective magnetic field does not significantly affect the d.c. transport. For stronger modulation $((\Delta n/n)k_F l \sim 1)$, the conductivity component σ_{xx} is appreciably changed.

The magnetic field dependence of the electron conductivity is determined by the value of the parameter gR. When $gR \gg 1$, correction to the conductivity is small

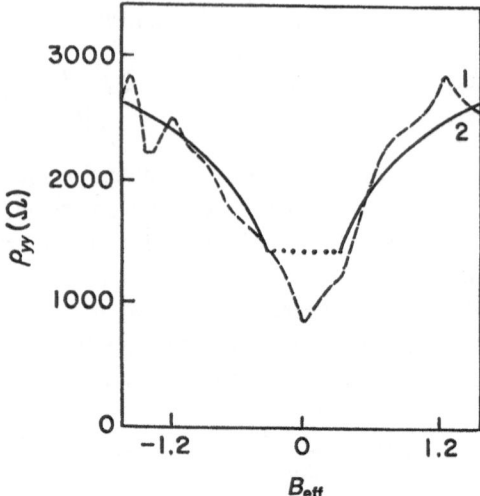

FIGURE 8.8. Direct current magnetoresistivity as a function of the effective magnetic field. Dashed line (1): experiment of [204]; solid line (2): theory for the parameters: $n = 1.8 \cdot 10^{11}$ cm^{-2}, $\Delta n/n = 0.02$, $g = 10^5$ cm^{-1}, $l = 1.4 \cdot 10^{-4}$ cm, $\sigma_0^{-1} = 270\,\Omega$. The dotted part of the curve 2 corresponds to the region of the values of B_{eff} where (8.99) cannot be applied.

in magnitude (because of the small parameter $(gR)^{-1}$) and describes the Weiss oscillations' effect in low magnetic fields (see, e.g., [207]–[208]). For $gR < 1$, the function $J_1^2(gR)$ takes values of the order of unity and can increase upon increasing B_{eff} (it corresponds to the decrease of the parameter gR). Hence, for $gR < 1$ and $(\Delta n/n)k_F l \sim 1$, the expression (8.99) can describe a large minimum in the magnetic field dependence of the electron conductivity σ_{xx} (or magnetoresistivity ρ_{yy}) around $\nu = \frac{1}{2}$ similar to the corresponding result of [205]. Such a minimum of the longitudinal magnetoresistivity in a modulated two-dimensional electron gas, for a weak modulation perpendicular to the current direction, was observed recently [204]. According to the conditions of these experiments $\left(\Delta n/n \sim 10^{-2}, n \sim 1, 8 \cdot 10^{11}\right.$ cm^{-2}, $l \sim 1, 4 \cdot 10^{-4}$ cm, $g \sim 10^5$ cm$^{-1}\right)$ the effect occurs when the parameters $(\Delta n/n)k_F l$ and gR are of the order of unity, which agrees with the theoretical estimates. The estimation based on the expression (8.99) also gives reasonable agreement with these experimental results (see Figure 8.8).

Straightforward calculations starting from (8.95), (8.96) show that for the un-deformed composite fermions' Fermi surfaces, with large q ($ql > 1$) in the same way as for small q, only σ_{xx}^{cf} changes, due to the inhomogeneity of the magnetic field along the "y"-direction. However, the correction to σ_{xx}^{cf} does not signifi-cantly influence the composite fermions' resistivity ρ_{yy}^{cf} (and, consequently, the corresponding component of the electron conductivity σ_{xx}) close to $\nu = \frac{1}{2}$. This follows from the expression (8.88) for σ_{xx}. The ratio $(\sigma_{xy}^{\text{cf}})^2/\sigma_{xx}^{\text{cf}}\sigma_{yy}^{\text{cf}}$ for $ql \gg 1$ is of the order of $(\Omega\tau)^2/ql$. Therefore, when $\Omega\tau < \sqrt{ql}$, we can use the approxima-tion $\rho_{yy}^{\text{cf}} \approx (\sigma_{yy}^{\text{cf}})^{-1}$, and σ_{xx} does not depend on σ_{xx}^{cf}. Hence near one half-filling, where $\Delta \nu = |\nu - \frac{1}{2}|$ is less than $\sqrt{(q/k_{\mathrm{F}})(k_{\mathrm{F}}l)^{-1}}$, the correction arising due to the inhomogeneity of the magnetic field does not contribute to the magnetoacoustic response of the modulated electron system to the surface acoustic waves propagating perpendicular to the direction of the density modulation.

At larger $\Delta \nu$, when $\Omega \tau \sim \sqrt{ql}$, the influence of the density modulation on the electron conductivity becomes significant. When B_{eff} is large enough to satisfy the condition

$$\frac{\Delta B}{B_{\mathrm{eff}}}ql = \frac{\Delta n}{n}k_{\mathrm{F}}lqR \ll 1,$$

we obtain the following asymptotic expression for σ_{xx}:

$$\sigma_{xx} = \frac{e^4}{(4\pi\,\hbar)^2}\rho_{yy}^{\mathrm{cf}}(\Delta n = 0)\left\{1 - \left(\frac{\Delta n}{n}k_{\mathrm{F}}l\right)^2 (qR)^2 J_1^2(gR)\right\}. \tag{8.101}$$

It is clear from the comparison of (8.99) and (8.101) that the corrections to the conductivity, due to the inhomogeneity of the effective magnetic field in the local $(ql \ll 1)$ and nonlocal $(ql > 1)$ limits, are of opposite signs. This difference arises due to the distinctions between contributions, due to the guiding center drift under the local and nonlocal regimes.

In the region of small q the diffusion due to the guiding center drift enhances the conductivity. In the opposite limit $(ql \gg 1)$, the main term in the conductivity arises from the effective parts of the Fermi surface where $\mathbf{q} \cdot \mathbf{v} \approx 0$ (at $\omega\tau \ll 1$). The existence of a nonzero average of the velocity component v_x over the cyclotron orbit prevents satisfying this condition and leads to a decrease of the conductivity. Positive corrections to the conductivity, which arise from the diffusion, are smaller in magnitude than negative. The ratio of magnitudes is of the order of $(ql)^{-1}$ for the two-dimensional electron gas system with the undeformed composite fermions' Fermi surface.

To analyze semiquantitatively the effect of the inhomogeneity of the magnetic field on the two-dimensional electron gas conductivity, taking into account the distortion of the composite fermions' Fermi surface due to the density modulation, we will assume at first that $(\Delta B/B_{\mathrm{eff}})ql \ll 1$. Also we will assume here that the velocity of the guiding center drift is of the same order as in the system whose Fermi surface is undeformed, because a small deformation of the effective parts of the Fermi surface may not result in significant changes of the averages over the composite fermions' Fermi surface for this particular case.

When the Fermi surface is deformed the component of the conductivity $\sigma_{yy}^{\mathrm{cf}}$ changes in the presence of the inhomogeneous magnetic field. Using (8.95), (8.96) we obtain

$$\sigma_{(1)yy}^{\mathrm{cf}} \approx \tfrac{1}{2}\sigma_{(1)yy}^{\mathrm{cf}}\left(\nu = \tfrac{1}{2}\right)\frac{g}{2\pi}\int_{-\pi/g}^{\pi/g} dY\left(S_\mu^+(\Omega\tau, Y) + S_\mu^-(\Omega\tau, Y)\right), \tag{8.102}$$

where

$$S_\mu^{\pm}(\Omega\tau, Y) = \int_{-\infty}^{0} e^\eta\left(\kappa_{\pm}(\eta) + iql\frac{\Delta B}{B_{\mathrm{eff}}}\varphi(Y)\right)^{\mu-1} d\eta \tag{8.103}$$

and

$$\varphi(Y) = \sin(gY)\Phi(g, B_{\text{eff}}). \tag{8.104}$$

The function Φ is of the order or less than unity. For the undeformed composite fermions' Fermi surface this function coincides with the Bessel function (see (8.93)).

For $\Omega\tau < 1$ we have

$$\sigma^{\text{cf}}_{(1)yy} \approx \sigma^{\text{cf}}_{(1)yy}\left(\nu = \tfrac{1}{2}\right)\left\{1 - a^2\left[(\Omega\tau)^2 + \frac{1}{2}\left(ql\frac{\Delta B}{B_{\text{eff}}}\right)^2\Phi^2\right] + \xi(\Omega\tau)^2\right\}. \tag{8.105}$$

Substituting this expression into (8.88) and using (8.55), (8.56) we arrive at the result

$$\frac{\Delta s}{s} = \frac{\alpha^2}{2}\frac{1}{1+\tilde\sigma^2}\left\{1 - \frac{2\tilde\sigma^2}{1+\tilde\sigma^2}\left[(a^2 - \xi)(\Omega\tau)^2 + \frac{a^2}{2}\left(\frac{q}{g}\right)^2\right.\right.$$
$$\left.\left.\times\left(\frac{\Delta n}{n}k_F l\right)^2(gR)^2\Phi^2\right]\right\}. \tag{8.106}$$

Hence the effect of the field $\Delta B(r)$ leads to the appearance of an additional term in the expression for the composite fermions' conductivity. This term enhances the decrease of the surface acoustic wave velocity shift upon increasing the effective magnetic field and thus strengthens the surface acoustic waves' anomaly. The ratio of the magnitudes of the two terms enclosed in square brackets in (8.106) depends on the magnitude of the density modulation and on the magnetic field B_{eff}. When B_{eff} increases the first term becomes larger than the second. Both terms can influence the composite fermions' conductivity and, consequently, the surface acoustic waves' velocity shift only when the composite fermions' Fermi surface is deformed because a^2 goes to zero for $\mu = 1$ (undeformed Fermi surface). Hence the deformation of the composite fermions' Fermi surface due to the periodic electric field, applied to the two-dimensional electron gas, is the most important factor determining the two-dimensional electron gas response to the surface acoustic waves in the nonlocal regime ($ql \gg 1$).

In metals it is known that the variation in the crystalline field by some external effect and/or in electron density, and consequently a change of the Fermi energy, can cause a change in the Fermi surface topology [209], [210]. Usually these electronic topological transitions represent some changes in the Fermi surface connectivity, i.e., when a number of voids increases or decreases. Such changes in the Fermi surface topology are accompanied by variations in its local geometry such as forming or losing lines of parabolic points, points of flattening, or conic points. In the case under consideration, the factor which could possibly cause a topological transition in the composite fermions' system is an increase in the modulating field magnitude. According to the symmetry of the system the transition could occur by means of the formation of two additional small voids arranged along the

p_y-direction. At present, such a topological change is a conjecture which needs theoretical investigation separately for the case of the composite fermions.

One can expect a change in the composite fermions' Fermi surface connectivity to be accompanied by the disappearance of the flattened part on the main void. In this case, the anomalous maximum in the surface acoustic waves' velocity shift magnetic field dependence has to be replaced by a minimum again. Thus assuming the possible relevance of the electron topological transition, one can explain the disappearance of the anomalous peak in the surface acoustic waves' velocity shift under increasing the modulation strength. It was mentioned above that this effect was observed in the experiment [203].

8.5 Dynamical Kohn Anomaly in a Surface Acoustic Wave Response in Modulated Quantum Hall Systems

In this section we consider a new dynamical manifestation of the Kohn anomaly in the response of a two-dimensional electron gas, and in the fractional quantum Hall effect regime at the half-filling of the lowest Landau level to an interacting surface acoustic wave. This dynamical Kohn anomaly may be measurable from the surface wave-vector dependence of the velocity shift and the attenuation of the acoustic wave of its wave vector. The kinks can be enhanced by an order-of-magnitude when external periodic modulation is applied, similar to the recent experiments [203], [204].

The Kohn anomaly in the acoustic wave response is well known in conventional metals. It was noted in [199] that, in principle, a similar anomaly can also exhibit itself in the surface acoustic waves' response of the two-dimensional electron gas at one half-filling. However, it was concluded there that the effect would be very small. Here it is shown that in proper geometry and with modulation the effect can be sufficiently large to be measurable using the response of the system to the surface acoustic wave at $\hbar q \approx 2p_F$, and, if measured, this response would contain important information. In this text we show that the image of the composite fermions' Fermi surface appears in the structure of the surface acoustic waves' velocity shift $\Delta s/s$ and the attenuation rate Γ, as functions of the surface acoustic waves' wave vector $\mathbf{q}$ for large q ($\hbar q \sim 2p_F$), thus giving another means to determine the composite fermions' Fermi surface.

The surface acoustic waves propagating in a piezoelectric medium interacts with a two-dimensional electron gas in a nearby semiconductor. In the usual geometry the surface acoustic waves propagate along the [011] direction on the (100) surface of an AlGaAs crystal. Defining the x-axis as $x \equiv [011]$ we have [199]:

$$\frac{\Delta s}{s} - \frac{i\Gamma}{q} = \frac{\alpha^2/2}{1 + i\sigma_{xx}(q,\omega)/\sigma_m} = \frac{\alpha^2}{2}\varepsilon(q,\omega). \tag{8.107}$$

The dielectric function $\varepsilon(q, \omega)$ is

$$\varepsilon(q, \omega) = 1 - V(q)K_{00}(q, \omega), \tag{8.108}$$

where $V(q)$ is the Fourier component of the Coulomb interaction: $V(q) = 2\pi e^2/\varepsilon_b q$ and $K_{00}(q, \omega)$ is the composite fermions' density–density response function. Note that in (8.108) the dielectric function $\varepsilon(q, \omega)$ is in the numerator: this takes into account that the piezoelectric field generated by the surface acoustic waves is unscreened, while the two-dimensional electron gas interacts with the screened electric field.

In the random phase approximation the density–density response function is described by the following expression given in [199]:

$$K_{00}(q, \omega) = \frac{K_{00}^0(q, \omega)}{1 + V(q)K_{00}^0(q, \omega) - (4\pi \hbar/q)^2 K_{11}^0(q, \omega)K_{00}^0(q, \omega)}. \tag{8.109}$$

Below we write the unperturbed response functions $K_{11}^0(q, \omega)$ and $K_{00}^0(q, \omega)$ corresponding to the noninteracting composite fermions' system at $\nu = \frac{1}{2}$, which is the Fermi sea with Fermi momentum p_F. The functions $K_{00}^0(q, \omega)$ and $K_{11}^0(q, \omega)$ for the undeformed composite fermions' Fermi surface are

$$K_{00}^0(q, \omega) = \iint \frac{f(E(\mathbf{p} + \hbar\mathbf{q})) - f(E(\mathbf{p}))}{\hbar\omega - E(\mathbf{p} + \hbar\mathbf{q}) + E(\mathbf{p}) + i\eta} \frac{d^2 p}{(2\pi \hbar)^2}, \tag{8.110}$$

$$K_{11}^0(q, \omega) = -\frac{n}{m^*} + \iint v_y^2 \frac{f(E(\mathbf{p} + \hbar\mathbf{q})) - f(E(\mathbf{p}))}{\hbar\omega - E(\mathbf{p} + \hbar\mathbf{q}) + E(\mathbf{p}) + i\eta} \frac{d^2 p}{(2\pi \hbar)^2}. \tag{8.111}$$

Let us first calculate the response functions K_{00}^0, K_{11}^0, and the attenuation and velocity shift when the composite fermions' Fermi surface is a circle (undeformed case), the Fermi momentum is p_F, and the isotropic effective mass is m^*. After straightforward calculation we arrive at the result ($\omega \ll q v_F, v_F = p_F/m^*, \eta \to 0$):

$$K_{00}^0 = N \left\{ 1 - 2\delta \left(\sqrt{\left(\frac{\hbar q}{2p_F} - \delta\right)^2 - 1} + \sqrt{\left(\frac{\hbar q}{2p_F} + \delta\right)^2 - 1} \right)^{-1} \right\},$$
$$\tag{8.112}$$

where

$$\sqrt{\left(\frac{\hbar q}{2p_F} \pm \delta\right)^2 - 1} \equiv \sqrt{\left|\left(\frac{\hbar q}{2p_F} \pm \delta\right)^2 - 1\right|}$$
$$\times \left\{ \theta\left(\frac{\hbar q}{2p_F} \pm \delta - 1\right) + i\theta\left(1 - \frac{\hbar q}{2p_F} \mp \delta\right) \right\}$$

and

$$\theta(x) = \begin{cases} 0 & \text{for } x \leq 0, \\ 1 & \text{for } x > 0. \end{cases}$$

Here $N = 2\pi m^*/(2\pi\hbar)^2$ is the density of states at the composite fermions' Fermi surface, $\delta \equiv \omega/qv_F$. Thus the unperturbed density–density response function exhibits an anomaly near $\hbar q = 2p_F$ both in real and imaginary parts. The main part of the response function $K_{11}^0(q, \omega)$ at $\omega \ll qv_F$ ($\delta \ll 1$) equals

$$K_{11}^0 = -\frac{q^2}{24\pi m^*} - \frac{2}{3}\frac{n}{m^*}\frac{k_F}{q}\left\{\sqrt{\left[\left(\frac{\hbar q}{2p_F} - \delta\right)^2 - 1\right]^3} - \sqrt{\left[\left(\frac{\hbar q}{2p_F} + \delta\right)^2 - 1\right]^3}\right\}.$$

(8.113)

For small q ($\hbar q \ll p_F$) Im K_{11}^0 is large relative to the real part and gives the linear dependence of the conductivity σ_{xx} upon q. However for $\hbar q \sim p_F$ and $\delta \ll 1$, Im K_{11}^0 is small compared to Re K_{11}^0. The ratio of these magnitudes is of the order of δ.

Substituting the expressions (8.112) and (8.113) into (8.109) and using (8.107) we obtain the following result for the σ_{xx}-component of the electron conductivity tensor:

$$\sigma_{xx} = -i\sigma_0(q)\left(1 - \frac{3}{2}\frac{\delta}{\sqrt{\left(\frac{\hbar q}{2p_F} - \delta\right)^2 - 1} + \sqrt{\left(\frac{\hbar q}{2p_F} + \delta\right)^2 - 1 - \frac{1}{2}\delta}}\right)$$

$$+ \frac{3}{4}\delta\sigma_0(q)\left(1 + \frac{1}{2}\frac{\delta}{\sqrt{\left(\frac{\hbar q}{2p_F} - \delta\right)^2 - 1} + \sqrt{\left(\frac{\hbar q}{2p_F} + \delta\right)^2 - 1 - \frac{1}{2}\delta}}\right),$$

(8.114)

where $\sigma_0(q) = \frac{3}{4}(\omega w^2/q^2)N$.

In the vicinity of $\hbar q = 2p_F$ the denominators are of the order of δ, therefore the last two terms are smaller than the first two terms. Neglecting them we obtain for the real part (σ') of σ_{xx}:

$$\sigma' = \sigma_0(q)F(q).$$

(8.115)

The function $F(q)$ increases when the wave vector approaches $2p_F/\hbar$ (but $1 - \hbar q/2p_F \gg \delta$) as

$$F(q) = \frac{3}{8\sqrt{2}}\delta\left/\sqrt{1 - \frac{\hbar q}{2p_F}}\right.$$

(8.116)

and has its maximum value $F_{\max}$ ($F_{\max} = \frac{3}{16}\sqrt{\delta}$) when $\hbar q = 2p_F(1 - \delta)$. A further increase of the wave vector leads to a decrease of the function $F(q)$ which goes to zero when $\hbar q > 2p_F(1 + \delta)$.

The derivative of the function $F(q)$ tends to infinity when $\hbar q$ approaches the points $2p_F(1 - \delta)$ and $2p_F(1 + \delta)$ from the left side. These peculiarities in $F'(q)$

correspond to kinks in the dependence of the real part of the conductivity upon q near the "Kohn anomaly point" ($\hbar q = 2p_F$). Due to the low density of electrons in the heterostructures under consideration ($n \sim 10^{10}$–10^{11} cm^2) the composite fermions' Fermi velocity is comparatively small ($v_F \sim 10^6$ cm/s). So the parameter δ cannot be neglected ($\delta \sim 10^{-1}$, $s \approx 3 \cdot 10^5$ cm/s [214]) and the variations of σ' near $\hbar q = 2p_F$ can be important. The imaginary part σ'' of σ_{xx} is not very sensitive to the variation in q in the vicinity of the point $q = 2p_F/\hbar$, so

$$\sigma'' \approx -\sigma_0(q). \tag{8.117}$$

However, the derivative of this function diverges when $\hbar q$ approaches the points $2p_F(1 \mp \delta)$ from the right-hand side. This corresponds to the results obtained long ago [211] for a metal with a nearly cylindrical Fermi surface. The real and imaginary parts of the dielectric function are

$$\frac{\Delta s}{s} = \frac{\alpha^2}{2} \operatorname{Re} \varepsilon(q, \omega = sq) = \frac{\alpha^2}{2} \frac{1}{1+\bar{\sigma}^2} \frac{1}{1 + \dfrac{\bar{\sigma}^2}{1+\bar{\sigma}^2} F^2(q)}, \tag{8.118}$$

$$\Gamma = -q\frac{\alpha^2}{2} \operatorname{Im} \varepsilon(q, \omega = sq) = q\frac{\alpha^2}{2} \frac{\bar{\sigma}}{1+\bar{\sigma}^2} \frac{F(q)}{1 + \dfrac{\bar{\sigma}^2}{1+\bar{\sigma}^2} F^2(q)}, \tag{8.119}$$

where $\bar{\sigma} = \sigma_0(q)/\sigma_m$. These formulas predict a minimum in the velocity shift and a maximum in the attenuation at $\hbar q = 2p_F(1 - \delta)$. A slight modification of this statement is needed. The actual location of the minimum (maximum) in $\Delta s/s$ (Γ) is slightly closer to $\hbar q = 2p_F$. This point can be found by taking the q-dependence of σ'' into account more carefully, near $\hbar q = 2p_F$. Thus, when the composite fermions' Fermi surface is a circle, the Kohn anomaly appears as nonmonotonic behavior of the surface acoustic waves' velocity shift and attenuation dependencies of q near $q = 2p_F/\hbar$. Equations (8.118) and (8.119) are illustrated in Figure 8.9. Note that the location of the anomalies are shifted from $2p_F$.

The ratio $\sigma_0(q)/\sigma_m$ for GaAs at $q \approx 2p_F/\hbar$ and $n \sim 10^{10}$ cm^{-2} is of the order of unity and the piezoelectric constant α^2 is of the order of 10^{-3} [213], [214]. So the magnitude of the minimum of the surface acoustic waves' velocity shift can be of the order of 10^{-4}–10^{-5}. According to [199], a more systematic calculation of the response functions at large q ($q \sim 2p_F/\hbar$), including contributions beyond the random phase approximation, gives a significant suppression of the Kohn anomaly due to strong interaction between electrons in the two-dimensional system. Estimations of [199] gave an estimate of 0.1 reduction. Taking this estimate into account we conclude that probably the dynamical Kohn anomaly in the surface acoustic waves' response of the two-dimensional electron gas with the undeformed (circular) composite fermions' Fermi surface is not large enough for observation in experiments.

We can expect a much stronger Kohn anomaly in the surface acoustic waves' spectrum when the composite fermions' Fermi surface is deformed due to some external effect. For instance, the composite fermions' Fermi surface can be flat-

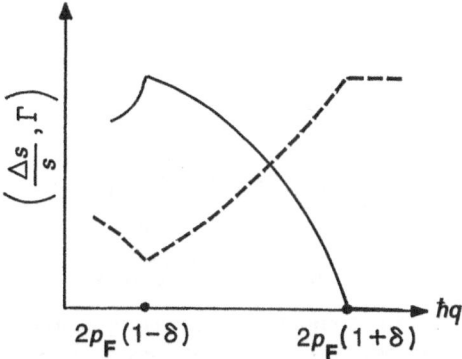

FIGURE 8.9. The dependence of the surface acoustic wave velocity shift $\Delta s/s$ (dashed curve) and the attenuation rate Γ (solid curve) upon q near the Kohn anomaly point. The composite fermions' Fermi surface is supposed to be circular.

tened as a result of the application of a periodic modulation potential, as in the experiments [197]. We showed in a previous section that this deformation of the composite fermions' Fermi surface resulting from the modulating potential could be at the origin of the anomalies in the magnetotransport observed in [197]. Here we show that the distortion of the composite fermions' Fermi surface can essentially strengthen the Kohn anomaly and make it large enough for observation in experiments.

Assume that the modulation potential wave vector $\mathbf{g}$ is parallel to the surface acoustic waves' wave vector $\mathbf{q}$. This parallel geometry is the most favorable for a new experiment under deformation because, as in a conventional metal, it provides the "flattened" segments of the composite fermions' Fermi surface which are its effective parts for the absorption of phonons with large q of the order of $2p_F/\hbar$. To analyze the Kohn effect under these conditions we parametrize the energy–momentum relation for composite fermions in the modulated system

$$E(\mathbf{p}) = \frac{p_x^2}{2m_1} + \frac{p_0^2}{2m_2} \left| \frac{p_y}{p_0} \right|^\gamma, \tag{8.120}$$

where p_0 is a constant with the dimension of momentum, the m_i are effective masses, and γ is a dimensionless parameter which determines the shape of the composite fermions' Fermi surface ($\gamma > 1$). When $\gamma > 2$ the composite fermions' Fermi surface is flattened near the vertices ($\pm p_0, 0$). To calculate the response functions K_{00}^0 and K_{11}^0 we introduce

$$p_x = p_0 \sqrt{\frac{m_1}{m_2}} \cos t, \qquad p_y = \pm p_0 |\sin t|^{2/\gamma}, \tag{8.121}$$

where $0 \leq t \leq 2\pi$ and the $+$ and $-$ signs are chosen corresponding to normal domains of the positive and negative values of the sine. To proceed we convert the integral in (8.110) to an integral over the composite fermions' Fermi surface and over energy.

As a result, the expression (8.110) becomes

$$K_{00}^0(q, \omega) = \Phi_+(q, \omega) - \Phi_-(q, \omega), \tag{8.122}$$

where

$$\Phi_\pm(q, \omega) = \frac{8b_\pm}{\gamma} \frac{m_1 m_2}{\hbar q p_0} \frac{1}{(2\pi \hbar)^2} \int dE f(E) \int_0^{\pi/2} \frac{\sin t^{2/\gamma-1} dt}{(b_\pm)^2 - \cos^2 t}, \tag{8.123}$$

$$b_\pm = \frac{\hbar q}{2p_0^*} \mp \delta + \frac{i}{ql}, \qquad p_0^* = \sqrt{\frac{m_1}{m_2}} p_0,$$

is the maximum value of the component of the composite fermions' momentum in the direction of the surface acoustic waves' propagation: $\delta = \omega/qv_0$, $v_0 = p_0/\sqrt{m_1 m_2}$, and $l = v_0 \tau$.

The inner integrals in (8.123) can be transformed to

$$\frac{1}{2b_\pm^2} \int_0^\infty z^{1/\gamma-1}(1+z)^{1/\gamma-1/2} \left(z + \frac{b_\pm^2 - 1}{b_\pm^2} \right)^{-1} dz$$

$$= \frac{1}{2b_\pm^2} B\left(\frac{1}{\gamma}; \frac{3}{2} - \frac{2}{\gamma} \right) {}_2F_1\left(1; \frac{1}{\gamma}; \frac{3}{2} - \frac{1}{\gamma}; 1 - \frac{b_\pm^2 - 1}{b_\pm^2} \right). \tag{8.124}$$

Here, $\beta(x, y)$ is the beta function and ${}_2F_1(\alpha; \beta; \rho; x)$ is the hypergeometric function. Using the asymptotic expression for the hypergeometric function in the limit $|b_\pm^2 - 1| \ll 1$, we obtain the following expression for the response function $K_{00}^0(q, \omega)$ for the deformed composite fermions' Fermi surface near $q = 2p_0^*(\delta \ll 1; \eta \to 0)$:

$$K_{00}^0(q, \omega) = N \left\{ 1 + \left(\frac{p_0^*}{\hbar q} \right)^{2/\gamma} \left[\left(\frac{\hbar q}{2p_0^*} - \delta \right)^2 - 1 \right]^{1/\gamma} \right.$$

$$\left. - \left(\frac{p_0^*}{\hbar q} \right)^{2/\gamma} \left[\left(\frac{\hbar q}{2p_0^*} + \delta \right)^2 - 1 \right]^{1/\gamma} \right\} \equiv N\{1 + Y_\gamma(q, \delta)\}, \tag{8.125}$$

where the density of states N equals $2\pi \sqrt{m_1 m_2} 2^{2/\gamma}/\gamma \sin(\pi/\gamma)(2\pi \hbar)^2$.

Similarly, we can obtain the asymptotic expression for the main term in the response function $K_{11}^0(q, \omega)$ near the point of the anomaly. The expression for the σ_{xx}-component of the electron conductivity tensor for the flattened composite fermions' Fermi surface is

$$\sigma_{xx} = -i\sigma_0(q) \left\{ 1 - \frac{Y_\gamma(q, \delta)}{1 + a + Y_\gamma(q, \delta)} \right\}, \tag{8.126}$$

where $\sigma_0(q) = (\omega e^2/q^2)(N/1 + a)$, the dimensionless constant "a" equals $4^{1/\gamma-1}\gamma/2\gamma - 1$, and "a" carries the information about the "flattening" of the initially circular composite fermions' Fermi surface.

From (8.126) we obtain important features of the Kohn anomaly for the deformed composite fermions' Fermi surface. When $\gamma = 2$ and $m_1 = m_2 = m^*$ (the

undeformed composite fermions' Fermi surface) this formula will coincide with
(8.116). Near $\hbar q = 2p_0^*$ (p_0^* is defined below) (8.123) for strong flattening of the
composite fermions' Fermi surface ($\gamma \gg 1$) the real part σ' of the conductivity
σ_{xx} is very small compared to $\sigma_0(q)$ and can be neglected. Its imaginary part (σ''),
near but not too close to $\hbar q = 2p_0^*$ ($\hbar q < 2p_0^*$), equals $-\sigma_0(q)$ as in the case
when the composite fermions' Fermi surface is undeformed.

The imaginary part of the conductivity at $\hbar q \approx 2p_0^*(1 - \delta)$ is now

$$\sigma'' = -\sigma_0(q) \left\{ 1 - \delta^{1/\gamma} \frac{\cos(\pi/\gamma)}{1 + a + \delta^{1/\gamma} \cos(\pi/\gamma)} \right\}. \tag{8.127}$$

Its magnitude decreases significantly in the vicinity of the Kohn anomaly point
when the composite fermions' Fermi surface flattening is strong ($\gamma \gg 1$). When
$\hbar q$ tends to $2p_0^*(1 + \delta)$ ($\hbar q < 2p_0^*(1 + \delta)$) the quantity σ'' increases sharply, and

$$\sigma'' \approx -\sigma_0(q) \left\{ 1 + \delta^{1/\gamma} \left[1 + a - \delta^{1/\gamma} \right]^{-1} \right\}. \tag{8.128}$$

When $\gamma \gg 1$ the parameter $\delta^{1/\gamma}$ is of the order of unity and $\sigma' \sim -\sigma_0(q)(1+a)/a$.
Then for strong flattening of the composite fermions Fermi surface the factor
$(1 + a)/a$ varies from 1 to 10.

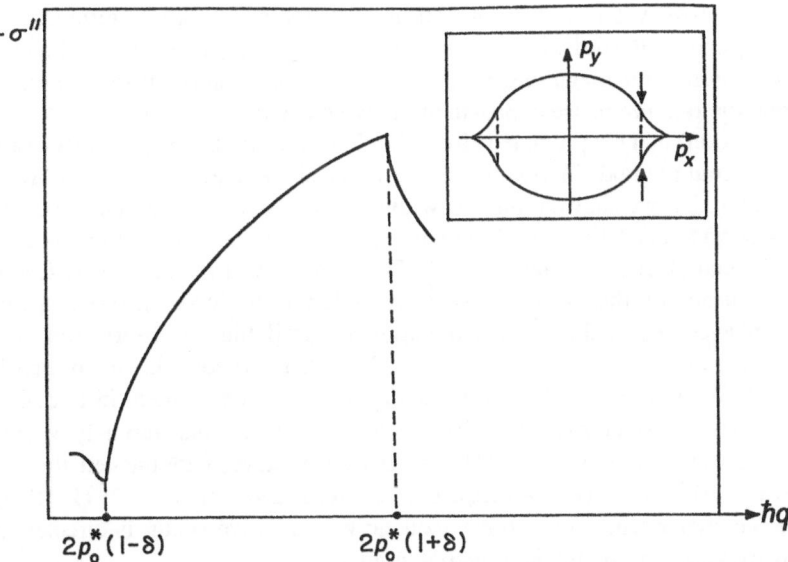

FIGURE 8.10. The dependence of σ'' on q in the vicinity of the Kohn anomaly point
for a strongly flattened composite fermion Fermi surface. The inset shows schematically
the shape of the composite fermions' Fermi surface deformed due to an external periodic
modulation in the "x"-direction in the nearly free electron approximation (solid line) and
in the Fermi surface corresponding to (8.120) (dashed line). The locations of regions of
particularly high density of states are noted by arrows.

The wave vector dependence of the function $\sigma''(q)$ near the point $\hbar q = 2p_0^*$ is shown in Figure 8.10. We observe kinks at points $\hbar q = 2p_0^*(1 \mp \delta)$ corresponding to discontinuities in the derivative of the imaginary part of the conductivity. Thus, flattening the composite fermions' Fermi surface will critically change the dependence of σ on q in the vicinity of the point $\hbar q = 2p_0^*$. In this case, the most important contribution to the Re $\varepsilon(q, \omega = qs)$ is

$$\mathrm{Re}\,\varepsilon(q, \omega = qs) = \frac{1}{1 + |\sigma''/\sigma_m|} \tag{8.129}$$

and, correspondingly,

$$\frac{\Delta s}{s} = \frac{\alpha^2}{2} \frac{1}{1 + |\sigma''/\sigma_m|}. \tag{8.130}$$

For $\gamma \gg 1$, $\Delta s/s$ will decrease significantly in the region $2p_0^*(1 - \delta) < \hbar q < 2p_0^*(1 + \delta)$. The amount of decrease in Re $\varepsilon(q, \omega)$ may be of the order of 10. Thus the anomaly produces a significant supression of the piezoelectric coupling (i.e., enhanced screening of the electric field) by the electrons at $2p_0^*(1 - \delta) < \hbar q < 2p_0^*(1 + \delta)$. A further increase of q leads to a sharp increase in $\Delta s/s$ which corresponds to the weakening of the electron system's coupling to the surface acoustic waves. This follows because the electrons cannot absorb phonons with wave vectors more than $2p_F(1 + \delta)/\hbar$. The magnitude of this dip in the surface acoustic waves' velocity shift for a strongly flattened composite fermions' Fermi surface can be of the order of 10^{-3}. Even estimating as it was performed did in [199] the reduction of the magnitude of the effect due to the contribution from the terms beyond the random phase approximation, we can conclude (using the estimate of 0.1 reduction) that the predicted effect should be observable by presently available experimental methods in modulated two-dimensional electron gas systems.

In summary, we showed that a new dynamical Kohn anomaly can be observed in the surface acoustic waves' velocity shift and attenuation in modulated two-dimensional electron gas near $\nu = \frac{1}{2}$. The distortion of the composite fermions' Fermi surface by the external density modulation produces an order of magnitude enhancement and can make the anomaly available for observation, in spite of the suppression due to the electron–electron interaction in two-dimensional systems. We propose a new surface acoustic waves' experiment in a modulated two-dimensional electron gas to observe this dynamical Kohn anomaly. We expect it to be observable in the GaAl/AlGaAs heterostructures with the density of electrons $n \sim 10^{10}$ cm^{-2} at the surface acoustic waves' frequency $\omega \sim 2\pi(1-10)$ GHZ in the geometry when the surface acoustic waves' wave vector is directed along the wave vector of the modulating potential.

We have to remark that the results of this and previous sections of work is based on the charged composite fermions' picture for the fractional quantum Hall effect, for example, as derived at $\nu = \frac{1}{2}$ in [199]. We followed the previous work by assuming that the composite fermions Fermi surface exists. An alternate picture for the fractional quantum Hall effect, also derived from a Chern–Simons approach, gives the quasi particles at $\nu = \frac{1}{2}$ as neutral dipolar objects, with the Hall current being carried by a set of collective magnetoplasmon oscillators [215], [216]. A magnetotransport theory based on this second picture was recently proposed in [219]. It is shown there that these two approaches are equivalent and lead to similar results for observables.

Appendix 1

To calculate the integrals $\Phi_m(u)$:

$$\Phi_m(u) = \int_{-1}^{1} \frac{\bar{s}(x)x^m}{u\chi - \bar{v}(x)}\,dx, \tag{A1.1}$$

we have to take into account that, because of the bilateral symmetry of the considered Fermi surface, the longitudinal component of the velocity $\bar{v}(x)$ is an odd function, and the cross-sectional area $\bar{s}(x)$ an even function, of x. Combining the contributions from the symmetric segments of the Fermi surface we can carry out integration over the half-surface corresponding to positive x. We can make a change of variables in the integrals (A1.1) to transform them to integrals with regard to $\bar{v}$:

$$\Phi_0(u) = 2u\chi \sum_i \int \frac{\bar{s}_i(\bar{v})\,dx_i/d\bar{v}}{u^2\chi^2 - \bar{v}^2}\,d\bar{v}, \tag{A1.2}$$

$$\Phi_1(u) = 2 \sum_i \int \frac{\bar{s}_i(\bar{v})x_i(\bar{v})\bar{v}\,dx_i/d\bar{v}}{u^2\chi^2 - \bar{v}^2}\,d\bar{v}, \tag{A1.3}$$

$$\Phi_2(u) = 2u\chi \sum_i \int \frac{\bar{s}_i(\bar{v})x_i^2(\bar{v})\,dx_i/d\bar{v}}{u^2\chi^2 - \bar{v}^2}\,d\bar{v}. \tag{A1.4}$$

Summation over i has to be performed over segments of the Fermi surface supposing that there is a one-to-one correspondence of p_z and v_z for each segment. Some segments have points $\bar{v} = 0$ which correspond to the effective cross-sections of the Fermi surface. To proceed we separate the corresponding terms from the sums over "i" in (A1.2)–(A1.4) and we consider their contributions to the integrals $\Phi_m(u)$.

Suppose that the longitudinal velocity tends to zero at the point x_0. In the vicinity of this point we can use the following approximation:

$$\bar{v}(x) = D\nu \, \text{sign} \, (x - x_0)|x - x_0|^r, \qquad\qquad r > 0. \qquad (A1.5)$$

Here D is a dimensionless positive constant, and the factor ν equals 1 when $\bar{v}(x)$ increases and -1 when it descreases upon an increase in the variable x. Using this approximation we obtain

$$\bar{s}(\bar{v}) = \bar{s}(0) - G|\bar{v}|^{\beta+2}, \qquad\qquad (A1.6)$$

$$dx/d\bar{v} = L|\bar{v}|^\beta, \qquad\qquad (A1.7)$$

where $G = 2\nu(\beta + 1)/\rho(\beta + 2)D^{\beta+1}$ and $L = (\beta + 1)/D^{\beta+1}$ are dimensionless constants $\rho = S(0)/\pi m_\perp v_m p_m$, and the parameter β is expressed in terms of "r": $\beta = 1/r - 1$.

The curvature of an axisymmetric Fermi surface is

$$K(p_z) = \frac{m_\perp^2 v_z^2 + m_\perp p_\perp^2 \, dv_z/dp_z}{(p_\perp^2 + m_\perp^2 v_z^2)^2}. \qquad\qquad (A1.8)$$

Hence the curvature at the effective cross-sections is proportional to $d\bar{v}/dx$. When $\beta > 0$ the curvature tends to infinity in the vicinity of the corresponding cross-section, when $\beta < 0$ it tends to zero, and when $\beta = 0$ it corresponds to the effective cross-sections with a finite and nonzero curvature.

We can transform the integrals over the segments containing an effective cross-section as follows:

$$\int_{|\bar{v}_1|}^{|\bar{v}_2|} \left(\frac{\bar{s}_i(\bar{v}) \, dx_i/d\bar{v}}{u^2\chi^2 - v^2} - \frac{L(\bar{s}(0) - G|\bar{v}|^{\beta+2})|\bar{v}|^\beta}{u^2\chi^2 - \bar{v}^2} \right) d\bar{v}$$

$$+ \int_{|\bar{v}_1|}^{|\bar{v}_2|} \frac{L(\bar{s}(0) - G|\bar{v}|^{\beta+2})|\bar{v}|^\beta}{u^2\chi^2 - \bar{v}^2} \, dv. \qquad (A1.9)$$

Here the second term represents a singular part of the integral which gives the contribution from the neighborhood of the effective cross-section. When $\beta < 1$ this contribution equals

$$2L \frac{u^{\beta-1}}{\chi^2} [\bar{s}(0) - G(u\chi)^{\beta+2}]$$

$$\times \left[\int_0^\infty \frac{t^\beta dt}{1 - (t/\chi)^2} + \frac{\chi^2}{2} \int_{|\bar{v}_1|/u}^\infty t^{\beta-2} dt + \frac{\chi^2}{2} \int_{|\bar{v}_2|/u}^\infty t^{\beta-2} dt \right]$$

$$= L[\bar{s}(0) - G(u\chi)^{\beta+2}] \left[-i\pi(u\chi)^{\beta-1} \left(1 - i \tan \frac{\pi\beta}{2} \right) - \frac{|\bar{v}_2|^{\beta-1} + |\bar{v}_1|^{\beta-1}}{\beta - 1} \right]. \qquad (A1.10)$$

This result remains valid for $1 < \beta < 2$. When $\beta > 2$ we can neglect the contribution from the vicinity of the effective cross-section, because under this condition it is too small to be considered. For $\beta = 1$ this contribution is

$$- L\bar{s}(0)\left[\pi i + \ln\left(\frac{|\bar{v}_1|}{|u\chi|}\right) + \ln\left(\frac{|\bar{v}_2|}{|u\chi|}\right)\right]. \tag{A1.11}$$

We can expand the integrand in the first term of (A1.9) in powers of the parameter $(u\chi)$. This expansion has the form

$$\sum_{n=0}^{\infty} \mu_n(u\chi)^n, \tag{A1.12}$$

where

$$\mu_n = -\left[\int_0^{|\bar{v}_1|}\left(\bar{s}(\bar{v})\frac{dx}{d\bar{v}} - L\bar{s}v^\beta\right)\frac{d\bar{v}}{\bar{v}^{2n+2}} + \int_0^{|\bar{v}_2|}\left(\bar{s}(\bar{v})\frac{dx}{d\bar{v}} - L\bar{s}(0)\bar{v}^\beta\right)\frac{d\bar{v}}{\bar{v}^{2n+2}}\right]. \tag{A1.13}$$

For the center cross-section of the Fermi surface the lower limit in the integrals (A1.9) equals zero. Therefore, instead of (A1.10), we obtain

$$\frac{L}{2}[\bar{s}(0) - G(u\chi)^{\beta+2}]\left[-i\pi(u\chi)^{\beta-1}\left(1 - i\tan\frac{\pi\beta}{2}\right) - \frac{|\bar{v}_0|^{\beta-1}}{\beta-1}\right], \tag{A1.14}$$

and the expression for the coefficient μ_n now contains only the first term. As a result, keeping terms in an order of magnitude no less than $(u\chi)^2$, we arrive at the formula

$$\Phi_0(u) = -2i\pi\sum_j\left(1 - \tfrac{1}{2}\delta_{j0}\right)L_j\left(\bar{s}(0) - G_j(u\chi)^{\beta+2}\right)$$

$$\times (u\chi)^{\beta_j}\left(1 - i\tan\frac{\pi\beta_j}{2}\right) - au\chi. \tag{A1.15}$$

Here summation over j is carried out over the effective cross-sections and the coefficient "a" equals

$$a = 2\sum_j\left[\int_0^{|\bar{v}_{1j}|}\frac{\gamma_j(\bar{v})}{\bar{v}^2}d\bar{v} + \int_0^{|\bar{v}_{2j}|}\frac{\gamma_j(\bar{v})}{\bar{v}^2}d\bar{v} + \frac{|\bar{v}_{1j}| + |\bar{v}_{2j}|}{\beta_j-1}\right](1 - \delta_{j0})$$

$$+ 2\sum_k\left[\int_0^{|\bar{v}_{1k}|}\frac{s_k(\bar{v})dx_k/d\bar{v}}{\bar{v}^2}d\bar{v} + \int_0^{|\bar{v}_{2k}|}\frac{s_k(\bar{v})dx_k/d\bar{v}}{\bar{v}^2}d\bar{v}\right]$$

$$+ 2\left[\int_0^{|\bar{v}_0|}\frac{\gamma_0(\bar{v})}{\bar{v}^2}d\bar{v} + \frac{|\bar{v}|^{\beta_0-1}}{\beta_0-1}\right]\delta_{j0}. \tag{A1.16}$$

Performing summation over k we take into account contributions from those segments of the Fermi surface which do not contain any effective cross-section. In (A1.16) we introduced functions $\chi_j(\bar{v})$ defined as follows:

$$\chi_j(\bar{v}) = \bar{s}_j(\bar{v})\frac{dx_j}{d\bar{v}} - L_j\bar{s}_j(0)\bar{v}^{\beta_j}. \tag{A1.17}$$

When the Fermi surface everywhere has a finite and nonzero curvature we can significantly simplify the first term in (A1.15):

$$-\frac{2\pi i}{\rho p_m^2}\sum_j\left(1-\tfrac{1}{2}\delta_{j0}\right)\left(1-\frac{v_j(u\chi)^2}{\rho p_m^2 K_j(0)\bar{s}_j(0)}\right)\frac{1}{K_j(0)}=-\frac{i\pi p_0^2}{\rho p_m^2}(1+\rho(u\chi)^2 d),$$

$$(A1.18)$$

where

$$p_0^2=2\sum_j\left(1-\tfrac{1}{2}\delta_{j0}\right)\frac{1}{K_j(0)},\qquad (A1.19)$$

$$d=\frac{2}{\rho^2 p_m^2 p_0^2}\sum_j\left(1-\tfrac{1}{2}\delta_{j0}\right)\frac{v_j}{\bar{s}_j(0)K_j^2(0)},\qquad (A1.20)$$

and $K_j(0)$ is the curvature at the j-th effective cross-section.

When any of the effective cross-sections of the Fermi surface has a curvature corresponding to $\beta=1$, its contribution to the first term of (A1.15) equals the expression (A1.11) multiplied by $2u\chi$.

The desired asymptotics for the integrals $\Phi_1(u)$ and $\Phi_2(u)$ can be obtained in a similar way. The main contribution to the integral $\Phi_1(u)$ originates from the vicinity of the center cross-section of the Fermi surface and from those segments which do not contain effective lines:

$$\Phi_1(u)\approx -i\pi(\beta_0+1)D_0^{-2(\beta_0+1)}(1-i\tan\pi\beta_0)(u\chi)^{2\beta_0+1}-b(u\chi)^2,\quad (A1.21)$$

where

$$b=2\int_0^{|\bar{v}_0|}\frac{d\bar{v}}{\bar{v}}\left(\bar{s}_0(\bar{v})x_0(\bar{v})\frac{dx_0}{d\bar{v}}-L_0 x_0(\bar{v})\frac{\bar{v}^{2\beta_0}}{D_0^{\beta_0+1}}\right)+\frac{2L_0}{2\beta_0+1}\frac{|\bar{v}_0|^{2\beta+1}}{D_0^{\beta_0+1}}$$

$$+2\sum_k\left[\int_0^{|\bar{v}_{1k}|}\frac{d\bar{v}}{\bar{v}}\bar{s}_k(\bar{v})x_k(\bar{v})\frac{dx_k}{d\bar{v}}+\int_0^{|\bar{v}_{2k}|}\frac{d\bar{v}}{\bar{v}}\bar{s}_k(\bar{v})x_k(\bar{v})\frac{dx_k}{d\bar{v}}\right].\quad (A1.22)$$

All the terms in the expression for the transverse conductivity, which include the integral $\Phi_2(u)$, are less than u^2 in the order of their magnitude. Therefore we can neglect them. Supposing that the Fermi surface of a metal everywhere has a finite and nonzero curvature and, using the formulas (A1.15)–(A1.20), we can derive the following asymptotic for the transverse conductivity

$$\sigma=\sigma_0(1+\Lambda_1 u+\Lambda_2 u+\cdots),\qquad (A1.23)$$

where

$$\sigma_0=\frac{e^2 p_0^2}{4\pi\hbar^3 q},\qquad (A1.24)$$

$$\Lambda_1=-\frac{ig}{\pi}\left(a\chi+\frac{\pi^2}{g^2}\alpha_2-b^2\alpha_2\right),\qquad (A1.25)$$

$$\Lambda_2=d\chi^2-2a\alpha_1\chi-2\alpha_2\frac{b}{c}g\chi-2\alpha_1\alpha_2 b^2-\frac{\pi^2}{g^2}\alpha_1^2.\qquad (A1.26)$$

The constants a, b, d are defined by the relations (A1.16), (A1.19), and (A1.20) and the constants c, g equal

$$c = \left[\rho p_0^2 p_m^2 K_0^2(0)\right]^{-1}, \qquad g = \rho p_m^2 / p_0^2. \tag{A1.27}$$

For a spherical Fermi surface this gives $c = d = g = 1$, $a = -4$, $b = \frac{4}{3}$.

To derive the asymptotic expression for the transverse conductivity of a compensated metal we can proceed in a similar way. Within the framework of the model assumed here we obtain the result

$$\sigma \approx \sigma_0 \left[1 + \Lambda_1 u + \Lambda_2 u^2 + R\left(1 - i \tan \frac{\pi\beta}{2}\right)(u\chi)^\beta \right], \tag{A1.28}$$

where

$$\Lambda_1 = -\frac{i}{\pi}(g_e a_e \chi_e + \bar{g}_e \bar{a}_e \bar{\chi}_e + g_h a_h \chi_h), \tag{A1.29}$$

$$\Lambda_2 = \bar{d}_e \chi_e^2 + d_h \chi_h^2, \tag{A1.30}$$

and

$$g_e = \frac{S_e(0) p_{me}^2}{\pi p_0^2 m_{\perp e} v_{me}} \equiv \frac{p_{me}^2}{p_{0e}^2} \rho_e,$$

$$\bar{g}_e = \frac{\bar{S}_e(0) p_{me}^2}{\pi p_0^2 m_{\perp e} v_{me}} \equiv \frac{\bar{p}_{me}^2}{p_0^2} \rho_e,$$

$$g_h = \frac{S_h(0) p_{mh}^2}{\pi p_0^2 m_{\perp h} v_{mh}} \equiv \frac{p_{mh}^2}{p_0^2} \rho_h.$$

Here v_{me}, $\bar{v}_{me}$, v_{mh} are the maximum values of the longitudinal velocity for charge carriers belonging to the three assigned groups: the electrons associated with the center lens, the remaining electrons and holes, p_{me}, $\bar{p}_{me}$, p_{mh} and $S_e(0)$, $\bar{S}_e(0)$, $S_h(0)$ are, respectively, the maximum projections of their momenta in the direction of the magnetic field and the areas of center cross-sections of the corresponding parts of the Fermi surface. To calculate the constants a_e, $\bar{a}_e$, a_h, $\bar{d}_e$, d_h we can use (A1.16), (A1.19).

Appendix 2

Here we analyze the magnetic field dependence of the integrals Y_1 and Y_2 included in the expression for the surface impedance (4.11). These integrals have the form

$$Y_1 = \frac{1}{\xi^2} \int_1^\infty dy \left\{ \frac{y}{\overline{\sigma}(i/y)} - y + \frac{g}{\pi} \left(a\chi + \frac{\pi^2}{g^2}\alpha_1 - b^2\alpha_2 \right) \right.$$
$$\left. + \frac{1}{y} \left[-\left(\frac{a^2 g^2}{\pi^2} + d \right) \chi^2 + \left(\frac{abg}{\pi^2} + c \right) \chi - \frac{g^2 b^4 \alpha_2^2}{\pi^2} \right] \right\}, \quad \text{(A2.1)}$$

$$Y_2 = \frac{1}{\xi^2} \int_0^1 dy \frac{y}{\overline{\sigma}(i/y)}. \quad \text{(A2.2)}$$

We make change of variables in these integrals introducing a new variable $z(\chi z = y)$ and transform them to the form

$$Y_1 = \frac{1}{2\xi^2}(1 - \chi^2) - \frac{9}{\pi\xi^2} \left(a\chi + \frac{\pi^2}{g^2}\alpha_1 - b^2\alpha_2 \right)$$
$$+ \frac{1}{2\xi^2} \ln\chi \left[-\left(\frac{a^2 g^2}{\pi^2} + d \right) \chi^2 + 2\alpha_2 bg \left(\frac{abg}{\pi^2} + c \right) \chi + \frac{g^2 b^4 \alpha_2^2}{\pi^2} \right]$$
$$- \frac{i\pi}{g} \frac{\chi^2}{\xi} F_0 - \frac{\pi}{g} \frac{\chi\alpha_2}{\xi^2} F_1 - \frac{i\pi}{g} \frac{\alpha_2^2}{\xi^2} F_2 + \frac{\pi}{g} \frac{1}{\xi^2} \frac{\alpha_2^3}{\chi} F_3. \quad \text{(A2.3)}$$

Here

$$F_0 = \int_1^\infty \left\{ \frac{z}{\Phi_0(z)} - \frac{ig}{\pi}z + \frac{ig^2}{\pi^2}a - \frac{ig}{\pi z} \left(\frac{a^2 g^2}{\pi^2} + d \right) \right\} dz, \quad \text{(A2.4)}$$

$$F_1 = \int_1^\infty \left\{ P(z) + \frac{g^2}{\pi^2}b^2 - \frac{2}{z}\frac{g^2}{\pi} \left(\frac{abg}{\pi^2} + c \right) \right\} dz, \quad \text{(A2.5)}$$

$$F_2 = \int_1^\infty \left(\frac{P(z)}{zR(z)} - \frac{i}{z}\frac{g^3}{\pi^3}b^4 \right) dz, \quad \text{(A2.6)}$$

$$F_3 = \int_1^\infty \frac{P(z)}{z^2 R(z)} \frac{dz}{1 - i\alpha_2/\chi z R(z)}. \tag{A2.7}$$

Besides, we used the notations

$$P(z) = \frac{\Phi_1^2(z)}{\Phi_0^2(z)},$$

$$R(z) = \frac{\Phi_0(z)}{\Phi_0(z)\Phi_2(z) - \Phi_1^2(z)}, \tag{A2.8}$$

Here $\Phi_m(z)$ are the integrals defined by (A1.1), where the quantity $u\chi$ in each integrand is replaced by i/z.

The integral Y_2 can be transformed to the form

$$Y_2 = -\frac{\chi\pi}{8}\frac{\alpha_1}{\xi^2} - \frac{i\pi}{8}\frac{\chi^2}{\xi^2}X_0 - \frac{\pi}{8}\frac{\chi}{\xi^2}\alpha_2 X_1 - \frac{i\pi}{8}\frac{\alpha_2^2}{\xi^2}X_2 + \frac{\alpha_2^3}{\chi}\frac{\pi}{8}\frac{1}{\xi^2}X_3, \tag{A2.9}$$

$$X_0 = \int_0^1 \frac{z\,dz}{\Phi_0(z)}, \tag{A2.10}$$

$$X_1 = \int_0^1 P(z)\,dz, \tag{A2.11}$$

$$X_2 = \int_0^1 \frac{P(z)}{zR(z)}\,dz, \tag{A2.12}$$

$$X_3 = \int_0^1 \frac{P(z)}{z^2 R(z)} \frac{dz}{1 - i\alpha_2/\chi z R(z)}. \tag{A2.13}$$

Suppose that $\omega\tau \to \infty$ and the quantity χ is real. We can write the expression for the Mellin transform of the integral F_3 with respect to the variable $\eta (\eta = |\alpha_2/\chi|)$:

$$\psi(s) = \int_0^\infty d\eta \eta^{s-1} F_3(\eta) = \frac{\pi}{\sin \pi s} \exp\left[i\pi s\left(\theta(\eta) + \frac{1}{2}\right)\right]$$

$$\times \int_0^\infty P(z)(zR(z))^{s-2}\,dz. \tag{A2.14}$$

Here $\theta(\eta)$ is a step function:

$$\begin{cases} \theta(\eta) = 1, & \eta \geq 0, \\ \theta(\eta) = 0, & \eta < 0. \end{cases}$$

We can expand the integrand in series in the inverse powers of the variable z:

$$P(z)(zR(z))^{s-2} = z^{s-2} \sum_{n=0}^\infty \frac{\Delta_n(s-2)}{z^n}. \tag{A2.15}$$

Substituting this expansion into (A2.14) we can rewrite the expression for $\psi(s)$ in the form

$$\psi(s) = \frac{\pi}{\sin \pi s} \exp\left[i\pi s\left(\theta(\eta) + \frac{1}{2}\right)\right] \sum_{n=0}^\infty \frac{\Delta_n(s-2)}{n-s+1}, \quad s < n+1. \tag{A2.16}$$

The inverse Mellin transformation gives

$$F_3(\eta) = \frac{1}{2\pi i} \int_{c-i\infty}^{c+i\infty} \eta^{-s} \psi(s)\, ds$$

$$= -\frac{i}{2} \sum_{n=0}^{\infty} \int_{c-i\infty}^{c+i\infty} \exp\left[i\pi s \left(\theta(\eta) + \tfrac{1}{2}\right)\right] \frac{\eta^{-s}}{\sin \pi s} \frac{\Delta_n(s-2)}{n-s+1}\, ds. \quad (A2.17)$$

The integral included in (A2.17) equals the sum of residues of the poles of the integrand arranged to the right from the line $\operatorname{Re} s = c (0 < c < n+1)$. The contribution from the simple pole corresponding to $s = 1, n \neq 0$ is

$$\frac{i\chi}{\alpha_2^2} \int_1^{\infty} \left[\frac{P(z)}{zR(z)} - \frac{\Delta_0(-1)}{z}\right] dz. \quad (A2.18)$$

Using the expressions (A2.8) for the functions $P(z)$ and $R(z)$, and the asymptotics (A1.15) and (A1.21) for the integrals Φ_0 and Φ_1 (we have to replace $u\chi$ by i/z in the integrands), we obtain

$$\Delta_0(-1) = \frac{ig^3}{\pi^3} b^4. \quad (A2.19)$$

Thus the desired residue equals $i\pi\alpha_2^2 F_2/g\xi^2$.

The contribution from the double pole corresponding to $s = 1, n = 0$ is

$$-\Delta_0(-1)\lim_{s\to 1}\frac{d}{ds}\left(\exp\left[i\pi s\left(\theta(\eta)+\tfrac{1}{2}\right)\right]\eta^{-s}\right)$$

$$= \frac{g^3}{\pi^3}\frac{\chi}{\alpha_2} b^4\left[\ln\left|\frac{\alpha_2}{\chi}\right| - i\pi\theta\left(\frac{\alpha^2}{\chi}\right) - i\frac{\pi}{2}\right]. \quad (A2.20)$$

We also have to take into account the contribution from the simple pole ($s = 2, n \neq 1$) which equals

$$\frac{\chi^2}{\alpha_2^2}\int_1^{\infty}\left[P(z) - \Delta_0(0) - \frac{\Delta_1(0)}{z}\right] dz = \frac{\chi^2}{\alpha_2^2}F_1. \quad (A2.21)$$

Here we used the expressions for $\Delta_0(0)$ and $\Delta_1(0)$:

$$\Delta_0(0) = -\frac{g^2}{\pi^2} b^2, \qquad \Delta_1(0) = \frac{2bg^2}{\pi}\left(\frac{abg}{\pi^2} + c\right).$$

At last we have to consider the double pole ($s = 2, n = 1$), which gives the contribution

$$\frac{2bg^2}{\pi}\left(\frac{abg}{\pi^2} + c\right)\frac{\chi^2}{\alpha_2^2}\left\{2\pi i\theta\left(\frac{\alpha_2}{\chi}\right) + i\pi - \ln\left|\frac{\alpha_2}{\chi}\right|\right\}, \quad (A2.22)$$

and the simple pole at $s = 3, n \neq 2$ and the double pole at $s = 3, n = 2$. The simple pole gives us the following contribution:

$$-\frac{i\chi^3}{\alpha_2^3}\int_1^{\infty} dz\left(\frac{z}{\Phi_0(z)} + z\Delta_0(1) + \Delta_1(1) + \frac{\Delta_2(1)}{z}\right) = -i\frac{\chi^3}{\alpha_2^3}F_0, \quad (A2.23)$$

and the contribution from the double pole equals

$$- \frac{\chi^2}{\xi^2} \left(\frac{a^2 g^2}{\pi^2} + d \right) \left\{ 3\pi i \theta \left(\frac{\alpha_2}{\chi} \right) + 3\pi \frac{i}{2} - \ln \left| \frac{\alpha_2}{\chi} \right| \right\}. \tag{A2.24}$$

We do not need poles corresponding to larger values of s because their contributions are smaller in the order of magnitude than χ^2. Thus we obtain the following asymptotic expression for the integral Y_1 in the vicinity of the cyclotron resonance $(\chi \to 0)$:

$$\begin{aligned} Y_1 &= \frac{1}{2\xi^2} - \frac{9}{\pi \xi^2} \left(\frac{\pi^2}{g^2} \alpha_1 - b^2 \alpha_2 \right) \\ &\quad + \frac{1}{\xi^2} \frac{g^2}{\pi^2} b^4 \alpha_2^2 \left[\ln |\alpha_2| - \pi i \theta \left(\frac{\alpha_2}{\chi} \right) - i\pi \left(1 - \theta(\chi) - \frac{i\pi}{2} \right) \right] \\ &\quad + o \left(\frac{\chi}{\xi^2} \right). \end{aligned} \tag{A2.25}$$

To consider the integral Y_2 near the resonance we can expand X_3 in powers of the small parameter χ / α_2:

$$X_3 = \frac{i\chi}{\alpha_2} X_2 + \frac{\chi^2}{\alpha_2^2} X_1 + \frac{\chi^3}{\alpha_2^3} X_0 + o \left(\frac{\chi^4}{\alpha_2^4} \right). \tag{A2.26}$$

Substituting these results into (A2.9) we arrive at the formula

$$Y_2 = - \frac{\chi}{\xi^2} \frac{\pi}{g} \alpha_1 + \frac{\alpha_2^2}{\xi^2} + o \left(\frac{\chi^3}{\alpha_2^3} \right). \tag{A2.27}$$

It follows from (A2.25) and (A2.27) that the surface impedance of a metal whose Fermi surface everywhere has a finite and nonzero curvature, in chosen geometry does not exhibit any resonance features corresponding to the cyclotron resonance in a normal magnetic field.

Appendix 3

The procedure of calculation of the quantity

$$F_q = \frac{|e|B}{cg} \frac{1}{(2\pi\hbar)^2} \sum_{n,\sigma} \int_{-\infty}^{\infty} dp_z \frac{f_{np_z}^{\sigma} - f_{np_z - \hbar q}^{\sigma}}{E_{np_z - \hbar q}^{\sigma} - E_{np_z}^{\sigma} + \hbar\omega + i\hbar/\tau} \tag{A3.1}$$

is expounded in [161] for a spherical Fermi surface. Here we extend this procedure to include all axisymmetric Fermi surfaces. We assume that the external magnetic field is directed along the symmetry axis. Transforming the sum over n into the integral we arrive at the following result for the oscillating part of this quantity F_q:

$$\tilde{F}_q = \frac{|e|B}{cg} \frac{1}{(2\pi\hbar)^2} \sum_{\sigma} \sum_{r=1}^{\infty} \int_{-\infty}^{\infty} dp_z \int_{\delta}^{\infty} dn \frac{f_{np_z}^{\sigma} - f_{np_z - \hbar q}^{\sigma}}{E_{np_z - \hbar q}^{\sigma} - E_{np_z}^{\sigma} + \hbar\omega + i\hbar/\tau}$$
$$\times (\exp(2\pi irn) + \exp(-2\pi irn)). \tag{A3.2}$$

Here δ takes on values from the interval $(0, 1)$. In the following calculations we assume that the magnetic field is strong enough to satisfy the inequality $qR \ll 1$. Under this condition we can rewrite (A3.2) in the form

$$\tilde{F}_q = -\frac{|e|B}{cg} \frac{1}{(2\pi\hbar)^2} \sum_{\sigma} \sum_{r=1}^{\infty} \int_{-\infty}^{\infty} dp_z \int_{\delta}^{\infty} dn \frac{df_{np_z}^{\sigma}}{dE_{np_z}^{\sigma}} \frac{v_z}{v_z - s - i/\tau}$$
$$\times (\exp(2\pi irn) + \exp(-2\pi irn)). \tag{A3.3}$$

here $s = \omega/q$. Using the well-known relation

$$S(E, p_z) = \frac{2\pi\hbar|e|B}{c}(n + \delta), \tag{A3.4}$$

we can make a change of variables in the integral (A3.3) and convert from the integration with respect to n to the integration with respect to energy:

$$
\tilde{F}_q = -\frac{1}{4\pi^2\hbar^3 g} \sum_\sigma \sum_{r=1}^\infty \int_{E_{\min}}^\infty dE \frac{df}{dE} \int_{p_{\min}}^{p_{\max}} \frac{m_\perp(E, p_z)v_z(E, p_z)}{v_z(E, p_z) - s - i/\tau}
$$
$$
\times \left\{ \exp\left[\frac{ircS(E, p_z)}{\hbar|e|B} - 2\pi ir\delta\right] + \exp\left[-\frac{ircS(E, p_z)}{\hbar|e|B} + 2\pi ir\delta\right] \right\}. \quad (A3.5)
$$

To simplify the following calculations we consider a Fermi surface which has the unique extremum cross-section. Due to the reflection symmetry of the surface this extremum cross-section has to be its center cross-section. We also conclude that $p_{\min}(E) = -p_{\max}(E)$. The main contribution to the integral over energies originates from the vicinity of the point $E = \zeta$. Therefore we can replace $p_{\max}(E)$ by $p_m(\zeta)$ which we denote p_m. In our model the longitudinal velocity of the quasi particles v_z is a monotonous function of p_z and the maximum value of the velocity v_m corresponds to the maximum value of the longitudinal component of the momentum: $v_m = v_z(p_m)$. Under these assumptions we can represent the oscillating part of F_q in the form

$$
\tilde{F}_q = \Delta_1 + \Delta_2, \quad (A3.6)
$$

where

$$
\Delta_1 = -\frac{p_m}{4\pi^2\hbar^3 g} \sum_\sigma \sum_{r=1}^\infty \exp(-2\pi ir\delta) \int_{E_{\min}}^\infty dE \frac{df}{dE} I_1(E),
$$
$$
\Delta_2 = -\frac{p_m}{4\pi^2\hbar^3 g} \sum_\sigma \sum_{r=1}^\infty \exp(2\pi ir\delta) \int_{E_{\min}}^\infty dE \frac{df}{dE} I_2(E),
$$

$$(A3.7)$$

and $u = s/v_m$ and $\beta = 1/qv_m\tau$:

$$
I_1(E) = \int_{-1}^1 \frac{m_\perp(E, x) \exp\left[ircS(E, x)/\hbar|e|B\right]}{x - u - i\beta} \, dx;
$$
$$
I_2(E) = \int_{-1}^1 \frac{m_\perp(E, x) \exp\left[-ircS(E, x)/\hbar|e|B\right]}{x - u - i\beta} \, dx.
$$

$$(A3.8)$$

To calculate the integrals $I_1(E)$ and $I_2(E)$ we use the Laplace transform

$$
\frac{1}{\omega} = \int_0^\infty \exp[-\omega t] \, dt, \quad (\Re w > 0). \quad (A3.9)
$$

As a result we obtain

$$
I_1(E) = -\int_0^\infty dt \exp[-(u + i\beta)t] \int_{-1}^u m_\perp(E, x) \exp\left[-\frac{ircS(E, x)}{\hbar|e|B} + tx\right] dx
$$
$$
+ \int_0^\infty dt \exp[(u + i\beta)t] \int_u^1 m_\perp(E, x) \exp\left[-\frac{ircS(E, x)}{\hbar|e|B} - tx\right] dx,
$$

$$(A3.10)$$

$$I_2(E) = -\int_0^\infty dt \exp[-(u+i\beta)t] \int_{-1}^u m_\perp(E,x) \exp\left[\frac{ircS(E,x)}{\hbar|e|B} + tx\right] dx$$

$$+ \int_0^\infty dt \exp[(u+i\beta)t] \int_u^1 m_\perp(E,x) \exp\left[\frac{ircS(E,x)}{\hbar|e|B} - tx\right] dx.$$

$$(A3.11)$$

We have to remark that the quantity u is of the order of s/v_F and $\beta \sim (ql)^{-1}$ (l is the mean free path). Therefore, under the considered conditions, both parameters are small in magnitude compared to unity and the main contributions to the integrals with respect to x in (A3.10), (A3.11) originate from the interval where x is close to zero. For small x we can replace the function $S(E,x)$ by the approximation

$$S(E,x) = S_{ex}(E) + \tfrac{1}{2}\left(\frac{d^2 S}{dp_z^2}\right)_0 p_m^2 x^2 + \cdots. \qquad (A3.12)$$

Using this approximation we can apply the method of steepest descents to evaluate the integrals over x. Alongside the contributions from saddle points we have to consider the contribution from the point $x = u$. As a result we have

$$I_1(E) \approx \exp\left[-ir\gamma^2 \frac{S_{ex}(E)}{p_m^2}\right] \left\{ \exp(iau^2)um_\perp(E,0) \int_0^\infty \frac{\exp(-i\beta t)\,dt}{2iau + t} \right.$$

$$+ \exp(iau^2)um_\perp(E,0) \int_0^\infty \frac{\exp(i\beta t)\,dt}{2iau - t}$$

$$\left. + i\left(\frac{i\pi}{a}\right)^{1/2} m_\perp(E,0) \int_0^\infty \exp[iy - 2(u+i\beta)\sqrt{ay}]dy \right\}, \qquad (A3.13)$$

Here $a = 1/(2r\gamma^2|d^2 S/dp_z^2|_0)$. A similar calculation gives the following result for $I_2(E)$:

$$I_2(E) \approx -\exp\left[ir\gamma^2 \frac{S_{ex}(E)}{p_m^2}\right] \left\{ \exp(-iau^2)um_\perp(E,0) \int_0^\infty \frac{\exp(-i\beta t)\,dt}{t - 2iau} \right.$$

$$- \exp(-iau^2)um_\perp(E,0) \int_0^\infty \frac{\exp(i\beta t)\,dt}{t + 2iau}$$

$$\left. + i\left(-\frac{i\pi}{a}\right)^{1/2} m_\perp(E,0) \int_0^\infty \exp[-iy - 2(u+i\beta)\sqrt{ay}]dy \right\}. \qquad (A3.14)$$

The last terms included in (A3.13), (A3.14) can be expressed in terms of the Fresnel integrals. For $u\sqrt{a} \gg 1$ we can use asymptotic expressions for these Fresnel integrals to simplify (A3.13), (A3.14):

$$I_1(E) \approx \exp\left[-ir\gamma^2 \frac{S_{ex}(E)}{p_m^2}\right]$$

$$\times \left[2\pi i \exp(iau^2 - 2au\beta)m_\perp(E,0)u - m_\perp(E,0)\left(\frac{i\pi}{a}\right)^{1/2}\right], \qquad (A3.15)$$

$$I_2(E) \approx -\exp\left[ir\gamma^2 \frac{S_{\text{ex}}(E)}{p_m^2}\right] m_\perp(E,0) \left(-\frac{i\pi}{a}\right)^{1/2}. \tag{A3.16}$$

To evaluate integrals with respect to energy in (A3.6), (A3.7) we can replace all the smooth functions of E in the integrands by their values at $E = \zeta$. Performing integration with respect to E, we obtain the following expression for the oscillating part of the function F:

$$\tilde{F} = \Delta' + \Delta. \tag{A3.17}$$

Here

$$\Delta' = \mu i\pi \left(\frac{\omega}{qv_m}\right) \sum_{r=1}^{\infty} \psi_r(\theta) \exp\left[-r\gamma^2 \left(\frac{2\pi m_\perp \omega}{qp_m}\right)^2 \frac{1}{\omega\tau} \left|\frac{d^2S}{dp_z^2}\right|_0^{-1}\right]$$

$$\times \exp\left[-ir\left(\gamma^2 \frac{S_{\text{ex}}}{p_m^2} - \gamma^2 \left(\frac{2\pi m_\perp \omega}{qp_m}\right)^2 \left|\frac{d^2S}{dp_z^2}\right|_0 + \pi\delta\right)\right] \cos\left(\pi r \frac{m_\perp}{m}\right), \tag{A3.18}$$

$$\Delta = \frac{\mu}{2} \frac{1}{\gamma} \left|\frac{d^2S}{dp_z^2}\right|_0^{-1/2} \sum_{r=1}^{\infty} \frac{\psi_r(\theta)}{r^{1/2}} \cos\left(\pi r \frac{m_\perp}{m}\right)$$

$$\times \left\{\exp\left[-ir\frac{cS_{\text{ex}}}{\hbar|e|B} + i\frac{\pi}{4}\right] G_r^+ + \exp\left[ir\frac{cS_{\text{ex}}}{\hbar|e|B} - i\frac{\pi}{4}\right] G_r^-\right\}, \tag{A3.19}$$

where $\mu = m_\perp p_m/\pi^2\hbar^3 g$ is a dimensionless constant:

$$G_r^\pm = \pm\frac{i}{2} \int_0^\infty \exp\left[\pm iy - \left(\frac{\omega + i/\tau}{qv_m}\right) \left|\frac{d^2S}{dp_z^2}\right|_0^{-1/2} \left(\frac{2\hbar|e|B}{cp_m^2}\right)^{1/2} \sqrt{ry}\right] dy. \tag{A3.20}$$

When $\omega(cp_m^2/2\hbar|e|B)^{1/2}/qv_m \gg 1$ we can simplify the expression for the function Δ and transform it to the form

$$\Delta = \mu \left(\frac{2\hbar|e|B}{cp_m^2}\right)^{1/2} \left|\frac{d^2S}{dp_z^2}\right|_0^{-1/2} \sum_{r=1}^{\infty} \frac{\psi_r(\theta)}{r^{1/2}}$$

$$\times \cos\left(\pi r \frac{m_\perp}{m}\right) \cos\left(\frac{rcS_{\text{ex}}}{\hbar|e|B} - \frac{\pi}{4} - \pi r\delta\right). \tag{A3.21}$$

When the curvature of the Fermi surface turns to zero at the extremum cross-section we have to use a different approximation for $S(E,x)$. This approximation is

$$S(E,x) = S_{\text{ex}}(E) + \frac{1}{(2l)!} \left(\frac{d^{2l}S}{dp_z^{2l}}\right)_0 p_m^{2l} x^{2l} + \cdots, \qquad l > 1. \tag{A3.22}$$

In this case the integrals $I_1(E)$ and $I_2(E)$ can be transformed to the form

$$I_1(E) \approx -\exp\left[-irc\frac{S_{\text{ex}}(E)}{\hbar|e|B}\right]$$

$$\times \left\{\exp[ia(u)^{2l/(2l-1)}]m_\perp(E,0)u\left[\int_0^\infty \frac{\exp(-i\beta t)\,dt}{2liau + (2l-1)u^{2(l-1)/(2l-1)}t}\right.\right.$$

$$+ \int_0^\infty \frac{\exp(i\beta t)\,dt}{2liau - (2l-1)u^{2(l-1)/(2l-1)}t}\Bigg]$$

$$+ m_\perp(E,0)\frac{(a)^{-1/2l}}{l}\Gamma\left(\frac{1}{2l}\right)\exp\left(-\frac{i\pi}{4l}\right)\Bigg\}, \tag{A3.23}$$

$$I_2(E) \approx -\frac{(a)^{-1/2l}}{l}\Gamma\left(\frac{1}{2l}\right)\exp\left(\frac{i\pi}{4l}\right)\exp\left(irc\frac{S_{ex}(E)}{\hbar|e|B}\right), \tag{A3.24}$$

here $a = rcp_m^{2l}\left|d^{2l}S/dp_z^{2l}\right|_0/(2l)\hbar|e|B$ and $\Gamma(x)$ is the gamma function.

As well as before the first two terms in (A3.23) correspond to the contributions from the point $x = u$ and the last term gives the contribution from the saddle point. We do not write the contributions from the point $x = u$ in the expression for $I_2(E)$ because these terms cancel out after performing integration with respect to t. The resulting asymptotics for the quantities $I_1(E)$ and $I_2(E)$ describe them rather well when $a(u)^{2l/(2l-1)} \gg 1$. These asymptotic expressions are

$$I_1(E) \approx 2\pi i m_\perp(E,0)u^{1/(2l-1)}$$

$$\times \exp\left[-irc\frac{S_{ex}(E)}{\hbar|e|B}\right]\exp\left[ia(u)^{2l/(2l-1)}\right]\exp\left[-\frac{2la\beta u^{1/(2l-1)}}{2l-1}\right]$$

$$+ m_\perp(E,0)\frac{(a)^{-1/2l}}{l}\Gamma\left(\frac{1}{2l}\right)\exp\left[-irc\frac{S_{ex}(E)}{\hbar|e|B} + \frac{i\pi}{4l}\right]. \tag{A3.25}$$

Substitution of (A3.25) into (A3.7) gives:

$$\Delta' = i\pi\mu\left(\frac{\omega}{qv_m}\right)^{1/(2l-1)}\sum_{r=1}^\infty \psi_r(\theta)\cos\left(\pi r\frac{m_\perp}{m}\right)$$

$$\times \exp\left[-\frac{2l}{2l-1}a\left(\frac{\omega}{qv_m}\right)^{1/(2l-1)}\frac{1}{qv_m\tau}\right]$$

$$\times \exp\left[-\frac{ircS_{ex}}{\hbar|e|B} + \frac{ircp_m^{2l}}{\hbar|e|B}\left|\frac{d^{2l}S}{dp_z^{2l}}\right|_0\left(\frac{\omega}{qv_m}\right)^{2l/(2l-1)}\frac{1}{(2l)!} - \pi ir\delta\right], \tag{A3.26}$$

$$\Delta = \mu\left(\Gamma(2l+1)\frac{\hbar|e|B}{cp_m^{2l}}\right)^{1/2l}\Gamma\left(\frac{1}{2l}\right)\left|\frac{d^{2l}S}{dp_z^{2l}}\right|_0^{1/2l}\frac{1}{l}$$

$$\times \sum_{r=1}^\infty \frac{\psi_r(\theta)}{r^{1/2l}}\cos\left(\pi r\frac{m_\perp}{m}\right)\cos\left(\frac{rcS_{ex}}{\hbar|e|B} - \frac{\pi}{4l}\right). \tag{A3.27}$$

The resulting expression for the function Δ coincides with (7.9).

Appendix 4

We consider the method of calculation of the quantity

$$N_\zeta^* = -\sum_{\nu\nu'} \frac{f_\nu - f_{\nu'}}{E_\nu - E_{\nu'}} n_{\nu\nu'}^*(-\mathbf{q}) n_{\nu'\nu}(\mathbf{q})|_{q=0}. \tag{A4.1}$$

Matrix elements of the effective operators $n_{\nu\nu'}^*(\mathbf{q})$ are defined by a relation

$$n_{\nu\nu'}^*(-\mathbf{q}) = n_{\nu\nu'}(-\mathbf{q}) + \sum_{\nu_1\nu_2} \frac{f_{\nu_1} - f_{\nu_2}}{E_{\nu_1} - E_{\nu_2}} F_{\nu\nu'}^{\nu_1\nu_2} n_{\nu\nu'}^*(-\mathbf{q}) \tag{A4.2}$$

and matrix elements of the Fermi-liquid kernel have a form

$$F_{\nu\nu'}^{\nu_1\nu_2}(-\mathbf{q}) = \varphi_{\alpha\alpha'}^{\alpha_1\alpha_2}(-\mathbf{q})\delta_{\sigma\sigma'}\delta_{\sigma_1\sigma_2} + 4(\mathbf{s}_{\sigma\sigma'}\mathbf{s}_{\sigma_1\sigma_2})\psi_{\alpha\alpha'}^{\alpha_1\alpha_2}(-\mathbf{q}). \tag{A4.3}$$

In the case when $(\mathbf{q} \perp \mathbf{B})$ the matrix elements $\varphi_{\alpha\alpha'}^{\alpha_1\alpha_2}$ are nonzero at $\alpha = n$, p_z, x_0 and $\alpha' = n'$, p_z, $x_0 + (\hbar cq/|e|B)$. Similar relations should connect the corresponding quantum numbers from the sets α_1 and α_2. One can represent the nonzero matrix elements in the form

$$
\begin{aligned}
\varphi(\mathbf{q}, Nlp_z; N'l'p_z') = \frac{1}{g_0} \sum_{s=-\infty}^{\infty} &\Bigg\{ \left[P_{00}^{|2s|}(p_z)P_{00}^{|2s|}(p_z') + p_z p_z' P_{01}^{|2s|}(p_z)P_{01}^{|2s|}(p_z') \right] \\
&\times J_{l-2s}\left(\frac{\hbar cq}{|e|B}\sqrt{2N+1} \right) J_{l'-2s}\left(\frac{\hbar cq}{|e|B}\sqrt{2N'+1} \right) \\
&+ \left[P_{10}^{|2s|}(p_z)P_{10}^{|2s|}(p_z') + p_z p_z' P_{11}^{|2s|}(p_z)P_{11}^{|2s|}(p_z') \right] \\
&\times \Bigg[J_{l-2s-1}\left(\frac{\hbar cq}{|e|B}\sqrt{2N+1} \right) J_{l'-2s-1}\left(\frac{\hbar cq}{|e|B}\sqrt{2N'+1} \right) \\
&+ J_{l-2s+1}\left(\frac{\hbar cq}{|e|B}\sqrt{2N+1} \right) J_{l'-2s+1}\left(\frac{\hbar cq}{|e|B}\sqrt{2N'+1} \right) \Bigg] \Bigg\}.
\end{aligned}
$$

$$\tag{A4.4}$$

Here $2N = n + n'$, $2N' = n_1 + n_2$, $l = n - n'$, $l' = n_1 - n_2$ and g_0 is the density of states of the quasi particles at the Fermi surface in the absence of the external magnetic field. One can receive the expressions for the matrix elements $\psi_{\alpha\alpha'}^{\alpha_1\alpha_2}$ replacing the functions $P_{ik}^{|2s|}(p_z)$ and $P_{ik}^{|2s|}(p'_z)$ in (A4.2) with the functions $Q_{ik}^{|2s|}(p_z)$ and $Q_{ik}^{|2s|}(p'_z)$.

The functions $P_{ik}^{|2s|}(p_z)$ and $Q_{ik}^{|2s|}(p_z)$ are even and the matrix elements $n_{vv'}(-\mathbf{q})$ are diagonal in spin number σ. Because of this, we can take into account only the first term in the first square brackets in (A4.4) in calculating of the quantity N_ζ^*. The contributions from the other terms will be equal to zero. Thus in the following calculations, we can assume that $\varphi(\mathbf{q}; Nlp_z; N'l'p'_z)$ and $\psi(\mathbf{q}; Nlp_z; N'l'p'_z)$ have the form

$$\varphi(\mathbf{q}, Nlp_z; N'l'p'_z) = \frac{1}{g_0} \sum_{s=-\infty}^{\infty} P_{00}^{|2s|}(p_z) P_{00}^{|2s|}(p'_z) J_{l-2s}$$

$$\times \left(\frac{\hbar cq}{|e|B} \sqrt{2N+1}\right) J_{l'-2s}\left(\frac{\hbar cq}{|e|B}\sqrt{2N'+1}\right),$$

$$\psi(\mathbf{q}, Nlp_z; N'l'p'_z) = \frac{1}{g_0} \sum_{s=-\infty}^{\infty} Q_{00}^{|2s|}(p_z) Q_{00}^{|2s|}(p'_z) J_{l-2s}$$

$$\times \left(\frac{\hbar cq}{|e|B} \sqrt{2N+1}\right) J_{l'-2s}\left(\frac{\hbar cq}{|e|B}\sqrt{2N'+1}\right). \quad (A4.5)$$

Further, we omit the lower indices of the functions $P_{00}^{|2s|}(p_z)$ and $Q_{00}^{|2s|}(p_z)$.

We enter the notations

$$I_s = \sum_{vv'} \frac{f_v - f_{v'}}{E_v - E_{v'}} P^{|2s|}(p_z) J_{l-2s}\left(\frac{\hbar cq}{|e|B} \sqrt{2N+1}\right) n_{vv'}^*(-\mathbf{q}) \delta_{\sigma\sigma'} \delta_{x_0;x'_0 - \hbar cq/|e|B}, \quad (A4.6)$$

$$I'_s = \sum_{vv'} \frac{f_v - f_{v'}}{E_v - E_{v'}} Q^{|2s|}(p_z) J_{l-2s}\left(\frac{\hbar cq}{|e|B} \sqrt{2N+1}\right) n_{vv'}^*(-\mathbf{q}) \delta_{\sigma\sigma'} \delta_{x_0;x'_0 - \hbar cq/|e|B}, \quad (A4.7)$$

$$g_{ss'} = -\frac{1}{g_0} \sum_{vv'} \frac{f_v - f_{v'}}{E_v - E_{v'}} P^{|2s|}(p_z) P^{|2s'|}(p_z) J_{l-2s}\left(\frac{\hbar cq}{|e|B} \sqrt{2N+1}\right)$$

$$\times J_{l-2s'}\left(\frac{\hbar cq}{|e|B} \sqrt{2N+1}\right) \delta_{\sigma\sigma'} \delta_{x_0;x'_0 - \hbar cq/|e|B}, \quad (A4.8)$$

$$q_{ss'} = -\frac{1}{g_0} \sum_{vv'} \frac{f_v - f_{v'}}{E_v - E_{v'}} Q^{|2s|}(p_z) Q^{|2s'|}(p_z) J_{l-2s}\left(\frac{\hbar cq}{|e|B} \sqrt{2N+1}\right)$$

$$\times J_{l-2s'}\left(\frac{\hbar cq}{|e|B} \sqrt{2N+1}\right) \delta_{\sigma\sigma'} \delta_{x_0;x'_0 - \hbar cq/|e|B}, \quad (A4.9)$$

$$r_{ss'} = -\frac{1}{g_0} \sum_{vv'} \frac{f_v - f_{v'}}{E_v - E_{v'}} P^{|2s|}(p_z) Q^{|2s'|}(p_z) J_{l-2s}\left(\frac{\hbar cq}{|e|B} \sqrt{2N+1}\right)$$

$$\times J_{l-2s'}\left(\frac{\hbar cq}{|e|B} \sqrt{2N+1}\right) \delta_{\sigma\sigma'} \delta_{x_0;x'_0 - \hbar cq/|e|B}, \quad (A4.10)$$

$$N_s = \sum_{vv'} \frac{f_v - f_{v'}}{E_v - E_{v'}} P^{|2s|}(p_z)J_{l-2s}\left(\frac{\hbar c q}{|e|B}\sqrt{2N+1}\right)n_{v'v}(\mathbf{q}), \tag{A4.11}$$

$$N_s' = \sum_{vv'} \frac{f_v - f_{v'}}{E_v - E_{v'}} P^{|2s|}(p_z)J_{l-2s}\left(\frac{\hbar c q}{|e|B}\sqrt{2N+1}\right)\sigma n_{v'v}(\mathbf{q}). \tag{A4.12}$$

The expressions for the renormalized matrix elements $n^*_{vv'}(-\mathbf{q})$ have the form

$$n^*_{vv'}(-\mathbf{q}) = n^*_{Nlp_z}(-\mathbf{q})\delta_{\sigma\sigma'}\delta_{x_0;x_0'-\hbar c q/|e|B}\delta_{p_z p_z'},$$

where the quantity $n^*_{Nlp_z}(-q)$ satisfies the relation, following from the definition (A4.2):

$$n^*_{Nlp_z}(-\mathbf{q}) = n_{Nlp_z}(-\mathbf{q}) + \frac{1}{g_0}\sum_s P^{|2s|}(p_z)J_{l-2s}\left(\frac{\hbar c q}{|e|B}\sqrt{2N+1}\right)I_s$$

$$+ \frac{\sigma}{g_0}\sum_s Q^{|2s|}(p_z)J_{l-2s}\left(\frac{\hbar c q}{|e|B}\sqrt{2N+1}\right)I_s'. \tag{A4.13}$$

When the Fermi surface is axisymmetric the matrix elements $n^*_{Nlp_z}(-q)$ do not depend on p_z:

$$n_{Nlp_z}(-\mathbf{q}) = J_l\left(\frac{\hbar c q}{|e|B}\sqrt{2N+1}\right). \tag{A4.14}$$

The average quantities I_s and I_s' satisfy a system of linear equations which follows from the (A4.13):

$$\begin{cases} I_s + \sum_{s'}(g_{ss'}I_{s'} + r_{ss'}I_{s'}') = N_s, \\ I_s' + \sum_{s'}(r_{s's}I_{s'} + q_{ss'}I_{s'}') = N_s'. \end{cases} \tag{A4.15}$$

Under conditions of the semiclassical quantization when $\gamma \gg 1$ ($\gamma = \sqrt{2\zeta/\hbar\Omega}$) we have

$$\sum_{vv'} \frac{f_v - f_{v'}}{E_v - E_{v'}} \Phi_{vv'} = -\frac{1}{4\pi^2\hbar^3}\sum_l\sum_{\sigma\sigma'}\int dp_z[m_\perp(p_z)\Phi_l^{\sigma\sigma'}(p_z)$$

$$+ \delta_{l_0}\delta_{\sigma\sigma'}\sum_m(\Delta_m + \sigma\Delta_m')m_\perp(p_m)\Phi_l^{\sigma\sigma'}(p_m)]. \tag{A4.16}$$

We can obtain the expression for the function Δ' after replacing cosines by sines in the expression for the function Δ. Here $m_\perp(p_z)$ is the cyclotron mass which is assumed to be a constant in further calculations; p_m is the value of the longitudinal component of the quasi momentum, corresponding to the m-th extremal cross-section of the Fermi surface. The form of the oscillating functions Δ and Δ' depends on the particular character of the energy-momentum relation for the quasi particles. Integration with respect to p_z in (A4.14) is carried out within the limits determined by the shape and size of the Fermi surface.

Using the asymptotic formula (A4.16), we can obtain the expressions for the coefficients of the system (A4.15):

$$g_{ss'} = \tilde{A}_{2s}\delta_{ss'} + \sum_m \Delta_m P^{|2s|}(p_m)P^{|2s'|}(p_m)J_{-2s}(qR_m)J_{-2s'}(qR_m), \quad \text{(A4.17)}$$

$$q_{ss'} = \tilde{B}_{2s}\delta_{ss'} + \sum_m \Delta_m Q^{|2s|}(p_m)Q^{|2s'|}(p_m)J_{-2s}(qR_m)J_{-2s'}(qR_m), \quad \text{(A4.18)}$$

$$r_{ss'} = \sum_m \Delta'_m P^{|2s|}(p_m)Q^{|2s'|}(p_m)J_{-2s}(qR_m)J_{-2s'}(qR_m), \quad \text{(A4.19)}$$

where R_m in the radius of the cyclotron orbit corresponding to the mth extremal cross-section of the Fermi surface

$$\tilde{A}_{2s} = \frac{m_\perp}{2\pi^2\hbar^3} \frac{1}{g_0} \int [P^{|2s|}(p_z)]^2 dp_z,$$

$$\tilde{B}_{2s} = \frac{m_\perp}{2\pi^2\hbar^3} \frac{1}{g_0} \int [Q^{|2s|}(p_z)]^2 dp_z. \quad \text{(A4.20)}$$

In obtaining (A4.16)–(A4.18) we used the identity concerning the Bessel functions

$$\sum_{l=-\infty}^{\infty} J_{l-m}(x)J_l(x) = \delta_{l0}. \quad \text{(A4.21)}$$

In the case when the Fermi surface has the unique extremal cross-section (at $p_z = 0$), equations (A4.15) take the form

$$\begin{cases} I_s + \Delta\dfrac{P^{|2s|}(0)J_{-2s}(qR_{ex})}{1+\tilde{A}_{2s}}X + \Delta'\dfrac{P^{|2s|}(0)J_{-2s}(qR_{ex})}{1+\tilde{A}_{2s}}Y = \dfrac{N_s}{1+\tilde{A}_{2s}}, \\[2ex] I'_s + \Delta'\dfrac{Q^{|2s|}(0)J_{-2s}(qR_{ex})}{1+\tilde{B}_{2s}}X + \Delta\dfrac{Q^{|2s|}(0)J_{-2s}(qR_{ex})}{1+\tilde{B}_{2s}}Y = \dfrac{N'_s}{1+\tilde{B}_{2s}}, \end{cases}$$
$$\text{(A4.22)}$$

here

$$X = \sum_s P^{|2s|}(0)J_{-2s}(qR_{ex})I_s, \quad \text{(A4.23)}$$

$$Y = \sum_s Q^{|2s|}(0)J_{-2s}(qR_{ex})I'_s. \quad \text{(A4.24)}$$

We can find the quantities X and Y solving the system of equations

$$\begin{cases} X(1+\alpha_q\Delta) + Y\alpha_q\Delta' = \sum_s N_s \dfrac{P^{|2s|}(0)}{1+\tilde{A}_{2s}} J_{-2s}(qR_{ex}), \\[2ex] X\beta_q\Delta' + Y(1+\beta_q\Delta) = \sum_s N'_s \dfrac{Q^{|2s|}(0)}{1+\tilde{B}_{2s}} J_{-2s}(qR_{ex}). \end{cases}$$
$$\text{(A4.25)}$$

here

$$\alpha_q = \sum_s \frac{A^*_{2s}}{1+\tilde{A}_{2s}} J^2_{-2s}(qR_{ex}), \qquad \beta_q = \sum_s \frac{B^*_{2s}}{1+\tilde{B}_{2s}} J^2_{-2s}(qR_{ex}),$$

$$A^*_{2s} = [P^{|2s|}(0)]^2, \qquad\qquad B^*_{2s} = [Q^{|2s|}(0)]^2. \quad \text{(A4.26)}$$

As a result we arrive at the expressions

$$X = \frac{1+\beta\Delta}{D} \sum_s N_s \frac{P^{|2s|}(0)}{1+\tilde{A}_{2s}} J_{-2s}(q\,R_{\mathrm{ex}}) - \frac{\alpha\Delta'}{D} \sum_s N'_s \frac{Q^{|2s|}(0)}{1+\tilde{B}_{2s}} J_{-2s}(q\,R_{\mathrm{ex}}),$$

$$\text{(A4.27)}$$

$$X = \frac{1+\alpha\Delta}{D} \sum_s N'_s \frac{Q^{|2s|}(0)}{1+\tilde{B}_{2s}} J_{-2s}(q\,R_{\mathrm{ex}}) - \frac{\beta\Delta'}{D} \sum_s N_s \frac{P^{|2s|}(0)}{1+\tilde{A}_{2s}} J_{-2s}(q\,R_{\mathrm{ex}}),$$

where the determinant of the system equals

$$D = 1 + (\alpha_q + \beta_q)\Delta + \alpha_q\beta_q(\Delta^2 - \Delta'^2).$$

$$\text{(A4.28)}$$

Substituting the expressions (A4.27) into the system (A4.22), we can calculate the averages:

$$I_s = \frac{N_s}{1+\tilde{A}_{2s}} - \frac{\Delta + \beta_q(\Delta^2 - \Delta'^2)}{D} \frac{P^{|2s|}(0)}{1+\tilde{A}_{2s}} J_{-2s}(q\,R_{\mathrm{ex}})$$

$$\times \sum_{s'} N_{s'} \frac{P^{|2s'|}(0)}{1+\tilde{A}_{2s'}} J_{-2s'}(q\,R_{\mathrm{ex}}) - \frac{\Delta'}{D} \frac{P^{|2s|}(0)}{1+\tilde{A}_{2s}} J_{-2s}(q\,R_{\mathrm{ex}})$$

$$\times \sum_{s'} N'_{s'} \frac{Q^{|2s'|}(0)}{1+\tilde{B}_{2s'}} J_{-2s'}(q\,R_{\mathrm{ex}}),$$

$$\text{(A4.29)}$$

$$I'_s = \frac{N'_s}{1+\tilde{B}_{2s}} - \frac{\Delta + \alpha_q(\Delta^2 - \Delta'^2)}{D} \frac{Q^{|2s|}(0)}{1+\tilde{A}_{2s}} J_{-2s}(q\,R_{\mathrm{ex}})$$

$$\times \sum_{s'} N'_{s'} \frac{Q^{|2s'|}(0)}{1+\tilde{A}_{2s'}} J_{-2s'}(q\,R_{\mathrm{ex}}) - \frac{\Delta'}{D} \frac{Q^{|2s|}(0)}{1+\tilde{B}_{2s}} J_{-2s}(q\,R_{\mathrm{ex}})$$

$$\times \sum_{s'} N_{s'} \frac{P^{|2s'|}(0)}{1+\tilde{A}_{2s'}} J_{-2s'}(q\,R_{\mathrm{ex}}).$$

$$\text{(A4.30)}$$

The asymptotic expressions for the quantities N_s and N'_s can also be found by formula (A4.16). For the Fermi surface with the unique extremal cross-section we have

$$N_s = -\frac{m_\perp}{2\pi^2\hbar^3} \int dp_z\, P^0(p_z)\delta_{s0} - g_0\Delta J_0(q\,R_{\mathrm{ex}})P^{|2s|}(0)J_{-2s}(q\,R_{\mathrm{ex}}),$$

$$\text{(A4.31)}$$

$$N'_s = -g_0\Delta' J_0(q\,R_{\mathrm{ex}})Q^{|2s|}(0)J_{-2s}(q\,R_{\mathrm{ex}}).$$

Using the (A4.13) and (A4.29)–(A4.31), we obtain

$$-\sum_{\nu\nu'} \frac{f_\nu - f_{\nu'}}{E_\nu - E_{\nu'}} n^*_{\nu\nu'}(-q)n_{\nu'\nu}(q)$$

$$= g_0\left[1 - \alpha_0 + (1 - \bar{\alpha}_0)^2 J_0^2(q\,R_{\mathrm{ex}}) \frac{\Delta + \beta_q(\Delta^2 - \Delta'^2)}{1+(\alpha_q + \beta_q)\Delta + \alpha_q\beta_q(\Delta^2 - \Delta'^2)}\right]. \quad \text{(A4.32)}$$

Here we introduced additional notations

$$\alpha_0 = \frac{A_0}{1+\tilde{A}_0}, \qquad\qquad \bar{\alpha}_0 = \frac{\overline{A}_0}{1+\tilde{A}_0},$$

$$A_0 = \left(\frac{m_\perp}{2\pi^2\hbar^3 g_0}\int P^0(p_z)\,dp_z\right)^2, \qquad A_0 = \frac{P^0(0)m_\perp}{2\pi^2\hbar^3 g_0}\int P^0(p_z)\,dp_z. \quad (A4.33)$$

Passing to the limit $q \to 0$ in (A4.32) we obtain the final result

$$N_\zeta = g_0\left[1 - \alpha_0 + (1 - \overline{\alpha}_0)^2 \frac{\Delta + \beta(\Delta^2 - \Delta'^2)}{1 + (\alpha + \beta)\Delta + \alpha\beta(\Delta^2 - \Delta'^2)}\right]. \quad (A4.34)$$

Here

$$\alpha = \lim_{q\to 0}\alpha_q = \frac{A_0^*}{1 + \tilde{A}_0}, \qquad \beta = \lim_{q\to 0}\beta_q = \frac{B_0^*}{1 + \tilde{B}_0}. \quad (A4.35)$$

Neglecting terms proportional to the differences $\Delta^2 - \Delta'^2$ in the numerator and denominator of (A4.34) we can simplify the expression for the renormalized density of states

$$N_\zeta = g_0\left[1 - \alpha_0 + (1 - \overline{\alpha}_0)^2 \frac{\Delta}{1 + (\alpha + \beta)\Delta}\right]. \quad (A4.36)$$

The expression (A4.36) is generalized easily to cover the case when the Fermi surface has several extremal cross-sections. In this case we obtain

$$N_\zeta = g_0\left[1 - \alpha_0 + \frac{\sum_m (1 - \overline{\alpha}_m)^2 \Delta_m}{1 + \sum_m (\alpha_m + \beta_m)\Delta_m}\right]. \quad (A4.37)$$

Here summation over m is carried out over the extremal cross-sections:

$$\alpha_m = \frac{A_{0m}^*}{1 + \tilde{A}_0}, \qquad \overline{\alpha}_m = \frac{\overline{A}_m}{1 + \tilde{A}_0}, \qquad \beta_m = \frac{B_{0m}^*}{1 + \tilde{B}_0},$$

$$A_{0m}^* = [P^0(p_m)]^2, \qquad B_{0m}^* = [Q^0(p_m)]^2,$$

$$\overline{A}_0 = \frac{m_\perp}{2\pi^2\hbar^3 g_0} P^0(p_m)\int P^0(p_z)\,dp_z. \quad (A4.38)$$

In the case when there are two cross-sections with extremal areas we shall have

$$K = \frac{[1 - \overline{\alpha}_0(0)]^2\Delta(0) + [1 - \overline{\alpha}_0(p_0)]^2\Delta(p_0)}{1 + [\alpha_0(0) + \beta_0(0)]\Delta(0) + [\alpha_0(p_0) + \beta_0(p_0)]\Delta(p_0)}, \quad (A4.39)$$

where

$$\alpha_0 = \frac{A_0}{1 + \tilde{A}_0}, \qquad\qquad \overline{\alpha}_0(p_z) = \frac{\overline{A}_0(p_z)}{1 + \overline{A}_0},$$

$$\alpha_0(p_z) = \frac{[P^0(p_z)]^2}{1 + \tilde{A}_0}, \qquad \beta_0(p_z) = \frac{[Q^0(p_z)]^2}{1 + \tilde{B}_0},$$

$$A_0 = \frac{m_\perp}{2\pi^2\hbar^3 g_0}\left(\int_0^{p_0} P^0(p_z)\,dp_z\right)^2, \qquad \tilde{A}_0 = \frac{m_\perp}{2\pi^2\hbar^3 g_0}\int_0^{p_0}[P^0(p_z)]^2\,dp_z,$$

$$\tilde{B}_0 = \frac{m_\perp}{2\pi^2\hbar^3 g_0}\int_0^{p_0}[Q^0(p_z)]^2\,dp_z, \qquad \overline{A}_0 = \frac{m_\perp}{2\pi^2\hbar^3 g_0}P^0(p_z)\int_0^{p_0}P^0(p_z)\,dp_z.$$

$$(A4.40)$$

References

[1] A.P. Cracknell and K.C. Wong, *The Fermi Surface*, Clarendon Press, Oxford, 1973.

[2] G.E. Reuter and E.H. Sondheimer, *Proc. Roy. Soc. London, Ser.* A **195**, 336 (1948).

[3] M.Ya. Azbel and E.A. Kaner, *Zh. Èksper. Teoret. Fiz.* **32**, 896 (1957) [*Soviet Phys. JETP* **5**, 730 (1957)].

[4] V.P. Silin, *Zh. Èksper. Teoret. Fiz.* **33**, 1227 (1958) [*Soviet Phys. JETP* **6**, 1945 (1958)].

[5] O.V. Konstantinov and V.I. Perel, *Zh. Èksper. Teoret. Fiz.* **38**, 161 (1960) [*Soviet Phys. JETP* **11**, 117 (1960)].

[6] S.L. Bushbaum and J.K. Galt, *Phys. Fluids* **4**, 1514 (1961).

[7] V.F. Gantmakher, *Zh. Èksper. Teoret. Fiz.* **44**, 811 (1963) [*Soviet Phys. JETP* **17**, 549 (1963)].

[8] V.F. Gantmakher and E.A. Kaner, *Zh. Èksper. Teoret. Fiz.* **48**, 1572 (1965) [*Soviet Phys. JETP* **21**, 1053 (1965)].

[9] J.C. McGrody, J.A. Stanford, and E.A. Stern, *Phys. Rev.* **128**, 927 (1968).

[10] R.G. Chambers and V.G. Skobov, *J. Fluid Metal Phys.* **1**, 202 (1971).

[11] E.A. Kaner and V.G. Skobov, *Adv. in Phys.* **17**, 605 (1968).

[12] A.A. Abrikosov, *Introduction to the Theory of Normal Metals*, Academic Press, New York, 1972.

[13] I.M. Livshitz, M.Ya. Azbel, and M.I. Kaganov, *Electron Theory of Metals*, Consultants Bureau, New York, 1973.

[14] P.M. Platzman and P.A. Wolff, *Waves and Interaction in Solid State Plasmas*, Academic Press, New York, 1973.

[15] A.B. Pippard, *Phil. Mag.* **2**, 1147 (1957).

[16] J.M. Ziman, *Principles of the Theory of Solids*, Cambridge University Press, London, 1964.

[17] D. Shoenberg, *The Physics of Metals*, Vol. 1, edited by J.M. Ziman, Cambridge University Press, London, 1961.

[18] I. Shapira, *The Physical Acoustics*, Vol. V, edited by W.P. Mason, Academic Press, New York, 1968.

[19] J. Roberts, *The Physical Acoustics*, Vol. IV B, edited by W.P. Mason, Academic Press, New York, 1968.

[20] V.L. Gurevich, V.G. Skobov, and Yu.A. Firsov, *Zh. Èksper. Teoret. Fiz.* **40**, 787 (1961) [*Soviet Phys. JETP* **13**, 551 (1961)].

[21] V.P. Silin, *Fiz. Met. Metalloved.* **29**, 681 (1970) [in Russian].

[22] S. Schultz and G. Dunifer, *Phys. Rev. Lett.* **18**, 283 (1967).

[23] P.M. Platzman and W. Walsh, *Phys. Rev. Lett.* **19**, 519 (1967).

[24] P.M. Platzman, W. Walsh, and E-Ni Foo, *Phys. Rev.* **172**, 689 (1968).

[25] P.S. Zyryanov, V.I. Okulov, and V.P. Silin, *Zh. Èksper. Teoret. Fiz.*, Pis'ma **8**, 489 (1968) [*Soviet Phys. JETP Lett.* **8**, 432 (1968)].

[26] P.S. Zyryanov, V.I. Okulov, and V.P. Silin, *Zh. Èksper. Teoret. Fiz., Pis'ma*, **9**, 371 (1969) [*Soviet Phys. JETP Lett.* **9**, 283 (1969)].

[27] N.P. Zyryanova, V.I. Okulov, and V.P. Silin: *Problems in Solid State Physics*, Ural. Sci. Center, Sverdlovsk, 1975 [in Russian].

[28] V.I. Okulov and V.P. Silin, *Phys. Met. Metalloved.* **55**, 837 (1983) [in Russian].

[29] D. Pines and P. Nozieres, *The Theory of Quantum Liquids*, Benjamin, Boston, MA, 1966.

[30] A.A. Abrikosov, L.P. Gor'kov, and I.E. Dzialoshinskii, *Methods of Quantum Field Theory in Statistical Physics*, Prentice-Hall, Englewood Cliffs, NJ, 1963.

[31] G. Mahan, *Many-Particle Physics*, Plenum, New York, 1981.

[32] B.A. Tavger and V.Ya. Demikhovskii, *Uspekhi. Fiz. Nauk* **96**, 61 (1968) [*Soviet Phys. Uspekhi* **11**, 644 (1969)].

[33] M.S. Khaikin, in *Conduction Electrons*, Nauka, Moscow, 1986 [in Russian].

[34] Yu.A. Romanov and M.Sh. Erukhimov, *Zh. Èksper. Teoret. Fiz.* **55**, 1561 (1968) [*Soviet Phys. JETP* **28**, 817 (1969)].

[35] A.S. Kondrat'ev, A.E. Kuchma, and R.P. Meilanov, *Fiz. Met. Metalloved.* **41**, 742 (1976) [in Russian].

[36] E.I. Butikov, A.S. Kondrat'ev, and A.E. Kuchma, *Fiz. Met. Metalloved.* **36**, 485 (1973) [in Russian].

[37] A.S. Kondrat'ev, A.E. Kuchma, and R.P. Meilanov, *Fiz. Met. Metalloved.* **37**, 1138 (1974) [in Russian].

[38] T. Ando, A. Fowler, and F. Stern, *Rev. Mod. Phys.* **54**, 437 (1982).

[39] L.A. Falkovskii, *Zh. Èksper. Teoret. Fiz., Pis'ma* **11**, 138 (1970) [*Soviet Phys. JETP Lett.* **11**, 138 (1970)].

[40] A.F. Andreev, *Uspekhi. Fiz. Nauk* **105**, 113 (1971) [*Soviet Phys. Uspekhi* **14**, (1971–1972)].

[41] V.I. Okulov and V.V. Ustinov, *Zh. Èksper. Teoret. Fiz.* **67**, 1176 (1974) [*Soviet Phys. JETP* **40**, 584 (1974)].

[42] V.I. Okulov and V.V. Ustinov, *Fiz. Met. Metalloved.* **41**, 231 (1976) [in Russian].

[43] V.I. Okulov and V.V. Ustinov, *Fiz. Nizk. Temp.* **5**, 213 (1978) [*Soviet J. Low Temp. Phys.* **5**, 116 (1978)].

[44] A.M. Brodskii and M.I. Urbakh, *Uspekhi. Fiz. Nauk* **138**, 413 (1982) [*Soviet Phys. Uspekhi* **25**, 810 (1982)].

[45] J.K. Galf, F.R. Merrit, and J.R. Klauder, *Phys. Rev.* A **139**, 823 (1965).

[46] J.K. Galf, F.R. Merrit, and P.M. Schmidt, *Phys. Rev. Lett.* **6**, 458 (1961).

[47] V.P. Naberezhnykh and N.K. Dan'shin, *Zh. Èksper. Teoret. Fiz.* **56**, 1223 (1969) [*Soviet Phys. JETP* **29**, 658 (1969)].

[48] V.P. Naberezhnykh and V.L. Mel'nik, *Fiz. Tverd. Tela* **7**, 258 (1965) [Soviet Phys. Solid State **7**, 197 (1965)].

236   References

[49] G.A. Baraff, C.C. Grimes, and P.M. Platzman, *Phys. Rev. Lett.* **22**, 590 (1969).

[50] N.A. Zimbovskaya, V.I. Okulov, and E.A. Pamyatnykh, *Fiz. Met. Metalloved.* **54**, 224 (1982) [in Russian].

[51] N.A. Zimbovskaya, V.I. Okulov, A.Yu. Romanov, and V.P. Silin, *Fiz. Nizk. Temp.* **8**, 930 (1982) [*Soviet J. Low Temp. Phys.* **8**, 468 (1982)].

[52] N.A. Zimbovskaya, V.I. Okulov, A.Yu. Romanov, and V.P. Silin, *Fiz. Met. Metalloved.* **58**, 851 (1984) [in Russian].

[53] N.A. Zimbovskaya and V.I. Okulov, *Fiz. Met. Metalloved.* **61**, 230 (1986) [in Russian].

[54] N.A. Zimbovskaya, V.I. Okulov, A.Yu. Romanov, and V.P. Silin, *Fiz. Met. Metalloved.* **62**, 1095 (1986) [in Russian].

[55] N.A. Zimbovskaya and V.I. Okulov, *Zh. Èksper. Teoret. Fiz., Pis'ma* **46**, 102 (1987) [*JETP Lett.* **46**, 125 (1987)].

[56] N.A. Zimbovskaya and V.I. Okulov, *Fiz. Nizk. Temp.* **17**, 726 (1991) [*Soviet J. Low Temp. Phys.* **17**, 356 (1982)].

[57] N.A. Zimbovskaya, V.I. Okulov, N.G. Bebenin and N.S. Yartseva, *Fiz. Met. Metalloved.* **66**, 62 (1991) [in Russian].

[58] V.V. Gudkov, I.V. Zhevstovskich, N.A. Zimbovskaya, and V.I. Okulov, *Zh. Èksper. Teoret. Fiz.* **100**, 1286 (1991) [*Soviet Phys. JETP* **74**, 711 (1991)].

[59] N.A. Zimbovskaya, *Fiz. Met. Metalloved.* **73**, 72 (1992) [in Russian].

[60] N.A. Zimbovskaya, *Fiz. Nizk. Temp.* **18**, 1009 (1992) [*Soviet J. Low Temp. Phys.* **18**, 709 (1992)].

[61] N.A. Zimbovskaya, *Fiz. Nizk. Temp.* **18**, 1258 (1992) [*Soviet J. Low Temp. Phys.* **18**, 880 (1992)].

[62] N.A. Zimbovskaya, *Fiz. Met. Metalloved.* **74**, 35 (1992) [in Russian].

[63] N.A. Zimbovskaya, *Fiz. Met. Metalloved.* **76**, 43 (1993) [in Russsian].

[64] N.A. Zimbovskaya, *Fiz. Met. Metalloved.* **76**, 105 (1993) [in Russian].

[65] N.A. Zimbovskaya, *Fiz. Nizk. Temp.* **19**, 1337 (1993) [*Soviet J. Low Temp. Phys.* **19**, 949 (1993)].

[66] N.A. Zimbovskaya, Deposited article 1741–B93, All Union Institute for Scientific and Technological Information, 1993 [in Russian].

[67] N.A. Zimbovskaya. *Fiz. Met. Metalloved.* **77**, 36 (1994) [in Russian].

[68] N.A. Zimbovskaya. *Fiz. Nizk. Temp.* **21**, 286 (1995) [*Soviet J. Low Temp. Phys.* **21**, 217 (1995)].

[69] N.A. Zimbovskaya. *Fiz. Nizk. Temp.* **20**, 441 (1994) [*Soviet J. Low Temp. Phys.* **20**, 350 (1994)].

[70a] N.A. Zimbovskaya. *Fiz. Nizk. Temp.* **22**, 1137 (1996) [*Soviet J. Low Temp. Phys.* **22**, 869 (1996)].

[70b] N.A. Zimbovskaya, *Zh. Èksper. Teoret. Fiz.* **107**, 1672 (1995) [*Soviet Phys. JETP* **80**, 932 (1995)].

[71] N.A. Zimbovskaya, *Phys. Low-Dimen. Struct.* **11/12**, 29 (1996).

[72] N.A. Zimbovskaya, *Zh. Èksper. Teoret. Fiz.* **113**, 2229 (1995) [*Soviet Phys. JETP* **86**, 1220 (1998)].

[73] N.A. Zimbovskaya, *Phys. Low-Dimen. Struct.* **5/6**, 40 (1998).

[74] N.A. Zimbovskaya and J.L. Birman, *Internat. J. Mod. Phys.* **13**, 859 (1999).

[75] N.A. Zimbovskaya and J.L. Birman, *Phys. Rev. B*, **60**, 2864 (1999).

[76] N.A. Zimbovskaya and J.L. Birman, *Phys. Rev. B*, **60**, 12174 (1999).

[77] L.D. Landau, *Zh. Èksper. Teoret. Fiz.* **11**, 581 (1941) [in Russian].

[78] L.D. Landau, *Zh. Èksper. Teoret. Fiz.* **30**, 1058 (1956). [*Soviet Phys. JETP* **3**, 920 (1956)].

[79] V.M. Kontorovich and N.A. Sapogova, *Zh. Èksper. Teoret. Fiz., Pis'ma* **18**, 281 (1973) [Soviet Phys. JETP Lett. **18**, 165 (1973)].

[80] G.T. Avanesjan, M.I. Kaganov, and T.Yu. Lisovskaya, *Zh. Èksper. Teoret. Fiz.* **75**, 1786 (1978) [*Soviet Phys. JETP* **48**, 900 (1978)].

[81] V.M. Kontorovich and N.A. Stepanova, *Zh. Èksper. Teoret. Fiz.* **76**, 642 (1979) [*Soviet Phys. JETP* **49**, 321 (1979)].

[82] M.I. Kaganov and Yu.V. Gribkova, *Fiz. Nizk. Temp.* **17**, 907 (1991) [*Soviet J. Low Themp. Phys.* **17**, 473 (1991)].

[83] E.V. Bezuglyi, *Fiz. Nizk. Temp.* **9**, 543 (1983) [*Soviet J. Low Temp. Phys.* **9**, 277 (1983)].

[84] N.A. Stepanova, *Fiz. Nizk. Temp.* **3**, 1415 (1977) [*Soviet J. Low Temp. Phys.* **3**, 680 (1977)].

[85] V.D. Phil', V.I. Denisenko, and E.V. Bezuglyi, *Proc XXI All Union Conf. Low Temp. Phys.*, Part III, p. 98 (Khar'kov, 1980) [in Russian].

[86] I.M. Suslov, *Fiz. Tverd. Tela* **23**, 1652 (1981) [*Soviet Phys. Solid State* **23**, 1114 (1981)].

[87] G.T. Avanesjan, M.I. Kaganov, and T.Yu. Lisovskaya, *Zh. Èksper. Teoret. Fiz., Pis'ma* **25**, 381 (1977) [*Soviet Phys. JETP Lett.* **25**, 355 (1977)].

[88] V.G. Peschanskii, V.S. Lekhtzier, and A. Tada, *Fiz. Met. Metalloved.* **56**, 855 (1983) [in Russian].

[89] V.G. Peschanskii, M. Dasanaeka, and E.V. Tzibulina, *Fiz. Nizk. Temp.* **11**, 297 (1985) [*Soviet J. Low Temp. Phys.* **11**, 118 (1985)].

[90] S.V. Lavrova, V.T. Medvedev, V.G. Skobov, et al., *Zh. Èksper. Teoret. Fiz.* **64**, 1839 (1973) [*Soviet Phys. JETP* **37**, 929 (1973)].

[91] V.T. Medvedev, V.G. Skobov, L.M. Phisher, et al., *Zh. Èksper. Teoret. Fiz.* **69**, 2267 (1975) [*Soviet Phys. JETP* **42**, 1152 (1975)].

[92] C.R. Leavens and J.P. Carbotte, *Canad. J. Phys.* **51**, 398 (1974).

[93] F. London: *Superfluids*, Vol. 1, p. 152, Wiley, New York, 1950.

[94] V.I. Okulov and E.A. Pamyatnykh, *Phys. Stat. Sol. (B)*, **60**, 771 (1973).

[95] N.A. Zimbovskaya and V.I. Okulov, Deposited article 2750–77, All Union Institute for Scientific and Technological Information, Moscow, 1977 [in Russian].

[96] V.P. Silin, *Zh. Èksper. Teoret. Fiz.* **38**, 977 (1960) [*Soviet Phys. JETP* **11**, 977 (1960)].

[97] A.I. Akhiezer, M.I. Kaganov, and G.Ya. Liubarskii, *Zh. Èksper. Teoret. Fiz.* **32**, 837 (1957) [*Soviet Phys. JETP* **5**, (1964)].

[98] V.M. Kontorovich, *Zh. Èksper. Teoret. Fiz.* **45**, 1638 (1963) [*Soviet Phys. JETP* **18**, 1125 (1964)].

[99] K.B. Vlasov and B.H. Philippov, *Zh. Èksper. Teoret. Fiz.* **46**, 223 (1964) [*Soviet Phys. JETP* **19**, 156 (1964)].

[100] T.P. Alodzhantz, *Zh. Èksper. Teoret. Fiz.* **59**, 1429 (1970) [*Soviet Phys. JETP* **32**, 780 (1971)].

[101] J. Czervonko, *Acta Phys. Polon.* A **42**, 155 (1972).

[102] J. Czervonko, *Acta Phys. Polon.* A **43**, 381 (1973).

[103] J. Czervonko, *Acta Phys. Polon.* A **45**, 755 (1974).

[104] M. Hammermesh, *Group Theory and its Application to Physical Problems*, Pergamon Press, Oxford, 1962.

[105] A.P. Cracknell, *Applied Group Theory*, Pergamon Press, Oxford, 1968.

[106] A.B. Pippard, *The Dynamics of Condition Electrons*, Gordon and Breach, New York, 1965.

[107] A.B. Pippard, *Proc. Roy. Soc. A* **224**, 273 (1954).

[108] M.I. Kaganov and M.Ya. Azbel, *Dokl. Akad. Nauk USSR* **102**, 49 (1955) [in Russian].

[109] R.B. Dingle, *Physics* **19**, 311 (1953).

[110] V.V. Ustinov and D.T. Khusainov, *Fiz. Nizk. Temp.* **11**, 1156 (1985) [*Soviet J. Low Temp. Phys.* **11**, 617 (1985)].

[111] A.V. Kobelev and V.P. Silin, *Trudy Fig. Inst. Akad. Nauk USSR* **158**, 125 (1985) [in Russian].

[112] A.W. Overhauser, *Adv. in Phys.* **27**, 343 (1978).

[113] A.W. Overhauser, *Phys. Rev.* **167**, 691 (1968).

[114] G.F. Giuliany and A.W. Overhauser, *Phys. Rev. B* **20**, 1328 (1979).

[115] P.G. Coulder and W.R. Datars, *Canad. J. Phys.* **63**, 159 (1985).

[116] J. Jensen and E.W. Plummer, *Phys. Rev. Lett.* **55**, 1912 (1985).

[117] A.W. Overhauser, *Canad. J. Phys.* **60**, 687 (1982).

[118] A.W. Overhauser, *Phys. Rev. Lett.* **55**, 1916 (1985).

[119] A.W. Overhauser, *Phys. Rev. Lett.* **52**, 1966 (1987).

[120] G. Lacueva and A.W. Overhauser, *Phys. Rev. Lett.* **46**, 1273 (1992).

[121] G. Lacueva and A.W. Overhauser, *Phys. Rev. B* **33**, 37 (1986).

[122] G.A. Baraff, *Phys. Rev.* **187**, 851 (1969).

[123] V.L. Gurevich, *Zh. Èksper. Teoret. Fiz.* **31**, 71 (1959) [*Soviet Phys. JETP* **10**, 51 (1960)],

[124] M.H. Cohen, M.J. Harrison, and W.A. Harrison, *Phys. Rev.* **117**, 937 (1960).

[125] O.V. Kirichenko and V.G. Peschanskii, *Fiz. Nizk. Temp.* **20**, 574 (1994) [*Soviet J. Low Temp. Phys.* **20**, 453 (1994)].

[126] D. Shoenberg, *Magnetic Oscillations in Metals*, Cambridge University Press, New York, 1984.

[127] P.T. Coldridge, G.B. Scott, and I.M. Templeton, *Canad. J. Phys.* **50**, 1999 (1972).

[128] N.J. Micoshiba, *Phys. Soc. Japan* **13**, 759 (1958).

[129] E.A. Kaner, *Zh. Èksper. Teoret. Fiz.* **43**, 216 (1962) [*Soviet Phys. JETP* **16**, 154 (1963)].

[130] E.A. Kaner, V.L. Fal'ko, and V.Ya. Yampol'skii, *Zh. Èksper. Teoret. Fiz.* **71**, 2389 (1976) [*Soviet Phys. JETP* **44**, 1260 (1976)].

[131] G.A. Baraff, *Phys. Rev. B* **1**, 4307 (1970).

[132] H.C. Jones and K.H. Sondheimer, *Proc. Roy. Soc. London A*, **278**, 256 (1964).

[133] P.M. Platzman and K.S. Jacobs, *Phys. Rev.* **134**, 974 (1964).

[134] G.A. Baraff, *Phys. Rev. B* **2**, 637 (1970).

[135] Y.C. Cheng, J.S. Clarke, and D. Merminn, *Phys. Rev. Lett.* **20**, 1486 (1968).

[136] S.C. Ying and J.J. Queen, *Phys. Rev.* **180**, 193 (1969).

[137] J.J. Queen and S.C. Ying, *Phys. Rev.* **180**, 1283 (1970).

[138] D. Merminn and Y.C. Cheng, *Phys. Rev. Lett.* **20**, 838 (1968).

[139] V.P. Silin, *Zh. Èksper. Teoret. Fiz.* **54**, 1016 (1968) [*Soviet Phys. JETP* **27**, 541 (1968)].

[140] T.P. Alodzhantz and V.P. Silin, *Fiz. Met. Metalloved.* **39**, 674 (1975) [in Russian].

[141] A.V. Kobelev, *Zh. Èksper. Teoret. Fiz.* **61**, 1203 (1971) [*Soviet Phys. JETP* **34**, 641 (1972)].

[142] Y.C. Cheng, *Phys. Rev. B* **3**, 2287 (1973).

[143] A.Yu. Romanov and V.P. Silin, *Physics: Brief Reports*, Nauka, Moscow, 1980 [in Russian].

[144] E.N. Foo and P.M. Platzman, *Phys. Rev. Lett.* **27**, 1568 (1971).

[145] D.S. Folk, B. Gerson, and J.F. Carolan, *Phys. Rev. B* **1**, 406 (1970).

[146] M.I. Katznel'son, V.I. Okulov, and V.V. Ustinov, *Fiz. Nizk. Temp.* **6**, 1155 (1980) [*Soviet J. Low Temp. Phys.* **6**, 587 (1980)].

[147] V.I. Okulov and V.V. Ustinov, *Fiz. Met. Metalloved.* **56**, 421 (1983) [in Russian].

[148] I.Ph. Voloshin, V.T. Medvedev, V.G. Skobov, et al., *Zh. Èksper. Teoret. Fiz.* **71**, 1555 (1976) [*Soviet Phys. JETP* **44**, 814 (1976)].

[149] I.Ph. Voloshin, N.A. Podlevskikh, V.G. Skobov, et al., *Zh. Èksper. Teoret. Fiz.* **83**, 1348 (1982) [*Soviet Phys. JETP* **56**, 1130 (1982)].

[150] K. Fuchs, *Proc. Cambridge Phil. Soc.* **34**, 100 (1938).

[151] A.Yu. Romanov and V.P. Silin, *Fiz. Met. Metalloved.* **56**, 639 (1983) [in Russian].

[152] P.S. Zyryanov, *Zh. Èksper. Teoret. Fiz.* **40**, 1353 (1961) [*Soviet Phys. JETP* **13**, 953 (1961)].

[153] J.J. Queen and S. Rodrigez, *Phys. Rev.* **128**, 2467 (1962).

[154] S. Rodrigez, *Phys. Rev.* **132**, 535 (1963).

[155] A.Ya. Blank and E.A. Kaner, *Zh. Èksper. Teoret. Fiz.* **50**, 1013 (1966) [*Soviet Phys. JETP* **23**, 673 (1966)].

[156] P.S. Zyryanov and B.N. Philippov, *Fiz. Met. Metalloved.* **23**, 767 (1967); **24**, 18 (1967) [in Russian].

[157] L.R. Testardi and J.H. Condon, *Phys. Rev. B* **1**, 3928 (1970).

[158] V.N. Bagaev, V.I. Okulov, and E.A. Pamyatnykh, *Zh. Èksper. Teoret. Fiz. Pis'ma* **27**, 156 (1978) [*Soviet Phys. JETP Lett.* **27**, 144 (1978)].

[159] B.T. Lazarev, E.A. Kaner, and L.V. Chebotarev, *Fiz. Nizk. Temp.* **4**, 808 (1978) [*Soviet J. Low Temp. Phys.* **3**, 394 (1977)].

[160] V.N. Bagaev, V.I. Okulov, and E.A. Pamyatnykh, *Fiz. Nizk. Temp.* **4**, 742 (1978) [*Soviet Phys. J. Low Temp. Phys.* **4**, 316 (1978)].

[161] E.A. Kaner, L.V. Chebotarev, and E. Uvimana, *Fiz. Nizk. Temp.* **4**, 1218 (1978) [*Soviet Phys. J. Low Temp. Phys.* **4**, 789 (1978)].

[162] V.I. Okulov and V.P. Silin, *Trudy Fig. Inst. Akad. Nauk USSR* **158**, 3, (1985) [in Russian].

[163] O.V. Konstantiniv and V.I. Perel', *Zh. Èksper. Teoret. Fiz.* **53**, 2034 (1967) [*Soviet Phys. JETP* **26**, 1151 (1968)].

[164] E.A. Kaner and V.G. Skobov, *Phys. Stat. Sol.* **22**, 333 (1967).

[165] V.I. Okulov, and E.A. Pamyatnykh, *Fiz. Met. Mettalloved* **38**, 279 (1990) [in Russian].

[166] E.A. Kaner and V.G. Skobov, *Phys. Rev. Lett. A* **25**, 105 (1967).

[167] V.V. Gudkov and K.B. Vlasov, *Phys. Rev. Lett.* **103 A**, 129 (1984).

[168] V.V. Gudkov and I.V. Zhevstovskikh, *Fiz. Nizk. Temp.* **13**, 976 (1987) [*Soviet J. Low Temp. Phys.* **13**, 556 (1987)].

[169] T.Ph. Butenko, V.T. Vitchinkin, A.A. Galkin, et al., *Zh. Èksper. Teoret. Fiz.* **78**, 1811 (1980) [*Soviet Phys. JETP* **51**, 909 (1980)].

[170] R.F. Gurvan, A.V. Gold and R.A. Phillips, *J. Phys. Chem. Col.* **29**, 1485 (1968).

[171] M.V. Kartsovnic, V.N. Laukhin, V.I. Nizhanovskii, and A.I. Ignatjev, *Zh. Èksper. Teoret. Fiz. Pis'ma* **47**, 302 (1988) [*Soviet Phys. JETP Lett.* **47**, 363 (1988)].

[172] K. Oshima, T. Mori, H. Inokuchi, et al., *Phys. Rev. B* **38**, 938 (1988).

[173] M.V. Kartsovnic, P.A. Kononovich, V.N. Laukhin et al. *Zh. Èksper. Teoret. Fiz., Pis'ma*, **48**, 498 (1988) [*Soviet Phys. JETP Lett.* **48**, 541 (1988)].

[174] M.V. Kartsovnic, P.A. Kononovich, V.N. Laukhin, et al., *Zh. Èksper. Teoret. Fiz.* **97**, 1305 (1990) [*Soviet Phys. JETP* **70**, 735 (1990)].

[175] V.Z. Kresin, *Phys. Rev. B* **35**, 8716 (1987).

[176] W. Kang, G. Montambaux, J.R. Cooper, et al, *Phys. Rev. Lett.* **62**, 2559 (1989).

[177] F.L. Pratt, J. Singleton, M. Doporto, et al., *Phys. Rev. B* **45**, 13904 (1992).

[178] N. Harrison, A. House, I. Deckers, et al., *Phys. Rev. B* **52**, 5584 (1992).

[179] J. Wosnitza, *Fermi Surface of Low-Dimensional Organic Metals and Superconductors*, Springer-Verlag, Berlin, 1996.

[180] V.M. Gokhfeld, O.V. Kirichenko, and V.G. Peschanskii, *Fiz. Nizk. Temp.* **20**, 366 (1994) [*Soviet J. Low Temp. Phys.* **20**, 317 (1994)].

[181] V.M. Gokhfeld, O.V. Kirichenko and V.G. Peschanskii, *Zh. Èksper. Teoret. Fiz.* **108**, 2147 (1995) [*Soviet Phys. JETP* **81**, 1171 (1995)].

[182] V.M. Gokhfeld, M.I. Kanagov, and V.G. Peschanskii, *Fiz. Nizk. Temp.* **12**, 1173 (1986) [*Soviet J. Low Temp. Phys.* **12**, 661 (1986)].

[183] V.M. Gokhfeld and V.G. Peschanskii, *Ukrain. Fiz. Zh.* **37**, 1594 (1992) [in Ukraian].

[184] M.I. Kanagov and P. Coutreras, *Zh. Èksper. Teoret. Phys.* **106**, 1814 (1994) [JETP **79**, 360 (1994)].

[185] J. Singleton, F.L. Pratt, M. Doporto, et al., *Phys. Rev. Lett.* **68**, 2500 (1992).

[186] J. Singleton, F.L. Pratt, M. Doporto et al., *Physica B* **184**, 470 (1993).

[187] S.V. Demishev, A.V. Semeno, N.E. Sluchanko, and N.A. Samarin, *Zh. Èksper. Teoret. Fiz., Pis'ma* **61**, 299 (1995) [*Soviet Phys. JETP Lett.* **61**, 313 (1995)].

[188] S.V. Demishev, A.V. Semeno, N.E. Sluchanko et al., *Phys. Rev. B* **53**, 12794 (1996).

[189] S.V. Demishev, A.V. Semeno, N.E. Sluchanko, et al., *Zh. Èksper. Teoret. Fiz.* **111**, 979 (1997) [*Soviet Phys. JETP* **84**, 540 (1997)].

[190] I.F. Shchegolev, P.A. Kononovich, V.N. Laukhin, and M.V. Kartsovnic, *Phys. Scripta* **29**, 46 (1989).

[191] Ya.M. Blanter, M.I. Kaganov, and D.B. Posvyanskii, *Uspekhi. Fiz. Nauk* **165**, 213 (1995).

[192] I.F. Shchegolev, V.N. Laukhin, Yu.V. Sushko, and E.V. Kostuchenko, *Zh. Èksper. Teoret. Fiz., Pis'ma* **42**, 294 (1985) [*Soviet Phys. JETP Lett.* **42**, 194 (1985)].

[193] R.L. Willett, M.A. Paalanen, R.R. Ruel, K.W. West, L.N. Pfeiffer, and D.J. Bishop, *Phys. Rev. Lett.* **65**, 112 (1990)

[194] R.L. Willett, R.R. Ruel, K.W. West, and L.N. Pfeiffer, *Phys. Rev. B* **47**, 7344 (1993).

[195] R.L. Willett, *Surf. Sci.* **305**, 76 (1994).

[196] R.L. Willett, R.R. Ruel, K.W. West, and L.N. Pfeiffer, *Phys. Rev. Lett.* **71**, 3846 (1993).

[197] R.L. Willett, K.W. West, and L.N. Pfeiffer, *Phys. Rev. Lett.* **75**, 2988 (1995).

[198] R.L. Willett, *Adv. in Phys.* **46**, 447 (1997).

[199] B.I. Halperin, P.A. Lee, and N. Read, *Phys. Rev. B* **47**, 7312 (1993).

[200] B.I. Halperin, in *Low Dimensional Semiconductor Structures*, edited by S. Das Sarma and A. Pinczuk, Wiley, New York, 1996.

[201] K.A. Inbergrigsten, *J. Appl. Phys.* **40**, 2681 (1969).

[202] P. Bierbaum, *Appl. Phys. Lett.* **21**, 595 (1972).

[203] R.L. Willett, K.W. West, and L.N. Pfeiffer, *Phys. Rev. Lett.* **78**, 4478 (1997).

[204] J.H. Smet, K. von Klitzing, D. Weiss, and W. Wegscheider, *Phys. Rev. Lett.* **80**, 4538 (1998).

[205] A.D. Mirlin, P. Wolfle, Y. Levinson, and O. Entin-Wohlman, *Phys. Rev. Lett.* **81**, 1070 (1998).

[206] F.v. Oppen, A. Stern, and B.I. Halperin, *Phys. Rev. Lett.* **80**, 4494 (1998).

[207] R. Menne and R.R. Gerhardts, *Phys. Rev. B* **57**, 1707 (1998).

[208] C.W.J. Beenakher, *Phys. Rev. Lett.* **62**, 2020 (1989).

[209] Ya.M. Blanter, M.I. Kaganov, A.V. Pantsulava, and A.A. Varlamov, *Phys. Rep.* **245**, 159 (1994).

[210] I.M. Lifshitz, *Zh. Èksper. Teoret. Fiz.* **38**, 1569 (1960) [*Soviet Phys. JETP* **11**, 1130 (1960)].

[211] A. Afanas'ev and Yu.M. Kagan, *Soviet Phys. JETP* **16**, 1030 (1963).

[212] E.G. Brovman and Yu.M. Kagan, in *Dynamical Properties of Solids* Vol. 1, edited by G.K. Norton and A.A. Maradudin, North-Holland, Amsterdam, 1974, Chap. 4, pp. 194–301.

[213] A. Wixforth, J.P. Kotthaus, and G. Weiman, *Phys. Rev. Lett.* **56**, 2104 (1986).

[214] R.L. Willett, R.R. Ruel, M.A. Paalanen, K.W. West, and L.N. Pfeiffer, *Phys. Rev. B* **47**, 7344 (1993).

[215] D.-H. Lee, *Phys. Rev. Lett.* **80**, 4745 (1998).

[216] R. Shankar and G. Murthy, *Phys. Rev. Lett.* **79**, 4437 (1997).

[217] G. Knorr, F.R. Hansen, J.P. Lynov, H.L. Pecseli, and J.J. Rasmussen, *Phys. Scripta.* **38**, 829 (1988).

[218] S.H. Simon, *Phys. Rev. B* **54**, 13878 (1996).

[219] A. Stern, B.I. Halperin, F. von Oppen, and S.H. Simon, cond-mat/982135.

Index